10 06?1?55 6

D1423404

Genetic Glass Ceilings

Transgenics for Crop Biodiversity

Genetic Glass Ceilings

Transgenics for Crop Biodiversity

Jonathan Gressel

Foreword by
Klaus Ammann

UNIVERSITY OF NOTTINGHAM
JAMES CAMERON-GIFFORD LIBRARY

The Johns Hopkins University Press
Baltimore

© 2008 The Johns Hopkins University Press
All rights reserved. Published 2008
Printed in the United States of America on acid-free paper
9 8 7 6 5 4 3 2 1

The Johns Hopkins University Press
2715 North Charles Street
Baltimore, Maryland 21218-4363
www.press.jhu.edu

Library of Congress Cataloging-in-Publication Data
Gressel, Jonathan.
  Genetic glass ceilings : transgenics for crop biodiversity / Jonathan Gressel.
    p. cm.
  Includes bibliographical references and index.
  ISBN 13: 978-0-8018-8719-2 (hardcover : alk. paper)
  ISBN 10: 0-8018-8719-4 (hardcover : alk. paper)
    1. Crops—Genetic engineering. 2. Transgenic plants. 3. Plant
diversity. 4. Crop improvement. I. Title. II. Title: Transgenics for crop
biodiversity.
  SB123.57.G74 2008
  631.5'233—dc22   20007020365

A catalog record for this book is available from the British Library.

*Special discounts are available for bulk purchases of this book. For more
information, please contact Special Sales at 410-516-6936 or
specialsales@press.jhu.edu.*

1006310556

Dedicated to the memory of Professor Leroy (Whitey) Holm,
the person who stimulated me to think differently.

Contents

Foreword

The Needs for Plant Biodiversity: The General Case

Biological diversity (often contracted to *biodiversity*) has emerged in the past decade as a key area of concern for sustainable development, but crop biodiversity, the subject of this book, is rarely considered. Jonathan Gressel's important contribution to the discussion of crop biodiversity in this volume should be considered as part of the general case for biodiversity. Biodiversity provides a source of significant economic, aesthetic, health, and cultural benefits. The well-being of earth's ecological balance as well as the prosperity of human society directly depend on the extent and status of biological diversity. Biodiversity plays a crucial role in all the major biogeochemical cycles of the planet. Plant and animal diversity ensures a constant and varied source of food, medicine, and raw material of all sorts for human populations. Biodiversity in agriculture represents variety in the food supply, allowing choices for balanced human nutrition as well as a critical source of genetic material for the development of new and improved crop varieties. In addition to these direct-use benefits, enormous less tangible benefits can be derived from natural ecosystems and their components. These include the values attached to the persistence, locally or globally, of natural landscapes and wildlife, values that increase as such landscapes and wildlife become scarce. The relationships between biodiversity and ecological parameters, linking the value of biodiversity to human activities, are summarized in part in Table 1.

Biological diversity may refer to diversity in a gene, species, community of species, ecosystem, or even more broadly, the earth as a whole. Biodiversity comprises all living beings, from the most primitive forms of viruses to the most sophisticated and highly evolved animals and plants. According to the 1992 International Convention on Biological Diversity, biodiversity means "the variability among living organisms from all sources including, terrestrial,

Table 1. Primary Goods and Services Provided by Ecosystems

Ecosystem	Goods	Services

Agroecosystems	Food crops
	Fiber crops
	Crop genetic
	   resources
		Maintain limited watershed
		   functions (infiltration, flow
		   control, partial soil protection)
		Provide habitat for birds,
		   pollinators, soil organisms
		   important to agriculture
		Build soil organic matter
		Sequester atmospheric carbon
		Provide employment

Forest ecosystems	Timber
	Fuelwood
	Drinking and
	   irrigation water
	Fodder
	Nontimber products
	   (vines, bamboos,
	   leaves, etc.)
	Food (honey,
	   mushrooms, fruit,
	   and other edible
	   plants; game)
	Genetic resources
		Remove air pollutants, emit oxygen
		Cycle nutrients
		Maintain array of watershed
		   functions (infiltration,
		   purification, flow control, soil
		   stabilization)
		Maintain biodiversity
		Sequester atmospheric carbon
		Generate soil
		Provide employment
		Provide human and wildlife habitat
		Contribute aesthetic beauty and
		   provide recreation

Freshwater
   ecosystems
	Drinking and irrigation
	   water
	Fish
	Hydroelectricity
	Genetic resources
		Buffer water flow (control timing
		   and volume)
		Dilute and carry away wastes
		Cycle nutrients
		Maintain biodiversity
		Sequester atmospheric carbon
		Provide aquatic habitat
		Provide transportation corridor
		Provide employment
		Contribute aesthetic beauty and
		   provide recreation

Grassland
   ecosystems
	Livestock (food, game,
	   hides, fiber)
	
	Drinking and irrigation
	   water
	Genetic resources
		Maintain array of watershed
		   functions (infiltration,
		   purification, flow control, soil
		   stabilization)
		Cycle nutrients
		Remove air pollutants, emit oxygen
		Maintain biodiversity
		Generate soil
		Sequester atmospheric carbon
		Provide human and wildlife habitat
		Provide employment
		Contribute aesthetic beauty and
		   provide recreation

Table 1. Primary Goods and Services Provided by Ecosystems (continued)

Ecosystem	Goods	Services
Coastal and marine ecosystems	Fish and shellfish Fishmeal (animal feed) Seaweeds (for food and industrial use) Salt Genetic resources Petroleum, minerals	Moderate storm impacts (mangroves; barrier islands) Provide wildlife (marine and terrestrial) habitat Maintain biodiversity Dilute and treat wastes Sequester atmospheric carbon Provide harbors and transportation routes Provide human and wildlife habitat Provide employment Contribute aesthetic beauty and provide recreation
Desert ecosystems	Limited grazing, hunting Limited fuelwood Genetic resources Petroleum, minerals	Sequester atmospheric carbon Maintain biodiversity Provide human and wildlife habitat Provide employment Contribute aesthetic beauty and provide recreation
Urban ecosystems	Space	Provide housing and employment Provide transportation routes Contribute aesthetic beauty and provide recreation Maintain biodiversity

marine, and other aquatic ecosystems and the ecological complexes of which they are part."[197] It is not a simple task to evaluate the need for biodiversity, in particular, to quantify agroecosystem biodiversity versus total biodiversity.[39,874,1061]

Types, Distribution, and Loss of Biodiversity

Genetic Diversity

In many instances genetic sequences, the basic building blocks of life, that encode functions and proteins are almost identical (highly conserved) across all species. The small unconserved differences are important, as they often encode the ability to adapt to specific environments. Still, the greatest importance of genetic diversity is probably in the combination of genes within an organism (the genome), because the variability in phenotype produced confers resilience and survival under selection. Thus, it is widely accepted that natural ecosystems should be managed in a manner that protects the untapped resources of genes within the organisms needed to preserve the resilience of the ecosystem. Much work remains to be done to both characterize genetic diversity and understand how best to protect, preserve, and make wise use of genetic biodiversity.[880]

The number of metabolites found in one species exceeds the number of genes involved in their biosynthesis. The concept of one gene–one mRNA–one protein–one product needs modification. There are many more proteins than genes in cells because of posttranscriptional modification. This can in part explain the multitude of living organisms that differ in only a small portion of their genes. It also explains why the number of genes found in the few organisms sequenced is considerably lower than anticipated.

Species Diversity

For most practical purposes measuring species biodiversity is the most useful indicator of biodiversity, even though no single definition exists of what is a species. Nevertheless, a plant species is broadly understood to be a collection of populations that may differ genetically from one another to some extent, but whose members are usually able to mate and produce fertile offspring. These genetic differences manifest themselves as differences in morphology, physiology, behavior, and life histories; in other words, genetic characteristics affect expressed characteristics (phenotype). About 1.75 million species have been described and named but the majority remain unknown. The global total might be ten times greater, most being undescribed microorganisms and insects.

Ecosystem Diversity

At its highest level of organization, biodiversity is characterized as ecosystem diversity, which can be classified in the following three categories:

— *Natural ecosystems* (ecosystems free of human activities) are composed of what has been broadly defined as "native biodiversity." It is a matter of debate whether any truly natural ecosystem exists today, because human activity has had an impact on most regions on the earth. It is unclear why so many ecologists seem to classify humans as being "unnatural."

— *seminatural ecosystems* (in which human activity is limited) are subject to some level of low-intensity human disturbance. These areas are typically adjacent to managed ecosystems.

— *Managed ecosystems* can be managed by humans at varying degrees of intensity from the most intensive, conventional agriculture and urbanized areas, to less intensive systems, including some forms of agriculture in emerging economies or sustainably harvested forests.

Beyond simple models of how ecosystems appear to operate, we remain largely ignorant of how ecosystems function, how they might interact with each other, and which ecosystems are critical to the services most vital to life on earth. For example, the forests have a role in water management that is crucial to urban drinking water supply, flood management, and even shipping.

Because we know so little about the ecosystems that provide our life support, we should be cautious and work to preserve the broadest possible range of ecosystems, with the broadest range of species having the greatest spectrum of genetic diversity within the ecosystems. Nevertheless, we know enough about the threat to, and the value of, the main ecosystems to set priorities in conservation and better management. We have not yet learned enough about the threat to crop biodiversity, other than to construct gene banks. Even here we have much to learn, as the vast majority of the deposits in gene banks are varieties and landraces of the four major crops. The theory behind patterns of general biodiversity related to ecological factors such as productivity is rapidly evolving, but many phenomena are still enigmatic and far from understood.[950,1062]

The Global Distribution of Plant Biodiversity

Biodiversity is not distributed evenly over the planet. Species richness is highest in warmer, wetter, topographically varied, less seasonal, and lower elevation areas. Far more species live in temperate regions (per unit area) than in polar ones, and yet far more are in the tropics than in temperate regions. Latin America, the Caribbean, the tropical parts of Asia and the Pacific host 80 percent of the ecological megadiversity of the world.[1100]

Within each region, every specific type of ecosystem supports its own unique suite of species, with their diverse genotypes and phenotypes. In numerical terms, global species diversity is concentrated in tropical rain forests.

Amazon basin rain forests can contain up to three hundred different tree species per hectare. Species and genetic diversity within any agricultural field will be more limited than in a natural or seminatural ecosystem. Nevertheless, agricultural ecosystems can be dynamic in terms of species diversity over time due to management practices. This is often not understood by ecologists who involve themselves in biosafety issues related to transgenics. They still think about ecosystems close (or seemingly close) to nature. Biodiversity in agricultural settings can be considered to be important at the country level in areas where the proportion of land allocated to agriculture is high. This is the case in continental Europe, for example, where 45 percent of the land is dedicated to arable and permanent crops or permanent pasture.[329] In the United Kingdom, this figure is even higher, at 70 percent. Consequently, biodiversity has been heavily influenced by humans for centuries, and changes in agrobiological management will influence overall biodiversity in such countries. Innovative thinking about how to enhance biodiversity, in general, coupled with bold action is critical in dealing with the loss of biodiversity.

Centers of biodiversity are a controversial matter, and even the definition of centers of crop biodiversity is still debated. Harlan[468] proposed a theory that agriculture originated independently in three different areas and that, in each case, there was a system composed of a center of origin and a noncenter, in which activities of domestication were dispersed over a span of five- to ten-thousand kilometers. One system was in the Near East (the Fertile Crescent) with a noncenter in Africa, another center included a north Chinese center and a noncenter in southeast Asia and the south Pacific, and the third system included a Central American center and a South American noncenter.[468] He suggests that the centers and the noncenters interacted with each other.

It is widely believed that centers of crop origin should not be touched by modern breeding because these biodiversity treasures are so fragile that these centers should stay free of modern breeding. This is an erroneous opinion, based on the fact that regions of high biodiversity are particularly susceptible to invasive processes, which is wrong. On the contrary, studies show that a high biodiversity means more stability against invasive species, as well as against genetic introgression.[753,1062,1148] The introduction of new predators and pathogens has caused numerous well-documented extinctions of long-term resident species, in particular, in spatially restricted environments such as islands and lakes. However, surprisingly few instances of extinctions of resident species can be attributed to competition from new competing species.

This suggests either that competition-driven extinctions take longer to occur than those caused by predation or that biological invasions are much more likely to threaten species through intertrophic than through intratrophic interactions.[255] This also fits well with agricultural experience, which builds on much faster ecological processes. Many ecologists err by not taking the ephemeral nature of agricultural plant communities into account.[36]

Loss of Biodiversity

Biodiversity is being lost in many parts of the globe, often at a rapid pace. It can be measured by loss of individual species or groups of species, or by decreases in the numbers of individual organisms. In a given location, the loss will often reflect the degradation or destruction of a whole ecosystem. The unchecked rapid growth of any species can have dramatic effects on biodiversity. This is true of weeds, elephants, and especially humans, who, being at the top of the chain, can control the rate of proliferation of other species, as well as their own, when they put their minds to it.

Habitat loss due to the expansion of human urbanization and the increase in cultivated land surfaces is identified as a main threat to 85 percent of all species described as being close to extinction. This threat can increase as so-called marginal lands are planted to biofuel crops. These lands are the last habitats left for many species. The shift from natural habitats toward agricultural land paralleled population growth, often thoroughly and irreversibly changing habitats and landscapes, especially in the developed world. Many from the developed world are trying to prevent such changes from happening in developing nations, to the consternation of many of inhabitants of the developing world who consider this to be ecoimperialism, promulgated by those unable to correct their own mistakes.

Today, more than half of the human population lives in urban areas, a figure predicted to increase to 60 percent by 2020 when Europe and the Americas will have more than 80 percent of their population living in urban zones. Five thousand years ago, the amount of agricultural land in the world is believed to have been negligible. Now, arable and permanent cropland covers approximately 1.5 billion hectares of land, with some 3.5 billion hectares of additional land classed as permanent pasture. The sum represents approximately 38% of the total available land surface of 13 billion hectares.[329]

Habitat loss is of particular importance in tropical regions of high biological diversity where food security and poverty alleviation simultaneously are key priorities. The advance of the agricultural frontier has led to an overall decline in the world's forests. Although the area of forest in industrialized regions has remained fairly unchanged, natural forest cover has declined by 8 percent in developing regions. Ironically, the most biodiverse regions are also those of greatest poverty, highest population growth, and greatest dependence on local natural resources.

Introduced species are another threat to biodiversity. Unplanned or poorly planned introduction of nonnative ("exotic" or "alien") species and genetic stocks can be, in a worst case scenario, a major threat to terrestrial and aquatic biodiversity worldwide. Hundreds if not thousands of new and foreign genes are introduced with trees, shrubs, and herbs each year.[609,1025] Many of those survive and can, after years and even many decades of adaptation, begin to be invasive. This might be interpreted as increasing biodiversity, but the final effect is sometimes the opposite. The introduced species often displace native species such that many native species become extinct or severely limited.

Biodiversity should still act as biological insurance for ecosystem processes, except when mean trophic interaction strength increases strongly with diversity.[1053] The conclusion, which needs to be tested against field studies, is that in tropical environments with a natural high biodiversity, the interactions between potentially invasive hybrids of transgenic crops and their wild relatives should be buffered through the complexity of the surrounding ecosystems. This view is also confirmed by the results of Davis.[255] Taken together, theory and data suggest that compared with intertrophic interaction and habitat loss, competition from introduced species is not likely to be a common cause of extinctions in long-term resident species at global, metacommunity, and even most community levels.

This general case for understanding and enhancing biodiversity should teach us, as Gressel endeavors to do, that the overdependence on so few crop species could be disastrous to world food security. Humans have the capability and obligation to enlarge the cultivated gene pool within insufficiently cultivated species, so that they again can contribute to crop biodiversity.

Klaus Ammann
Department of Biotechnology
Delft University of Technology
The Netherlands

Preface

The present volume started out as a journey to write a book on lost crops of the world and how to revive them transgenically to greatly enhance crop biodiversity. The Rockefeller Foundation was kind enough to be host for a month for thought and work at their exquisite facility in Bellagio, Italy, where much of the first draft was outlined and part was written. It immediately became apparent that the object of the journey was naïve insofar as it could not be determined that any crops had really been lost. Conversely, it was discovered that even major crops were precarious in their ability to cope with an ever-changing planet. The journey took on a new direction—to select a representative variety of crops that had reached their genetic ceilings; that is, that have problems that seem intractable to standard breeding but are potentially amenable to further repair by genetic engineering. The choices made are not exhaustive; the aim is not meant to supply a blueprint of how to deal with each crop and constraint. The aim is to present a spectrum of problems using real examples and then to describe how they might be analyzed and dealt with. This is not a recipe for repair but a recipe on how to think about the issues in a book meant for an audience well beyond the molecular community. Purdue University and Professor Gebisa Ejeta provided the quiet necessary for finishing the manuscript.

The amount of literature on underutilized and neglected crops is considerable. Much of it is hyperbole meant to convince people of the potential importance of each such crop but lacking a good analysis of why the crops are underutilized and of the constraints to their cultivation, especially where modern breeding has been ineffective because of a lack of genetic variability in the species. One series of publications stands out, where the missing links are well analyzed, spelled out in a manner that does not require reading between the lines. This is the highly recommended series of more than twenty monographs published by the International Plant Genetic Resources Institution (IPGRI, now called Bioversity International) in Rome on "Promoting the conservation and use of underutilized and neglected crops," The IPGRI and their staff are thanked for providing a copy of this series, although each crop is not examined, as many suffer common problems.

Many of the insights to the problems of the species involved came from an earlier 4.5-day workshop held at the same Rockefeller Foundation facility in Bellagio on ferality in crops, on whether or how transgenic technologies would increase the possibilities of crops becoming feral (published as *Crop Ferality and Volunteerism*).[417] It contains a considerable amount of lore about many crops, some of which is recited again here in a different context, where appropriate. It provided essential information needed for the biosafety analyses described herein for crops and genes in each of the case studies presented.

The author is especially thankful for the discussions, advice, and comments on Chapters 1 through 5 by Klaus Ammann. He kindly wrote a foreword on the general need for biodiversity, to put this book within the larger context, and contributed a few sections (noted in the text) in Chapters 2 and 3 where his expertise far exceeds the author's.

The following people have assisted in gathering the material, analyzing or writing on the following subjects with the author, or providing unpublished information:

— Hani Al-Ahmad on gene mitigation
— Daniel Ben-Ghedalia on ruminant nutrition
— Rafael DePrado on silicon inclusions in rice
— Gebisa Ejeta on sorghum and a lot more
— Galo Jarrín on orchids
— Wally Marasas on mycotoxins
— James Ochanda on grain weevils
— Ziv Shani on cellulose modifications
— Bernal Valverde on rice
— Suzanne Warwick on domestication
— Sarit Weissmann and Moshe Feldman on gene flow from wheat
— Aviah Zilberstein on lignin modification

Without their input this book could not have been written, although they cannot be held responsible for the author's mistakes, interpretations, or opinions.

Useful discussions with Deborah Delmer of the Rockefeller Foundation and Coosje Hoogendoorn and Toby Hodgkin of the International Plant Genetic Resources Institution are also acknowledged.

Genetic Glass Ceilings

Transgenics for Crop Biodiversity

CHAPTER ONE

Why Crop Biodiversity?

The world now depends on four crops—rice, wheat, maize, and soybeans—for the vast majority of human and livestock caloric intake. These crops have spread globally, displacing native crops, as the "big four" were deemed superior by farmers and consumers. The question must be quickly asked as to whether this lack of crop biodiversity is good for the food security of a rapidly expanding human population, a population that displaces wildlife and agriculture by urbanization and that is now demanding that agricultural lands be used to also produce biofuels. This is especially cogent, because major climate changes are to be expected as this human population increases. Much can be said for the old clichés "variety is the spice of life" and "don't put too many eggs in one basket" as together they intelligently warned our ancestors against this overdependence on so few crops. So many crops have been cultivated, and it is time to look back and ask whether some of them should be brought back to lessen the risks. To do so, one must be sure that indeed there is a risk from this limited crop range, and then one must ascertain why so many ancient and recent crops have been abandoned. Without asking why a crop was dropped and dealing with its limitations, it would be futile to start cultivating it anew.

1.1 The Loss of Crop Biodiversity

There has been a rapid decline in the number of cultivated species and their genetic diversity brought about by the success of new varieties. Farmers tend, with growing success, to eliminate weeds in their fields to prevent yield loss due to competition and to prevent contamination of the crop. More than 80 percent of the crop varieties still cultivated, such as most varieties of apple, maize, tomato, wheat, and cabbage, has been lost worldwide. Population geneticists raise concern over genetic erosion, leading to efforts to collect germplasm in *ex situ* collections.

Initially, it was naively assumed that there were new major crops to be found in the biodiversity of nature and also that some crops that had been cultivated no longer existed in farmers' fields. Our ancestors clearly utilized a much wider variety of species than we do at present. This is apparent from the archeological history. For example, *Spergula arvensis*, now considered a weed, was eaten by the Tollund and Grauelle peoples in the third to fifth centuries CE in northwestern Europe, and until recently, by Scandinavians to make inferior bread in times of food shortage.[487] The use of pignut tubers (*Bunium bulbocastanum*) goes back to the Bronze Age.[737] This species still exists abundantly in the human-influenced dry meadows near villages in the Swiss Valais. Are these, or any of the many species listed in survival manuals or cookbooks using weed species, abandoned crops? Not likely; these are "emergency food" species that were never domesticated, left over from the prehistory of gathering. They did not have sufficiently desirable traits to be domesticated.

Major crops are defined in this book as crops that supply much of a population's daily caloric needs. Fiber crops are not considered herein, even though cotton has taken over from flax and other fibers, but there are synthetic alternatives should cotton crash, as it often does regionally when insects get out of hand due to the evolution of insecticide resistance. There are no truly synthetic foods, despite what we may think of some of the items on supermarket shelves. Previous generations did a rather good job of bioprospecting for food crops. Whereas some have gone into disuse (rye, oats) compared with previous generations, these crops still exist, although some landraces may have disappeared. So what kind of agrobiodiversity do we need with major crops?

There is still much bioprospecting to do with minor crops: the fruits and vegetables that make our diets so interesting and make up for the deficiencies of the major crops. Of course there are medicinal crops still to be found. New fruits, vegetables, and flowers from continuous standard domestication make it to the markets, and some succeed. The success of breeding the Chinese gooseberry into a larger, sweet, and flavorful kiwi fruit is a case with two points: the first is the success of this crop and the second is the double standard. If the kiwi fruit had been genetically engineered, it would not be on our tables. A very small proportion of the population develops severe allergies to kiwi fruit with a wide range of symptoms, from localized oral allergy syndrome to life-threatening anaphylaxis, which can occur within minutes after eating the fruit.[663] Some of the allergic reactions are coincident with allergies to latex, other fruits, or birch pollen, but most people do not have these other allergies, which were first recognized more than two decades ago.[26] Crops such as zucchini have back mutated to a feral form that produces cucurbitacins, and in one 1982 case sent twenty-two people to the hospital after they had collapsed.[725] Such allergic reactions and outright poisonings would preclude the release of kiwi or zucchini if they had been derived by genetic engineering, or their retraction from the market where zero risk now seems to be the requirement. Lawsuits would have ensued, and if allowed to remain on the market, labeling with warnings would be required if transgenes were involved. The philosophical question of whether regulators should keep kiwi, zucchini, or peanuts, or any other product off the shelves because a minuscule proportion of the populace is violently allergic, or if a crop can rarely mutate to a feral form, is an important one. The reciprocal question is: what "natural" foodstuff is on the shelves that someone, somewhere is not allergic to? Why the double standard?

In agrobiodiversity there is the question of whether a need exists to maintain all species eaten. There are berries and seeds that were gathered from the wild when all other crops failed. This diversity sustained some until the famine was over. Should these berries and seeds be brought into agriculture? Should their diversity be maintained? I would argue that the diversity must be maintained for the genes they contain, but not necessarily for them as a crop. A hypothetical case would be a small, wild, tasteless berry, low yielding, which is gathered in years of famine. The genetic engineer could remodel it; add the gene encoding thaumatin, an ultrasweet protein (more than a thousand times sweeter than sugar, by weight), but that would only be the beginning to the whole series of other genes required. Are we lacking fruits? Do we not have fruits that are closer to domestication (e.g., fruits that are bigger, sweeter, more flavorful, higher yielding, but just lacking shelf life, the only trait missing)? Clearly, priorities must be set in deciding which species are to be domesticated further. In the case studies (Chapters 5 and onward) it also becomes apparent that priorities must be set for augmenting major crops. Some tubers and legumes have been proposed for further domestication, but none of the proponents can say what needs to be done to make these crops acceptable to the consumer, who now shuns them. If the consumer does not pick rye bread off the shelf in the supermarket, or buy bambara nuts or mashua tubers in the market, why domesticate them further? There are commodities with similar properties on the market in competition, which the consumers prefer.

From a utilitarian point of view, agriculture needs general biodiversity for the variety of genes that the diverse species contain, and to maintain stability of the surrounding ecosystem, but not necessarily for the species that can conceivably be domesticated at great effort.

Real problems exist with the major crops now cultivated. First, the world now depends on too few major crops. Sometimes, those crops have proven to have a genetic base that is too narrow. Note the major large-scale failure of maize due to a disease epidemic because much of the hybrid maize had a susceptibility gene genetically linked to the male sterility gene being used to produce hybrids.[641] Maize breeders learned the lesson and have opted for a diversity of such genes, as well as manual detasseling. Still, we must expect that some crop species will approach extinction, just as some species become extinct in the most pristine ecosystems. That does not mean that crop biodiversity is unnecessary. Crop biodiversity is essential at several levels that are not mutually exclusive.

1.   *The agroecosystem.* The more and biodiverse the group of crops rotated in an agroecosystem, the healthier the system. Rotations typically lower weed, insect, and disease problems, necessitating fewer pesticides, and are better for soil fertility. Some pest species rely on specific crops as hosts, and rotating to another crop can reduce populations of such pests. Crop rotation has been applied in virtually all agricultural strategies, from classic and historic agriculture. The maize/soybean rotation in the United States as a means

of controlling corn rootworm is one example of such a rotation designed to aid in pest control efforts. A lesson can be learned from this rotation: pests can evolve methods to overcome strategies to eliminate them. The corn rootworm evolved biotypes with a two-year diapause, overcoming this rotational strategy. Thus, limited rotations too can be overcome by pest evolution.

2. *Monoculture.* The culturing of a single crop species year after year has been hard on agriculture. There are many agroecosystems, especially "at the edge" where monoculture has no alternative, for example, rice in very wet areas, wheat in dry areas, both where not too much else grows. In areas where monoculture is the norm, the rotation of different varieties of the crop, the more disparate the better, is better than no rotation at all.

Many agroecosystems are subject to frequent crop failures. More "emergency" crops are needed—short-season crops that can be planted after a failure so the farmer can harvest something. Short-season grass pea and tef are examples of crops planted after drought-induced crop failure.

3. *Food security.* Much of the agrolandscape is covered by four major crops. Large-scale failures due to El Nino, or a disease, or a pest pandemic can lead to major disruptions of food supply. It is a not a widely known economic/food security fact that, on any given day, the amount of wheat in storage, worldwide, is sufficient for only thirty days. If a crop failure somewhere drops this supply to twenty-seven days, the price of wheat skyrockets, and panic ensues. When overproduction puts a thirty-three-day supply of wheat into granaries, the market price plummets. The situation is similar with other commodity crops traded in world markets. The problem of low stocks will become more acute with high oil prices, as grains are siphoned off to produce bioethanol and biodiesel. This diversion of food to fuel will also increase crop prices. Agronomists and others worried about world food supply do not see the world as the economists do, and such low amounts in storage seem fearful, especially with the predicted increase in mouths to feed for the next few decades together with a climate that is expected to be more variable. Having more crop species in culture, and more varied varieties within each species, clearly will buffer the supply fluctuations, providing greater food security.

4. *Farmer economics.* When farmers have a greater choice of crop species and varieties, the economic risks of agriculture are diminished. Price fluctuations can be lessened or buffered when more species are grown. The better each crop is domesticated and adapted for regional needs, the more likely it will be cultivated, helping farmers spread their not inconsequential risks.

Poverty in developing countries can be alleviated in part by having diverse crops with diverse genes that meet the particular circumstances. Maize is loved in Africa, but maize farmers remain impoverished in many places, because maize internally lacks the biodiversity to cope with stem borers, grain weevils, and the parasitic weed *Striga*. Transgenic enhancement of the species to overcome this lack of genetic diversity is thus very important to the farmer.

5. *Consumers need and want crop biodiversity.* Studies have shown that despite southeast Asian children getting more calories per day than African children, there is more mal-nourishment (measured as distended bellies) in southeast Asia. Too many in southeast Asia receive only rice in their diet, while African children have a more varied diet, compensating for fewer calories. Most crops, including cereals, have inadequate levels of at least one critical nutrient (see section 2.5). The consumers in the developed world also need and mostly want (other than the junk food generation) a varied diet and will often pay dearly for rare fruits and vegetables.

1.2 Approaches to Preserving Crop Biodiversity

That there is a need to preserve such crop biodiversity is without doubt. The landraces and wild progenitors and relatives of crops contain genes that will surely be necessary in the future, especially the resistances to insects, diseases, and abiotic stresses. Even more genes that may be useful are sure to be found in species that are not related to our crops, and they too must be conserved. Pest populations will shift and proliferate, and the environment will change, necessitating new traits.

To accentuate their differences, the varying views on how to both preserve and enlarge the biodiversity of crops still cultivated around the world are referred to with obvious bias as: (1) misdirected romantic, (2) fatalistic anal-retentive, and (3) transgenic pragmatic.

1. The misdirected romantics demand that peasants continue to cultivate low-yielding landraces. There have often been calls for keeping the subsistence farmers poor and isolated or sometimes for supporting them with subsidies (as in the developed world). Most peasants either vote for such schemes with their feet by abandoning agriculture, or with their heads, abandoning the landraces for newer, higher yielding varieties, to provide food for their families. They do not want to

buy their food with subsidies, they want to grow it. Only in rare cases do the landraces have such high value that it is worth continuing with older low-yielding varieties or landraces. For example, basmati rice in India, which yields about half as much grain as green revolution varieties but commands more than double the price in the market. Another solution would be to involve farmers in participatory breeding programs, but these need to be carefully evaluated and lead to better crop plants for present-day markets. This might be possible in the case of the Mexican corn landraces, because they often meet precisely the special needs of the indigenous kitchen habits. It will be better yet when transgenes for desired traits can be added to the landraces.

2. Fatalistic anal-retentivists (formally termed "collectors or curators") realize that the romantic view is misdirected, so they collect the landraces and crop-related species either into botanical/agronomic gardens or into giant freezer facilities to preserve the germplasm for future generations. This approach is well justified when the material deposited is well catalogued and indexed, both by traits and by genomes, while reducing redundancies within the collections, such that breeders can easily access material. This is not the case with many "gene banks"; indeed, few gene banks yet have their catalogs on the internet.

3. The pragmatic molecular approach ascertains what genes can be added to the desired landraces, which are about to be or are already abandoned, and brings the landraces back into economic production. The rest of this book is about the pragmatic molecular approach, not only dealing with landraces or abandoned crops, but dealing with further domestication of present crops, where breeding cannot provide the genetic diversity.

CHAPTER TWO

Domestication

Reaching a Glass Ceiling

> Crop species ride bicycles—to keep their balance, they must
> keep moving.
>
> (PARAPHRASE OF EPIGRAM ATTRIBUTED
TO ALBERT EINSTEIN)

Humanity has clearly changed the face of this planet. Prior to the evolution of *Homo sapiens* from the joint ancestors of apes, there were neither crops nor weeds; both are the result of human activity. The history of how our ancestors went from hunters to hunter-gatherers to domesticators of crops, and the geopolitical effects this had at many centers on the globe was first formalized by Vavilov,[1106] retold many times by evolutionists (e.g., references 245 and 471), and popularized.[273]

The biological selection for a group of traits that did not require running through the woods or savanna but allowed the farmer to leave the species nearer home made life easier for the gatherer and led to the concept of home versus nomadic existence. The gathered and slowly domesticated plant species became coddled, wimpy, versions of their wild progenitors, losing many traits that had conferred the evolutionary advantages needed to compete with countless other species in the wild. The domesticator-coddled species were cultivated alone, or possibly with one other crop species, and soon molded into the lifestyle of the cultivator.

The changes that occurred are often referred to as the "syndrome of domestication." The neutral definition of syndrome is "a combination of issues that commonly go together, which might show the existence of a condition." This implies that not all the "issues" must be there together. Thus, one crop may have one group of traits constituting the syndrome, and another crop may have some other overlapping group of traits. As demonstrated below, domestication is never complete; it cannot be so in the changing world. This has been termed the "Red Queen Principle"[1098]; an organism must continue to evolve to maintain its position among other evolving organisms in an ever-changing ecosystem. This is based on the Red Queen in *Alice's Adventures in Wonderland* who said "in this place it takes all the running you can do to keep in the same place."[192]

Domestication provided sustenance to the domesticators, then the selectors. In the twentieth century, formal breeders, and later genetic engineers working with the breeders, gained their income from further domestication. Some crops are more domesticated than others. Some are no longer amenable to further domestication by traditional breeding because they have reached a genetic glass ceiling; they lack the genes within their genome and that of their interbreeding close relatives that will allow further domestication. Although genetic engineering can assist in domestication by suppressing contraindicated inherent genes, it is often imperative that the needed genes must come from other sources to breach the genetic glass ceiling.

2.1. Selection from Prehistoric Times through the Nineteenth Century

It is posited that over 10,000 years ago, the first gatherers of plants and grains observed that seeds of the wild species that they had collected and inadvertently dropped near their encampments produced the same plants that they had brought in from afar. Consciously planting those seeds precluded the necessity to go afar and look from them, competing with other humans or herbivores that might have reached the source earlier.

These protoagriculturalists learned how to store seeds and to prepare seeds and plant them. They quickly learned that weeds competed with the crops, and removed the weeds, often feeding them to animals that too were being domesticated. They discovered that herding is an easier and more secure source of food than hunting. The farmers learned to specifically save the seeds

Table 2. The Components of the Wild, Domestication, and Weediness Syndromes

Wild traits	Domestication traits	Weedy/invasive traits
Propagules that are not adapted to long-distance dispersal	Retention of the seed/fruit on the plant at maturity	Propagules that are adapted to long-distance dispersal and easily distributed
Seed dormancy	Loss of germination inhibitors	Seed dormancy
Discontinuous germination (secondary dormancy)	Synchrony in germination (loss of secondary dormancy)	Discontinuous germination (secondary dormancy)
Special germination requirements	Narrow germination requirements	Broad germination requirements
Reduced ability to germinate in a wide range of conditions	—	Ability to germinate in a wide range of conditions
Long-lived seeds (seedbank)	Short-lived seeds (no seedbank)	Long-lived seeds (seedbank)
Slow growth to flowering, perennial	Synchrony of flowering and fruit development	Rapid growth to flowering, annual
More determinate growth	More determinate growth	Continuous seed production for as long as growing conditions permit
Lower seed output	Smaller numbers of larger fruits or inflorescences	Very high seed output
Seed produced in a narrower range of environmental conditions	—	Seed produced in a wide range of environmental conditions
—	Increase in seed and/or fruit size	—
Propagule (seed) shattering	Reduction in seed dispersal (shattering)	Propagule (seed) shattering

No special adaptations for seed dispersal over either short or long distances	—	Special adaptations for seed dispersal over either short or long distances
Not vigorous vegetative reproduction, if perennial	Increase in vegetative vigor	Vigorous vegetative reproduction, if perennial
Reduced plasticity of growth	Increase in apical dominance	Plasticity of growth
Reduced competitive ability	Reduced competitive ability	Strong competitive ability
Selfing and/or self-incompatible, can be obligate selfer and/or apomictic	Selfing and/or self-incompatible	Selfing and/or self-incompatible but not obligate selfer or apomictic
Specialized pollinators	Unspecialized pollinators	Unspecialized pollinators
Not adapted to disturbed habitats	Adaptation to disturbed habitats	Adaptation to disturbed habitats
Diploid and/or polyploid	Polyploidy frequent	Polyploidy frequent
—	Increase in starch, sugar, or oil and decrease in protein content of the seed and/or fruit	—
—	Loss of bitter substances in the seed and/or fruit	Presence of bitter substances in the seed/fruit (increased pest resistance)

Source: Modified from Warwick and Stewart [1133] and other sources
Note: A change in a few traits is sufficient for a wild species to be domesticated or become weedy.

of the best plants and to select for desirable changes in the plant genotypes and phenotypes, without knowing those terms. The farmers were selecting for the domestication syndrome (Table 2).

More effort was invested in, or needed for, the domestication of some crops than for others. In some crops little if any genetic difference exists between the cultivated species and the species in the wild; for example, cranberry (*Vaccinium macrocarpon*) "cultivars" are still hardly domesticated. Most commercial cranberry varieties are essentially wild clones chosen by growers in the nineteenth century.[1133] The olive differs from the wild oleaster in pruning and some selection among natural variability for seed size and oil content.[151] Other domestications are far more complex, giving the taxonomists what is now considered unnecessary leeway to call the wild progenitor by a different Latin binomial based on morphological or other minor differences. The perennial *Oryza rufipogon*, or its wild annual form given a different name *Oryza nivara*, is the progenitor of annual rice *Oryza sativa*, without any major fertility barriers among them (Table 3). In some crops, a complete range of intermediates provides a link to the wild, either as weeds or as wild plants in their original range, whereas in other crops such links are no longer evident. Many other examples of conspecific (same species) wild species, crops, and weeds are summarized in Table 3.

Clearly the two most important factors in the domestication syndrome for agronomic crops (Table 2) are uniform germination after planting and the lack of shattering (the propensity of seeds to fall to the ground as they ripen, before a farmer can harvest). Lack of shattering was often selected concomitantly with uniform ripening, allowing a single harvest without losing seed. In general, these are now known to be recessively inherited,[384] often governed by single major genes with modifications by minor genes. Thus, the individual plants possessing these traits would have been exceedingly rare, but our domesticators must have had a keen eye for the desirable. Further, the early crops had to be diploid, as recessive traits are exceedingly hard to pick out in polyploids when all copies of a gene may be expressed. Polyploidization, increasing seed size, and enhancement of other desirable characters came later, often with the loss or silencing of genes, such that plants were functionally diploid for some traits and polyploid for others.[338] In some cases there may have been a genetic linkage between the domestication traits, facilitating such selection.

The number of major genes distinguishing between the wild progenitors and the present crop is often quite small. Maize (*Zea mays* ssp. *mays*) putatively arose once from a single teosinte strain (*Zea mays* ssp. *parviglumis*) in

Table 3. Conspecific Wild, Crop, and Weedy Species that Differ in Genotype, Phenotype, and Impact on Agriculture but Interbreed

Wild type	Crop type	Weedy type	Reference
Oryza rufipogon	Oryza sativa domestic rice	O. sativa Red or weedy rice	1102, 1103
Sorghum bicolor arundinaceum	S. bicolor bicolor[a] domestic sorghum	S. bicolor drummondii weedy sorghum	312
Beta vulgaris ssp. maritima	Beta vulgaris Red and sugar beets Swiss chard	B. vulgaris Weed beet	1026
Helianthus annuus	H. annuus sunflower	H. annuus Weedy sunflower	117
Avena sterilis	Avena sativa Domestic oats	A. sterilis and A. fatua[b] Wild oats	251
Olea europaea subsp. europaea oleaster	Olea europaea subsp. europaea Olive	O. europaea subsp. europaea feral olive	151
	Vigna unguiculata Cowpea	V. unguiculata Feral cowpea (Thailand)	1105
Lolium perenne[c]	L. perenne[c] Perennial ryegrass	L. perenne[c]	238
Carthamus palaestinus	C. tinctorius	C. oxyacanthus and C. persicus	64
Brassica rapa	B. rapa Turnip, rutabaga, etc.	B. rapa Wild turnip	451
Setaria viridis	Setaria italica Foxtail millet	S. viridis Green foxtail S. viridis var. major[d] Giant green foxtail	252
Secale cereale ancestrale	Secale cereale cereale Cultivated rye	S. cereale Feral or weedy rye	

[a]S. bicolor bicolor × S. propinquum = S. halepense (johnson grass), an even worse weed.
[b]Now considered conspecific.
[c]Interfertile with L. rigidum (annual ryegrass) and L. multiflorum as a ryegrass complex that can cross with some Festuca spp.
[d]Setaria viridis X Setaria italica = S. viridis var. major (=S. viridis subsp. pycnocoma).

southern Mexico. Apparently, people living in the proximity of the Mexican Balsas River valley selected a proto-maize about nine thousand years ago and began to cultivate it. Just six major traits, each controlled by a single gene, explain half of the variance between scrawny, wild teosinte with its traditional seed head and massive maize with its tassels and cobs (Fig. 1).[283, 284] Similar

Figure 1. A few genes can cause major differences. Teosinte (*left*) (*Zea mays* ssp *mexicana*) and maize (*right*) differ in five or six genes. An ear of their F₁ hybrid is in the center. *Source:* Photo by John Doebley, by permission: www.wisc.edu/teosinte/images.htm

numbers account for the domestication of beans.[600] The selectors chose these major effects and soon had crops quite different from the wild type, all the time choosing what we now know as genetic modifier genes that brought further nuances of change.[384]

Some crops are hybrids between more than one species; the prime example being bread wheat, a crop that did not evolve in the wild. Common (bread) wheat (*Triticum aestivum*) probably was formed by natural hybridization in farmers' fields. It is a hexaploid (made of three genomes from three different progenitors, commonly denoted as genome BBAADD) that is thought to have been produced by two sequential hybridization events.[337] The first hybridization was probably between an unknown diploid donor of the B genome crossed with diploid wheat *Triticum urartu* (genome AA), giving rise to tetraploid wheat *Triticum turgidum* (genome BBAA) (Fig. 2). Then, a domesticated form of *T. turgidum* similar to today's durum wheat migrated with the domesticators to the native habitat of wild diploid *Aegilops tauschii* (genome DD) to create hexaploid *T. aestivum* (Fig. 2).

In contrast, cultivated oats is also a hexaploid, but is conspecific to (i.e., the same species as) its weedy progenitor.[251] This crop was domesticated in

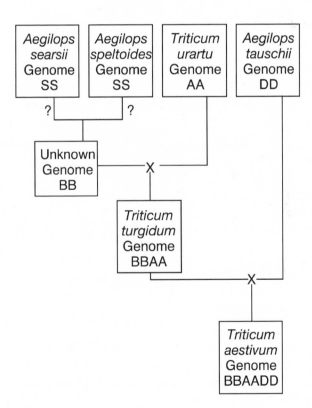

Figure 2. The presumed origins of hexaploid wheat, a species not existing in the wild. *Source:* Redrawn and condensed from Weissmann et al.[1139]

Europe two to three thousand years ago from the wild-weedy hexaploid *Avena sterilis* brought as a weed with wheat and barley from the Fertile Crescent. *A. sterilis* occurs wild in herbaceous plant communities and as a weed in the Middle East and throughout the Mediterranean basin, but is progressively replaced by another hexaploid weedy species, *Avena fatua*. *A. fatua* is confined to crops and could be considered as a specialized ecotype of *A. sterilis* or as a derivative of the crop.[251]

2.1.1. The Dedomestication of Crops as a Source of Weeds

Not all weeds come from the wild, as evident in Table 3. Some weeds probably evolved from the crops themselves by dedomestication, with the crops becoming feral, that is, evolving some but not necessarily all weedy or wild characteristics.[38, 1133] The differences between crop and weed are minimal (Table 2), and most of the domestication traits are recessively inherited in the crop. Thus, frequent back, dominant mutations can quickly confer some of the wild-type or weedy traits on the crop, allowing it to become feral. The ecological relationships between weed, crop, and wild species are summarized in Fig. 3. Ferality is quantitative, the more such genes accumulate, the weedier or wilder the progeny, depending on the ecosystem where the dedomesticating plant that is becoming feral resides. When the crop becomes a feral weed using only back mutations or variability within the crop, the process is referred to as endo-ferality.[415] At times, crops dedomesticate by hybridizing with their wild progenitors or with a wild or weedy relative to become "exoferal" weeds. The many rather different examples of these ferality processes are described in a recent book.[417]

Thus, ferality can affect biodiversity, especially transgene-enhanced biodiversity in the crop, in the realm of biosafety (Chapter 4). One does not want transgenes moving with impunity into feral forms.

2.2. Weeds—Codomesticated Camp Followers

Before there were crops, there were no weeds in the sense used in this book. Many argue over the definition of weeds, but the two major treatises on the worst weeds eschew definitions based on the concept that "you know a weed when you see one."[508,509] Here weeds are simply defined as plants competing with the crop being cultivated. Thus, if the crop drops seed to the ground before harvest, these seeds can give rise to "volunteer" weeds in the following crop.

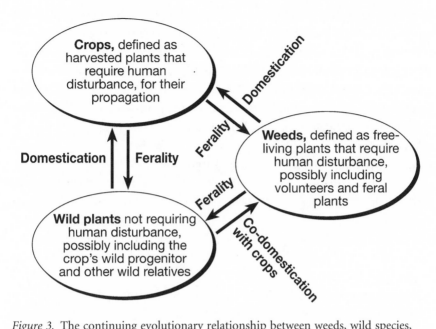

Figure 3. The continuing evolutionary relationship between weeds, wild species, and crops. Feral plants are partially dedomesticated crop plants that have evolved or introgressed some but not all traits of wild or weedy relatives and can exist without human intervention. Weeds as well as feral and wild species can be found in the somewhat human-disturbed but uncropped ruderal areas between the cultivated crop field and the more pristine undisturbed wild ecosystems. *Source:* Modified and redrawn from Warwick and Stewart.[1133]

As the gatherers became domesticating farmers in prehistory, propagules of some wild plants found fertile soil in the fields being tilled by the early farmers. In some cases volunteer crop seed may have gone feral, with genetic exchange among the wild, cultivated, and weedy forms (Fig. 3, Table 3). Most of the weeds that took hold were *not* the same genus and species as the crops. The early weeds further evolved in ways that mimicked the crop: similar phenologies (flowering and seed set at the same time as the crop or just before the crop) that allowed some of the seed to be spread with crop seed, similar seed size, shape, and color that made it harder for the farmer to differentiate between crop seed and weed seed; evolving similar seedling characters that made it harder for the farmer to remove the weeds.[92,401,656]

Some of the weeds became global, following the crop as contamination of crop seed. As selectors and breeders modified the crop to fit new climes, many of the weeds evolved, following suit. The selectors and breeders continually tried to obtain crops that were taller than weeds; here too the weeds followed suit. This trend was broken by the advent of selective herbicides and the sagacity of scientists who bucked the trend, that is, who brought about the green revolution (section 2.4). Thus, as many crops are quite different from their progenitors, so are the weeds different from their primeval forms.

In modern agriculture, some of the worst weeds are those that are closely related to crops. Modern herbicides that are selective between a crop and most of its weeds cannot typically distinguish between crop and a related weed, making a bad situation worse. Some grasses have extended their ability to mimic crops to the biochemical realm. They were initially susceptible to selective herbicides used for their control in cereal crops, but then they evolved resistance using metabolic pathways similar to that of the crop to detoxify the herbicides; examples are *Lolium* and *Alopecurus* in wheat, and *Echinochloa* in rice.[412]

Nothing is static in the evolution of weeds. Agronomic practices can change the spectrum of weeds in crops; for example, drainage and nitrogen fertilizer in wheat brought about a displacement of bog-loving species, along with those weed species that did not respond to nitrogen.[447] Many broad leaf weeds were eliminated from wheat fields by 2,4-dichlorophenoxyacetic acid (2,4-D). The ecological niches left by such practices were always filled by other weeds, or new evolution of plants from wild to weedy. The success of selective herbicides in the corn belt in the United States was recently shattered when a swamp-dwelling wild *Amaranthus rudis = A. tuberculatus*, previously unheard of by weed scientists but known to some botanists, emerged from being a wild swamp species in a sudden evolutionary swoop. It is not yet clear whether it did so by itself or by crossing with other *Amaranthus* species.

2.3. Breeding Following Mendel

The nineteenth century work of the monk Gregor Mendel on how traits were sexually inherited in crosses between plants was discovered at the beginning of the twentieth century, and modern breeding blossomed at universities, government research establishments, and seed companies. Yields began to soar, which was very important to agriculture as in the Northern Hemisphere, the industrial revolution was already in full swing. With farmers going to work in factories, farm sizes were growing, with fewer farmers having to produce more food for the growing urban populace. It was not breeding alone that allowed this. The Haber-Bosch process of fixing nitrogen from air to produce ammonia fertilizer allowed the augmentation of manure as the major mineral nitrogen source for plants. Plants were bred to be responsive to this fertilizer; plants lacking the internal genetic diversity to respond to fertilizer were left by the wayside, and breeders concentrated on crops that did respond. This intensified the erosion of crop biodiversity. They also bred taller crops, to better compete with weeds. Another major discovery was the finding that crosses between varieties can lead to hybrids that have greatly increased yields due to hybrid vigor (heterosis). These hybrids, especially in maize and some vegetable crops, must be performed every year because of genetic instability if replanted. This increases seed cost, but farmers happily bought the hybrid seed because of the enhanced yield.

2.4. The Green Revolution

As discussed above, the ancient selectors and the early twentieth century breeders both inadvertently and consciously bred for taller crops to compete with weeds. After World War II selective herbicides in grains obviated the need for height. Still, the impression that a correlation exists between height and yield continued, with devastating consequences in the developed world. The tall wheat and rice crops fertilized with the cheap nitrogen fertilizer that became available because of the Haber-Bosch process flopped over (lodged) at a hint of high wind, exacerbated by stem insect damage (section 9.2.1), leaving harder to harvest fields with inadequately dry, often moldy, grain. The breeders also knew that fields were near photosynthetic capacity. There is only so much sun, so the maximum yield per unit area is fixed. The only way then to get higher yield per unit area is to ensure that less photosynthate is partitioned into nongrain biomass. This could mean breeding for thinner straw or shorter straw, but considering the lodging problem, thinner was not an option. Shorter means more weed competition, but selective herbicides could cheaply deal with that, paid for with just a small part of the profit from the added grain.

Two ways were chosen to shorten cereals, genetically and chemically. Researchers in agrochemical companies were sure there must be some magical chemicals that would increase total yield, ignoring the limitations of photosynthesis. They found no chemicals that would reproducibly increase total biomass, but counterintuitively (at the time) noticed that plants with reduced height had larger yields of grain in weed-free environments. Most of these compounds inhibited the biosynthesis or action of gibberellic acid, a growth hormone that causes stems to elongate. Hormones are still used for that purpose in European cereal grain fields, as well as for shortening the stems of potted ornamental plants.

Norman Borlaug in wheat and other researchers in rice set out to genetically dwarf these cereal grains. They found dwarfing genes and bred them into the best, high-yield, disease-resistant, fertilizer-responsive backgrounds of the time. The task in wheat was especially onerous, as the chromosome carrying the dwarfing gene had yield-reducing genes closely linked to it. Crossing these away was not easy, as it requires rare chromosomal recombination (crossing over), meaning screening millions of plants in the field. The task was done and the varieties rapidly adopted by farmers in India and China. The tripled yield of green revolution crops led to food security in countries on the brink of war, which justified the awarding of the Nobel Peace Prize to Borlaug and colleagues. The success of the Green Revolution ran counter to the predictions of economists, sociologists, political scientists, agronomists, and the gurus from the pesticide and fertilizer industries. They were sure the populace would not be flexible enough to adopt, would not have the infrastructure, the desire or ability to pay, and on and on. It is surprising how the self-appointed experts on agriculture do not know farmers, an issue reappearing with the rapid adoption of transgenics by farmers, especially by small, resource-poor farmers, against predictions by a later generation of pseudoexperts.

The only ones not to utilize the genetic approach to increasing the harvest index are some Europeans. They prefer to lace their grain with the old dwarfing hormones to achieve the same results. The separation of the linked yield-reducing traits from dwarfing in wheat was too daunting in some European countries. Because weeds coevolved to be taller, mimicking the grain crops, the height has no advantage for weed control, and these European wheat farmers use more herbicides than most others, along with the hormones.

2.4.1. Is the Green Revolution "Problem Free"?

Nothing in agriculture is forever, other than continued evolution or extinction. The initial green revolution wheat and rice varieties with their huge yield increments were further improved by a continued breeding effort. As expected, the initial increments of yield increase were large and later per force slowly become lower. No matter how much you reshuffle the same gene pool, eventually a ceiling is reached. Initially governments subsidized fertilizers, the fuel to make them was cheap, and farmers used the maximum amounts to the detriment of logic and environment. As fertilizer prices rose, farmers rationalized to use less, and yields went down a bit, but greater profitability was attained. The yield going down in national statistics is the grist of many nay-sayers.

Other problems ensued. Much of the green revolution areas in India went into a rotation that was really two monocultures; paddy rice in the summer monsoons, irrigated wheat in winter. The paddying killed all winter-germinating weed seeds, except one species *Phalaris minor,* and a single herbicide was used for its control. After 15 continuous years of use, this weed inevitably evolved resistance,[429,683,987] forcing more logical rotations and agronomic procedures to deal with the resistance. These alternative procedures are more cumbersome, complex, and costly than a single herbicide, but agriculture rarely remains simple. The motto of many agricultural advisers is KISS—keep it simple stupid. Long-term sustainability requires that this abbreviation be redefined to keep it sophisticated, smarty.

In a similar vein, millions of hectares of Chinese rice have been overrun by *Echinochloa* that evolved resistance to the selective herbicides used in green revolution rice.[514] The solutions for both of these problems are in transgenic, herbicide-resistant crops, which will be successful for a period, as have other procedures developed in the ten-millennia history of agriculture. Evolution does not stop, but the rate could have been and can be modulated by more forethought on how procedures are used. Is/was the green revolution a failure? In balance, it was a resounding success by tripling yields and providing food security; it averted famine, wars, and death, fully justifying the Nobel Peace Prize to the legendary Norman Borlaug and his colleagues. It must be continued in new and different ways to keep up with evolution. The Red Queen must continue running.

2.5. Imbalanced Nutrition and Domestication

Conventional wisdom preaches that crops should provide a balanced nutrition and that first mutation breeding and now transgenics should be used to rectify what is missing in the major crops (Table 4). Conversely, many of the widely touted abandoned/underutilized crops are trumpeted as having balanced nutrition. Balanced nutrition is a necessity for insects, birds, and monogastric mammals such as humans, but not ruminants, that have bacteria that can bioconvert one amino acid to another, or create amino acids from carbohydrates and minerals. It was recognized long ago that wheat is nearly devoid of lysine in the seed storage proteins, whereas pea and bean proteins have nearly 5 percent of their amino acids as lysine in proteins.[827] Conversely beans are lacking in methionine and cysteine (Table 4). So how is it that each of our major crops is deficient in one or a few key nutrients? Were the human requirements ignored by human crop selectors/breeders? The question is especially cogent as the progenitors of the crops do typically have higher protein levels with better balanced amino acids, whereas the crops have less protein and imbalanced amino acids.

Were the selectors and breeders for the past ten thousand years uniformly stupid to select for nutritional imbalance? Were the few breeders who tried to improve nutritional quality in the twentieth century[782] off the mark? The breeders/selectors naturally went for higher yield, and higher yield naturally went for higher starch and fat contents, as it takes less photosynthate to produce starch than to produce protein. But why are amino acids typically imbalanced in domesticated crops? The answer may come from a contrarian view which posits that this is a direct result of domestication and that species with balanced amino acid content in their seeds are not domesticated, and will not be fully domesticated until an imbalance is achieved, or compensatory genes added.[751] Why was this imbalance part

Table 4. Imbalanced Nutrition in Grains

Crop	Imbalance
Maize	Low in tryptophan (and lysine)
Wheat	Low in tyrosine, leucine
Sorghum	Low in lysine
Rice	Low in leucine
Soybean	Low in sulfur amino acids[a]
Millet	Low in tyrosine, proline, methionine
Buckwheat	Sulfur amino acids,[a] threonine

Sources: Osborne,[827] Millward,[728] De Francischi et al.,[269] and Ejeta et al.[313]

[a]Sulfur amino acids: methionine, cysteine, and cystine.

of domestication? The early agriculturalists planted what they harvested and stored. They were able to harvest and store what was not eaten by insects, birds, and monogastric mammalian pests, that is, grain that had an inherent lack of nutrition for these animals, grain deficient in a critical amino acid. Is there any evidence to support this view that there was a directed (albeit not understood) selection for lack of a nutritional component? Oliver Nelson and colleagues spent decades selecting for maize having a higher lysine content, that is, was better balanced.[782] Their highly acclaimed success was maize that could be readily and selectively detected and decimated by rodents and insects (besides being lower yielding with shrunken grains), even when surrounded by conventional maize. This clearly supports the hypothesis that selection for pest tolerance resulted in nutritional imbalance. The pests preferred the nearby wild species in the past eras when farms covered a small part of the landscape.

Thus, in trangenically domesticating or further domesticating a new grain, there seem to be two (not mutually exclusive) choices of direction:

1. Transgenically cause an amino acid imbalance to mimic the domestication of other crops. This may seem antinutritional but one could envisage making different deletions in different crops, for example, a lysine-poor amaranth and a methionine-poor quinoa, and mixing the final grains to provide a balanced food.
2. Add compensating genes, genes that kill or deter the herbivores. This may seem daunting, because there are so many herbivores and possibly too many genes would be required, and herbivores continually evolve resistance. There is a problem with many such natural antinutritional factors; they often affect humans and livestock as well as pests, as will be discussed in later chapters.

A mixture of both strategies might be synergistic. Despite nutritional imbalances in our present grains, they still have pests, grain weevils, for example. A seed cannot be totally devoid of an amino acid, and thus a weevil can be inefficient, leaving much nutritious material behind. If there is sufficient humidity, microorganisms having enzymes not in herbivores, can further biotransform the spent grain to make the missing amino acids, providing additional food for the herbivores. This may explain the evolutionary phenomenon that grain insects are vectors of microorganisms that can transaminate the amino acids present in the grain to those that are missing.

2.6. Reaching the Genetic Glass Ceiling

Each species has its own genome, which in evolutionary time mutated, evolved, and recombined from other species to give the biodiversity now available within a crop. Our forefathers tasted, experimented, and were successful in domesticating a very small proportion of the plant kingdom. People still collect some food from the wild. Ethnobotanists have flocked to remote places to document what people eat that they gather and grow. Why can some species only be gathered, why have some species (e.g., wheat and maize) become very different from anything found in nature?

We must realize that a huge difference exists between evolutionary time and impatient-human time. Species evolution takes time, and as we know, most species stop evolving and at some point go extinct. Let us take an example from recent history. Roman miners in Wales contaminated pastures with heavy-metal-rich, toxic mine tailings. Of the species found on a pristine pasture, only about half had the genes to reshuffle to live on the nearby toxic mine tailings. Did the successful species add genes to their genomes that they did not have before? Probably not. They had genes that were always mutating and/or recombining such that a very minute part of the population had a combination of genes that could exist on mine tailings two thousand years ago. Without the genes in their inherent variability, they could not evolve to live on tailings. This constellation carried a "fitness" penalty on pristine ground, which kept resistant individuals at a low frequency in the population, except on the tailings, where other species and biotypes were poisoned.[144] In more recent times, young, only decades old tailings, far away in England carried all of the same species that are found on the distant Roman tailings. Both the pristine Welsh and the English pastures contained rare, metal-resistant individuals of the same species within their populations. There was no evolution of new genes in these species. It can take millions of years, if at all, to evolve new genes not found within the biodiversity of the genome of the species. The species that seem to evolve in human time are those that have the genes, previously unexpressed or at a very low frequency within their gene pool.

This implies that all crops can be domesticated to the extent that their pre-existing genes allow. Some have a broad spectrum of genes that allowed them to adapt to conditions around the world, for example, wheat from the subtropical fertile crescent, to evolve as spring or winter wheat in far more frigid climes, or maize from the American tropics to temperate areas. Other crops seem more localized, with an inability to adapt, no matter how the breeders try. Thus each crop has its own glass ceiling to advancement. Evolution in human time frames is far less flexible than management, where the "Peter Principle" is rampant.[845] The Peter principle states that workers are eventually typically promoted to one level above their inherent competence. Plants in nature go extinct when they need competences they do not possess, unlike human managers who get promoted.

Thus, the only way for a crop that has reached its glass ceiling to succeed further afield is to obtain genetic assistance from elsewhere. Some gathered species have a glass ceiling so low that no one has succeeded in cultivating them. *Dioscorea deltoidea,* described in Chapter 20, is a source of steroid precursors for the estrogens in birth control pills, and related compounds. This yam has not been domesticated despite its value, but it is not as if no one has tried.

Four ways to breach the genetic glass ceilings are: (1) Recombining with progenitors that have the genes (vertical gene transfer). Domestication has in general resulted in a loss of genetic diversity compared with the wild progenitor gene pool.[628] Going back to that wild gene pool to obtain needed genes can be futile if the genes needed for further evolution are not there. (2) Recombining with related but poorly interbreeding species that have the needed genes (diagonal gene transfer). Besides the possibility that the genes are not there, this is fraught with genetic complications, bringing deleterious genes with the needed ones (section 2.8). (3) Evolving the genes needed from other genes it already has (in million to billion year time frames). There is a strong possibility of going extinct before this happens. (4) Having the gene jockeys (genetic engineers) find the needed genes from wherever they can, and put them in the needy crop. The advantages of genetic engineering over waiting billions of years for internal evolution, or over interspecific or intergeneric crosses, is described in the following chapters of this book.

2.6.1. The Genetic Glass Ceiling Is Not the Only Reason for Crops to Be Forsaken

Barley is a case in point of a crop that is far from reaching its potential as a major feed grain, yet it has been forsaken except for brewing. Barley is a much more robust crop than its cousin wheat; it is far better competitor with weeds, its grain is of better quality for animal feeding than wheat, it is more disease and insect tolerant, and it has a modicum of salt tolerance. Indeed, one of its salt-tolerance genes has been isolated and transformed into oats, giving oats greater salt tolerance. More than a quarter of the wheat crop ends up in animal feed, and barley would be a better alternative for much of it. A half a century ago barley and wheat had similar yields on marginal lands. Considerable breeding efforts have increased the yields of wheat. Had barley reached its genetic glass ceiling fifty years ago? Most breeders doubt this. Wheat breeding continued intensively over this half century, but except for breeding better malting barleys for beer and whiskey (types with less and less protein, but enriched in specific starches), barley breeding stopped in comparison with the efforts with wheat. Most breeders believe that had more effort been put into barley breeding, the yields could still equal wheat, and the concomitant benefits would be there.

This lack of breeding effort is largely the result of ill-advised political meddling in agriculture. The cultivation of wheat was subsidized by the politicians; the cultivation of barley was not. Even if crop yields were similar, the easier to grow barley, with the somewhat better product for feeding animals, was at a vast economic disadvantage. Why breed a crop farmers will not grow? One advantage of a market economy without subsidies will be that the breeding effort might be exerted to render barley competitive with wheat, again. Molecular techniques such as marker-assisted breeding might help bridge the gap. In analyzing barley, no special instances could be found where genetic engineering traits would be imperative for barley used as a feed grain. Breeders should be able to successfully handle most of the further domestication of this crop. The only special need of barley may be transgenic herbicide resistance, so that related *Hordeum* and other grass species may be controlled.

The artificial economics of agricultural subsidies may have caused other crops to be forsaken, as well. Conversely, subsidies have kept other crops such as sugar beets from going by the wayside, because sugar beets are not economically competitive with sugar cane. Changing consumer tastes may cause crop abandonment as well. Unless these fashions can be attributed to specific genetic impairments (e.g., bitter flavor, allergies, texture), little rhyme or reason exists for the breeder or the genetic engineer to become involved.

2.7 Why Trangenics?-"Mutagenesis Is Fine"

The question mark in the preceding title is correctly placed. Some tell us that transgenics are unnecessary because we can use mutagens to increase the biodiversity necessary for our crops. Can we really?

Mutagenesis has been a useful yet very clumsy tool, but the utility that was proven in the past is misunderstood and incorrectly extrapolated to the future. Mutagens do not enhance genetic diversity, they increase the frequency of mutations already in the population, typically about tenfold. Thus, the haystack seems tenfold smaller, but the needle does not change. If you need a safety pin instead of a needle, and it is not in the proverbial haystack, you will not find it by reducing the size of the haystack because it was not there in the first place.

Mutagens are clumsy because they mutate the gene or few genes we want, but randomly mutate all the others in the crop, and genes that need not be mutated are the vast majority. Indeed most mutations are lethal, and to increase the mutation rate tenfold, the researcher typically applies a calibrated amount of chemical mutagen, UV light, gamma irradiation, and so on, to kill most of the population, and the nonlethal, but usually sickly mutations are found in the rest. Many remaining progeny will bear more than one mutation and one does not know whether poor growth is a function of the unfitness of the desired mutation or is subsidiary due to other mutations until they are separated by segregation breeding. If the mutations are near each other on the same chromosome, they may not easily segregate. If the required mutation is recessive, then the first-generation progeny will not bear the mutated phenotype, and the researcher must slowly select among large numbers of selfed progeny. With transgenics, the gene of choice is dominant and is typically inserted, linked to an easy-to-pick-out selectable marker as will be discussed later in further detail. This makes it much easier to find the elusive needle.

Care must be taken when working with mutagens, most are toxic and/or carcinogenic to humans (although the plant progeny are not). It is fascinating that many of those *against* transgenics per se as a process can be *for* mutagenesis,[635] in this case, ignoring the process. It is typical that dangerous and cumbersome processes that have been around for a few generations are feared much less than safer, more precise, yet novel processes. Just because this is human nature, and human nature can be irrational, does not mean that we must accept it, as some would like.

Mutagenesis has limitations even when there is a desire for a mutant of an endogenous gene existing in the crop. Assume that a mutation is needed to change a single amino acid A to a preferred one B in an enzyme to obtain a needed trait. In some cases, no mutagen can do this in human time: when two or more bases in the triplet codon must be changed, the possibilities of mutating two or three bases in the same codon are statistically remote (an understatement). Different mutagens cause different base substitutions, and if the wrong one is chosen, even a desired single-substitution mutation may not be increased in frequency. An additional problem with mutagenesis lies in the inability to usually achieve tissue-specific modulation of gene expression, a factor often needed. For example, a plant can make an insecticidal compound that is very useful in the leaves, but alas the same compound could be toxic to people eating the fruit. It is more likely that if one finds a mutant with low toxin expression in fruit, insects will devour its leaves as the toxin will not be there. Transgenics allow one to be more precise; tissue-specific promoters can be added to the transgene of choice to obtain expression or suppression when and where it is desired. In the preceding case, an RNAi or antisense construct can be made with a tissue-specific promoter to suppress expression of the toxin only in the fruit, as is discussed in more detail in Chapter 3.

Transgenic techniques allow replacing the target gene completely, with more than one base change in a codon triplet, and with many amino acid changes. Another engineering technique chimeraplasty allows surgically replacing specific codons in the endogenous gene to give rise to any desired amino acid. This is the process presently used in human gene therapy, but it is applied only to somatic cells, not those that will be gametes for sexual reproduction. This process has been applied to plant cells that become gametes,[1198] but is still highly inefficient in plants. This is exquisite genetic engineering, but the resultant plant is not transgenic by most definitions, because no outside DNA is present, as the carrier sequences that perform the transversion disappear and cannot be found. This is akin to finding the corpse of a person who had been murdered with an icicle; the weapon required for conviction has disappeared.

The most important point is that mutagenesis will not create genes that are absent in the crop (or are absent in the relatives that will hybridize with the crop). Bringing in new genes, or creating totally new genes can only be performed by the recombinant DNA techniques of genetic engineering. Thus, when crops need to breach a genetic glass ceiling because they have reached the end of their genetic tether with the endogenous alleles, mutagenesis will be of little avail.

2.8 Future Domestication—Toward a Continuing Decrease in Human Intervention during Cultivation

In many respects domestication can be seen as the selection of traits that require less human intervention in the culturing of a crop. For example, before the ancient selection for nonshattering, many cycles of harvesting were required, and in modern times selection for disease or insect resistance reduced the necessity for fungicide and insecticide sprays. There was often a balance: the selector selected for lack of off-flavors or toxicants, often due to compounds controlling diseases and pests, necessitating more pesticides. Dwarfing provided higher grain yield at the expense of long stems, but necessitated more herbicide use because the crops were less competitive. Apparently, with many crops, we have reached the end of the genetic tether; the genes are no longer available within the crop genome to deal with the constraints that nature continues to strew in its path. Thus, further domestication for insect and disease resistances is being achieved by transgenically finding the needed genes wherever they may be, and then inserting them into crops. Thus, the use of the bacterial (Bt) endotoxin gene in cotton reduces the need for nine or ten of the approximately twelve insecticide sprays previously applied. Newer Bt cotton containing more than one toxin gene may require even less human intervention.

Unfortunately, this type of continued domestication by transgenic means is being applied in a nonuniform manner to current crops. The genes are targeted to higher-value cash crops, in the developed countries, with developing countries getting mainly "hand me down" genes (e.g., Bt genes for cotton, rice, and maize, already being used, and little else). A recent study by sub-Saharan African agronomists and plant protection specialists pinpointed the constraints considered intractable to breeding in their commonly cultivated crops. Only a few key issues were being addressed by biotechnology in a limited manner (the parasitic weed *Striga* and stem borers in maize, but not sorghum and millet), and some constraints were not being dealt with at all (storage weevils and mycotoxins in grains and seed legumes).[431] Thus, there is ample room for transgenic domestication to continue in our currently cultivated crops.

One can posit that many crops were fully or partly abandoned because they arrived at their genetic limits earlier than the crops remaining in cultivation. Some of the species that have fallen into disrepute may require more transgenic intervention than others; probably those abandoned first are those needing the most help.

Whereas traditional selection and breeding are often for ill-defined or multicomponent traits, such as yield, quality, and storage, the use of transgenes requires a precise definition of the trait to ascertain the genes needed, even if more than one gene. Still, transgenic traits are typically dominant, whether gain of function, or RNAi-enforced loss of function, versus the recessiveness of most domestication traits selected for. The future, with genomics and whole-genome sequencing, together with bioinformatics, will probably soon allow transgenically picking and modifying polygenic traits.

2.9. Precision Design and Engineering versus Gross Recombination

2.9.1. Breeding, the Blunt, Crude Tool

Even though the wild progenitors of our crops typically have a broader genetic base with greater diversity than the crops, they should not be expected to have all the genes needed for further domestication. They will not have the herbicide resistance genes that provide selectivity between the crop and related weeds. Many diseases and insects that affect the crop also affect the progenitor species, but the related species can yield less and still replace themselves. They do not have to be high yielding for the farmer and unblemished for the market. The progenitors and related species may have more genetic diversity than the crops, but they too have a glass ceiling.

Many of the underdomesticated crops that have gone by the wayside suffer from the same problems: seed shatter, lodging, and secondary dormancy of seed. The reciprocal of these traits can be introduced by breeding, but in most cases in most species the mutations leading to them are recessive and occur at much lower frequencies in the population than dominant mutations. If a dominant mutation is found in one in a million plants, a recessive trait is there in one in a million millions, that is, one in a trillion. Our ancestors were pretty good at selecting for such rare mutants. These low rates are in diploid organisms. The frequency of a recessive trait can be exponentially lower in polyploid crops, if the genetic redundancy of a trait is expressed on each set of the chromosomes. We are told that polyploid crops have greater genetic diversity, especially the allopolyploid crops where the similar but identical chromosome sets come from different progenitors. This is true for dominant traits, but the diversity is hard to catch with recessive traits.

Going back to the wild progenitor or using wide crosses to diagonally transfer needed traits into a crop from distant relatives that rarely sexually intermingle with the crop, and bear mainly defective offspring, can entail great effort and brings with it great risks. The further the cross, the greater the fertility barriers, and the more laborious the effort. The related species with the desired trait often does not cross with the crop but does cross with another species that crosses with the crop. Thus, the transfer of a desired trait necessitates the use of such "bridge" species (or two, or three bridges). Often when hybrids are achieved, the embryos abort or the seeds do not germinate normally. This requires intervention through tissue culture (embryo rescue) with the hope of obtaining a wimpy plant with some fertile pollen to backcross to the crop. Perhaps another few backcross generations are required through embryo rescue. Problems due to different chromosome numbers, to chromosomal segment inversions, or to translocations to other chromosomes occur in many of these crosses between related species or genera.

Most interspecific hybridizations even within a genus do not work. Generations of researchers tried to cross black nightshade (*Solanum nigrum*) with potatoes (*Solanum tuberosum*) to transfer disease resistance, to no avail, because hybrids never formed. Their work was repeated in part by those fearful of transgene flow, who set out to try again, with no success.[310] In this case, the incompatibility between an old world and new world species in the same genus is at the nuclear level, as the plastome (plastid genome) can be transferred from one to the other by protoplast fusion after inactivation of the *S. nigrum* nuclear genome (cybridization).[425]

Even success with such interspecific and intergeneric crosses can have its limits. In exquisite pioneering work, Rick[897] brought numerous wild relatives of tomatoes from the wilds of South America to breeders' plots, and many wild traits were introduced, despite the excruciating breeding effort to rid the progeny of poisons and ill flavor and to return to progeny resembling tomatoes. Tomato breeders now feel they are reaching the genetic ceiling with this approach and must go elsewhere for the resistances to viruses that they need, and to find the genes to deal with parasitic *Orobanche* weed species and the *Bemesia* whitefly. Other tomato breeders are still convinced that recombining the limited gene pools of the crops and their interbreeding relatives ad nauseum will miraculously overcome all barriers. Surely they will need even more genes from outside the available gene pool, as pests and diseases seem to flock to this important crop. At least with genetic engineering you have a larger gene pool to select from, and finding the needed genes, while often daunting, becomes easier every day, unlike breeding, which becomes harder if you go further afield to find the genes in distant relatives even more reluctant to cross with the crop.

2.9.2. Precision Engineering

In this era where instant gratification is desirous, the time-consuming, risky, gargantuan efforts of the breeder to bring in novel traits through vertical and diagonal genetic breeding are increasingly been considered an anachronism. The molecular tools available (see Chapter 3) allow the breeder cum genetic engineer to specifically choose the genes and splice them with the elements that will control in which tissue, under what circumstances, and how much a gene will be expressed. This is done without introducing all the extraneous genetic baggage brought by crossing with the related species. Genetic engineering is like getting a spouse without in-laws, whereas breeding is like getting a spouse with a whole village. There is still considerable randomness (rapidly changing to become more refined) about where the gene construct will be inserted in the genome, with resulting position effects, necessitating generating many transformants and then screening for the best. Site-specific insertion methods for plants are becoming available,[967] lowering the required numbers.

Thus, genetic engineering allows the insertion of useful genes from anywhere in the living world without restrictions to relatedness, with the genes molded or remodeled and modulated as desired, without the baggage of extraneous genetic elements that must be bred out with repeated backcrossing.

2.10. The Rise and Fall of Crops

Can or should we preserve all abandoned crops just for the sake of crop biodiversity? Possibly, but only if we can overcome the reasons for their loss of acceptance. People stop eating foods they do not and never liked, but had to eat to stave off famine. There is no reason to cultivate a crop if it has no raison d'etre, that is, if it has no use, or market, or is too expensive compared with similar crops. If those issues can be overcome, the abandoned crops can be reconsidered.

Some major crops were abandoned in the twentieth century, and it is worth understanding why by looking at these four, nonexhaustive examples.

1. Oats once constituted about 20 percent of the grain area. This was in the days when the energy for agricultural production was produced on the farm. The horses had to be fed oats and the multitudes of workers to drive them were fed oatmeal porridge, on the farm. The need for this crop has disappeared.

2. Rye was grown for food and feed, as a diluent of wheat flour, lacking in gluten, so breads made with too much rye did not rise. Rye was not as responsive to nitrogen fertilizer as wheat, and the cheap fertilizer saw rye cultivation plummet. Even rye (=Canadian) whiskey is now made without much, if any, rye, and perhaps should be called wry whiskey. Thus, until it is known how to engineer fertilizer responsiveness, rye will remain forgotten, except in very poor areas.

3. Sugar beets are a crop requiring extensive cultivation and pesticide use. It is economically sustainable only against much cheaper-to-grow sugar cane because of the high subsidies offered by governments. This is yet another proof that government intervention in the free market is not agroecologically sustainable. Cane sugar has lost markets to high-fructose maize syrup, which has taken over much of the soft drink market. Much cane sugar is now used to make ethanol as a motor fuel. High-fructose sweetener is made by first digesting maize starch to glucose with transgenic amylase, and then transforming the glucose to fructose with a transgenically produced fructose isomerase. Per sweetening unit, the high-fructose maize syrup is cheaper than cane sugar, in part because of U.S. subsidies of maize production.

4. Safflower and other older, less productive sources of oil are being replaced by more efficient oil crops, and now any oil composition can be mimicked by transforming soybean, oil-palm, or oilseed rape with the requisite genes.

Let us analyze three crops that have expanded:

1. Wheat has expanded at the expense of most other small grains. Populations who not too many generations ago received much of their caloric sustenance from grain and tuber porridges and gruels, now

receive it from bread. Urbanization and rationalization of women's time from cooking and preparing food has resulted in this change. Lord Sandwich's invention has become a part of life, and bread preparation is mostly industrial. The adoption of even bad bread (Anglo-American sponge type) has become a status symbol of industrialization in the developing world, even in traditionally rice-eating nations.

2. Maize is the most efficient source of starch and protein for all non-bread products, the feedstock for sweeteners, and a major source of animal feed, especially with nonruminants. It has become the main nutrient source for the rural poor of Africa, heavily supplanting indigenous sorghum where water is sufficient. The yields are higher than sorghum, and the taste preferred. Many Africans think of maize as an African species, and in many respects it is. The white kernel, tropical maize varieties are resistant to indigenous fungal and viral diseases not found elsewhere (but the maize is not resistant to the parasitic weed *Striga* or to stem borers), but neither is sorghum.

3. Soybeans have become the major source of edible oils even though they are produced for the meal needed to augment cereal grains for monogastric animals, because soy contains lysine. While lower yielding than oil palm, the oil is of healthier quality and the meal is of great value as a protein source for animals. This nitrogen-fixing legume is nonresponsive to nitrogen fertilizer and does partially replenish soil fertility. It is typically rotated with maize (as one third of the rotation) in the developed world. One can only reflect on the consequences to agriculture if soybeans were not available for such rotations and more farmers cultivated monoculture maize.

The preceding examples, and others presented elsewhere, suggest that we must be cognizant of globalization, economics, agroecosystems, and consumer-driven fashions when we call for the reintroduction of a species.

We must be leery of reintroducing species that are inefficient under the banner of crop biodiversity as this will require putting more land under the plow. That could be at the expense of other ecosystems and biodiversities that we wish to preserve. It may be possible and desirable to reintroduce some of these desirous but less productive species in the latter half of this century, when demographers finally expect a contraction in the human population. There will be more limited resources available at that time due to present wastefulness, so lower yielding crops that also require lower inputs may then the desirable. Until then, we must put our minds to preventing famine among the expected multitudes. A large proportion of these famines is expected in Africa, which must deal with its own future (following section). African crop yields are one third to one half of world averages and, by being so low, lower the world averages. Bringing Africa to world average yields by dealing with the intractable constraints to crops that have reached their genetic ceiling on that continent[431] can easily supply the necessary food.

Thus, many factors must be considered before deciding if and how to transgenically raise a crop above its present genetic glass ceiling; these factors are discussed with each of the cases studied and should be considered at all times. Times will change, and today's "nay" may change to yes at some future date.

2.11. The Effect of the Price of Petroleum on Domestication

One need not be an economist to realize the rapid changes being wrought on agriculture because of the high fuel prices. A whole new economy is rapidly developing, with huge public and private investment, with promises of rapid profits, now that the economists are assuring that oil prices will not drop below fifty dollars a barrel. Fifty dollars is the magic threshold number that renders the technologies of turning quality grain into ethanol, and food oils to biodiesel profitable in the United States, where people pay only half for automotive fuels of what most of the developed world pays. The biofuel industry has hardly targeted waste substrates such as straw and stover; at fifty dollars a barrel they are happy to use quality grain, taking the food out of peoples' mouths. This is not just excess grain, this is whatever the market will bear in competition with petroleum, and has led to higher grain prices around the world.

This huge investment in factories to quickly reap a bonanza will clearly stabilize the bottom price for grain at a much higher price than at present. Subsidies will no longer be needed in the West, and the farmers in developing world will no longer have to compete with grain sold ("dumped" in economic jargon) below the actual production costs. However, the farmers in the developing world will have to gear up to production, instead of subsistence. Soon, stocks will no longer be available for famine relief in times of need as long as oil is more than fifty dollars a barrel. All excess grain going to biofuel production will only make a small dent in the total fuel needs of the West and the growing fuel needs in Asia. The magic fifty dollars a barrel also renders nuclear energy a viable alternative for much of the fuel, but it takes nearly a decade to build a nuclear power plant, and that is after the decision is made to build one. Such decisions are not fast in coming, and other alternative energy sources (e.g., wind, solar) cannot match the magnitude of the shortfall, no matter how appealing.

The only viable take-home message from this is that the developing world, especially Africa with its huge perennial food shortfalls, must quickly prepare itself to go it alone vis-à-vis its food security. The question "should we accept transgenic maize as food aid?" will be moot in a very short time, because such maize will no longer be available, it will be powering an automobile. Poor countries with sporadic shortfalls must quickly come to the realization that they must rapidly go from subsistence agriculture, with yields a third of global averages, to productive agriculture to feed their citizens. This can only come about by having good seed bred and available, fertilizer available at near international prices, and not an unjustifiable four times those prices. Extension services must get to the farmers and teach them the most sustainable, cost-effective practices. An infrastructure with good storage facilities is critical to ensure storage for times of need, as well as an equitable price to the farmer. If India could get such a storage infrastructure going decades ago, Africa has few excuses for not doing, as it will have to overcome its dependency on foreign grain.

The key needs described above started with good seed, not with the long ago discredited "farmer-saved seed." Seed deteriorates season after season in most farmers' hands. Only the very few best hands grow "certified" seed. The good seed must be of more crop species than presently grown to reduce risk, and it must be adapted to local conditions. It should come with as many built-in resistances as possible; resistance to abiotic stresses, high fertilizer use efficiency, resistance to local insect, rodent, and avian pests during cultivation and storage, resistance to indigenous diseases and the debilitating mycotoxins their pathogens produce, along with resistance to that scourge of much of Africa, the parasitic witchweeds (*Striga* spp.). Good breeding can surely help, but where decades of breeding have proven ineffectual, the biotechnology sector must be utilized. Biotechnology must deal with more species than maize, as crop biodiversity is also an essential element of food security. Biotech priorities should not be haphazard but based on evaluations of need.[431]

Biotechnology will play an important role in food security in the developing world, a role that will be useless if the other institutional and infrastructural issues are not addressed. And they must be dealt with quickly, as biofuel plants are quickly coming on line, sucking up the grain that went out as food aid. Politicians in the developing world may have thought that their countries do not need to produce and store grain for winter, but winter is on the way, even in the tropics.

2.12. Transgenic Organic Agriculture—Back to the Future (by Klaus Ammann)

Organic farming is a heterogeneous management method in agriculture having a multiplicity of origins. Certification of organic farming practices with follow-up inspection has been introduced over the past decades in many different places with different systems and requirements. Organic farming is rapidly moving from the realm of backward-thinking Luddites to becoming a veritable industry. Regulation has been imposed more or less strictly on all organic farms from local jurisdictions such as California[445] to whole countries such as Switzerland or the United States. Top-down regulation requires coming to terms with standards, including those of conventional agriculture such as defining levels of toxicity for biopesticides, which is often not easy. The main rules for the Swiss organic agriculture for plants follow.

- Natural cycles and processes are to be respected.
- The use of chemical substances is to be avoided.
- The use of transgenics or products derived from transgenics is not allowed, except in veterinary medicine.
- The products will not be treated with radiation, and no products having undergone irradiation will be used.

All agricultural systems must provide an economic return to the farmer; unprofitable agricultural systems will not continue unless they are heavily subsidized as in the United States or in Europe, which is problematic in the long run. Today's farming systems must produce more food on smaller areas. Efforts to maintain and enhance output, such as improving soil fertility and reducing losses to weeds and pests, are imperative. It is less easy to argue that a natural or diverse ecosystem is a critical input to a sustainable agricultural system. Although ecologists frequently stress the interrelationships between species, it is difficult to see how the existence of species such as the swallow tail butterfly or a rare orchid could contribute to the sustainability of a farming system.[1120] The degree of redundancy in ecological communities is largely unknown and remains a rich field for investigation. Agricultural systems can benefit from more biodiversity but these do not necessarily have to be in the production field. Biological networks hosting highly diverse arthropod populations can be near the production fields, which may make the whole region more resistant to pest invasions, or conversely, be a reservoir of pests.[783]

This is not to say that agriculture could continue in the absence of all non-farmed species. Rather, it has been suggested that only a subset of all existing species are essential for food and fiber production within the field.[309,1119]

2.12.1. Is Organic Farming More Sustainable than Conventional?

A thoughtful study concluded that organic farming systems are not sustainable in the strictest sense.[309] Considerable amounts of energy are put into organic farming systems, but most of the compounds utilized in the protection of conventional farming crops are derived from nonrenewable sources and incur processing and transport costs prior to application. Nevertheless, the long-term balance of input clearly favors organic farming systems,[675] except when compared with conventional no-till agriculture using herbicides. Nutrient inputs (N, P, K) in the organic systems are one third to one half lower than in the conventional systems, but crop yields have been only 20 percent lower over two decades, indicating an imbalance in efficient production, or an overuse of fertilizer in conventional systems, or both. The energy to produce a unit of crop dry matter was claimed to be 20 to 56 percent lower than in conventional or 36 to 53 percent lower per unit of land area [771] in one study, but this depends on the case. No-till conventional farming using herbicides uses far less nonrenewable energy than heavily tilled organic systems, with far more damage to the soil in the organic systems because of mechanical compaction and erosion, and organic agriculture should consider how to adopt such procedures.

Many of the "biopesticide" compounds used in organic agriculture are not without toxicological hazards to ecology or humans. For example, several research groups are working on the difficult question of how to avoid the input of copper sulfate as a fungicide in organic agriculture. It is clear from some studies, that copper deposited in high concentrations has a negative impact on soil microbes. Total microarthropod abundance was highest at intermediate copper concentrations, and was linearly related to grass biomass.[842] The continuous use of high copper concentrations has led to copper concentrations that are toxic to plants.[841] The lack of good organic fungicides has led to problems of mycotoxins in organic produce that are well above the allowable thresholds. When sustainability includes landscape quality as viewed in Europe, organic farming shows a positive influence,[226] but the results need to be verified elsewhere.

2.12.2. Is Organic Farming Better for Biodiversity?

Many studies demonstrate that organic farming has a definite advantage regarding biodiversity compared with traditional agriculture. An extensive review[506] of many field studies presents a wealth of evidence that agricultural intensification is the principal cause of the widespread decline in European farmland bird populations due to habitat loss and reductions in abundance and diversity of a host of plant and invertebrate taxa.[289,506,1156] Only a few studies have sought to integrate the changes in soil conditions, biodiversity, and socioeconomic welfare with the conversion from conventional to organic production.[227] The conclusions may not be representative for all organic conversions, but the findings are relevant to the debate over changing patterns of subsidies and other incentives in agricultural policy. The study demonstrated the differences between organic and nonorganic farming, showing evidence of increased species diversity and an eventual improvement in the profitability of the organic farming regime. Variations in farm management practices strongly influenced the on-farm and off-farm environmental consequences.[227] Similar positive effects of organic farming were seen in a two-decade study in Switzerland; root colonization by mycorrhizae in organic farming systems was 40 percent higher than in conventional systems.[355] Biomass and abundance of earthworms were higher by a factor of up to three in the organic plots as compared with conventional.[847] This all comes with a 20 percent yield loss compared with traditional farming. This triggers a debate about whether such a drop in yield is tolerable in the view of the protection of biodiversity, because it is imperative to produce more food on a shrinking amount of arable land.[395,675,771] Potato yields in the organic systems were 40 percent less than those in conventional plots, mainly because of low potassium supply and the incidence of *Phytophtora infestans*. Winter-wheat yields in the third year of a crop rotation were only 10 percent more in conventional systems. In an overall comparison, when energy input is lower, one can theoretically conclude that in some European conditions organic farming can be the more efficient production strategy. This is not the case when the conventional systems use reduced tillage with herbicides.

2.12.3. Should Future Organic Farming Include Transgenic Crops?

Organic farming has its most convincing answers as a niche structure with grassroots dynamics, with a diversity of management methods. The dilemma is clear: mainstream organic farming must go through a transition achieving more efficient production methods while maintaining its diversity of production methods on a local scale. In the past few years organic farming strategists have finally realized that breeding is of utmost importance to reach such goals. Organic farming systems aim at resilience and buffering capacity in the farm ecosystem by stimulating internal self-regulation through functional agrobiodiversity in and above the soil, instead of external regulation through chemical protectants. New varieties are required that are adapted to organic farming systems to further optimize organic product quality and yield stability. The desired variety traits include adaptation to organic soil fertility management, implying low(er) and organic inputs, a better root system, and ability to interact with beneficial soil microorganisms, ability to suppress weeds, contributing to soil, crop, and seed health, good product quality, high-yield level, and high-yield stability.[1092,1093]

Unfortunately, strictly excluding transgenics and their by-products has been an important part of an organic farming marketing strategy. Having this shopping list for breeding in mind, it would be wise to open the toolbox of breeding to the most modern methods. Although marker-gene-assisted breeding is accepted by many organic farmers, genetic engineering is excluded by almost everybody in this scene, except a few.[1034] The arguments against transgenics in organic farming are not convincing because they do not really show the whole picture of what happens in traditional breeding on a molecular level. Organic agriculture established a concept of "intrinsic naturalness of the genome," but this is pure fiction,[59,60] as most crops have thoroughly artificial genomic structures, for example, widespread durum (*Triticum durum*) wheat varieties, peanuts, etc.[1108] Many traits of well-known cultivars are the result of indiscriminate gamma radiation.[147,335,934] Radiation mutation quickly has done to the genomes what takes natural mutations millions of years. Natural mutations follow the same three strategies as genetic engineering:

- DNA acquisition through gene transfer
- DNA rearrangement through recombinational reshuffling of genomic DNA sequences
- Local change of DNA sequences through internal and environmental mutagens and replication infidelities.

Still, differences exist, but not related to the molecular mechanisms themselves. Whereas natural genetic variation is in general not directed (except for rare cases of adaptive mutations still to be studied more carefully), the genetic engineer plans the alteration and verifies its results. Under natural conditions, the pressure of natural selection eventually determines the direction taken by evolution, together with the available diversity of genetic variants. Natural selection also plays its decisive role in genetic engineering, as indeed not all preplanned sequence alterations withstand the power of natural selection. There is a notable difference: after successful transformation and biosafety certification the new crop will be widely sown, exceeding the natural distribution speed of any successful natural mutation.

In summary, there are no major reasons why organic farming strategies should exclude transgenic crops, except, of course, marketing reasons built on a totally misguided public perception in particular regions. Of course, the future of organic farming should also include some of the crops dealt with in this book; this could greatly add to the diversity of organic production in the future. The future belongs to "organotransgenic" crops that will be grown in knowledge-based precision agriculture, while respecting nature, its biodiversity, and at the same time opening perspectives for an intensification of agricultural production.

KLAUSS AMMANN

CHAPTER THREE

Transgenic Tools for Regaining Biodiversity

Breaching the Ceiling

Many excellent books have been written about the tools for genetically engineering plants: isolating the genes, making the constructs, and transforming and regenerating the transgenic plants.[220,371,936,1072] We will not concentrate on the "how" to do it, but on the why, when, and where to consider doing it. We will discuss what tools are available, except without many details on how they work, or protocols for using them, but we will try to discuss current limitations, even though they may soon be overcome. It is important to know what is available, and what can be done with it. It is wise not to go into the tools in detail, as they continually change, and in many instances, even the practitioners are engaged in using "black box" tools and kits. Just as decades ago people had to write their own computer software, biochemists and molecular biologists isolated the enzymes they used, made their own reagents, and had long and tedious protocols for getting the job done. In this age of instant gratification, most of us use the ready-made programs for the computers without understanding the inherent software, and use kits with reagents and enzymes in genetic engineering, often with very few clues as to their contents or how they extract, cut, or splice nucleic acids and proteins. Both computer mavens and gene jockeys use premade memory chips, one with electronic memories, the other with genetic memories by way of genomic sequences. Many users of either type of chip only vaguely understand the underpinning engineering or the logic. Using such chips saves vast amounts of time, making up for the loss of understanding about what is actually happening.

Gene sequencing and synthesis, as well as many aspects of genetic screening, are now performed by robots linked to computers. The productivity is so great (Fig. 4) that "all" the scientists need do is think and design, with much of the daily drudgery omitted. Indeed, the rate of increase in the efficiency of DNA sequencing is actually much greater than the parallel rate of increase in computer chip capacity governed by Moore's Law (Fig. 4). The cost of DNA

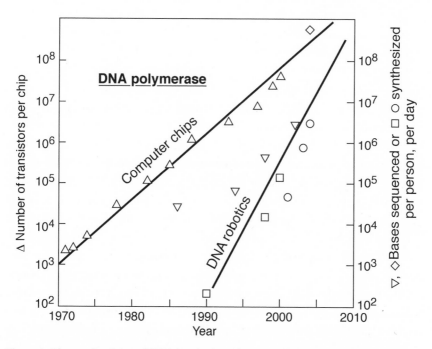

Figure 4. The productivity of DNA synthesis and sequencing increases faster than the predictions of Moore's Law for computer memories (Δ). Each of the points is the amount of DNA that can be processed for sequencing or gene synthesis by one person running multiple robots for one eight-hour day, defined by the time required for preprocessing and sample handling for each instrument; different symbols refer to different model DNA synthesizers. The approximate rate at which a single molecule of *E. coli* DNA polymerase III replicates DNA is shown (horizontal line), based on an eight-hour day. *Source:* Modified and redrawn from Carlson.[189]

sequencing and probe and gene synthesis has decreased by at least an order of magnitude in the past one and a half decades.[189] The equipment used a decade ago was also orders of magnitude heavier and costlier than now. The analytical ultracentrifuges and scintillation counters stand idle, and not many radioisotopes are used. As always, it is the bottom line that is important; the choice of research to perform and its importance and implications. The heavy equipment used at present has shifted to the more biochemical realm of metabolomics, not always needed as an adjunct to finding the necessary genes.

We are justified and obliged to stand on the shoulders of our predecessors and use these black boxes, but we must remember the hard work they performed. Just a few decades ago pessimistic tracts were written about the inability to transform soybeans, predicting that no legumes would ever be transformed because so many person-years had been invested in attempting to do so, to no avail. Though not easy, the intractable has proven tractable, and the largest share of the area devoted to transgenics in the world is planted with the legume that for decades could not be transformed, soybeans. Similarly, it was thought that gene repair (chimeraplasty) in plants, while easy in mammals and lower plants, was impossible,[871] and it remained at too low a frequency to become a method that could be used routinely in crop plants. Now, at least one plant biotechnology company is devoted to using this technology. One impediment has been found and then overcome by engineering plants with the yeast *RAD54* gene, a member of the *SWI2/SNF2* chromatin-remodeling gene family. Its expression enhanced gene targeting from a frequency of 10^{-4} to 10^{-3} in untransformed plants to a still low but acceptable 10^{-2} to 10^{-1}.[967]

Technicians can quickly be trained to perform tasks that were once relegated to the hands of only the most experienced scientists, then neophyte postdoctoral fellows, and later graduate students. All this is important because it means that more genes can be cloned, modified if necessary, and more species and varieties can be transformed in less well equipped laboratories by people without sophisticated training using these gene constructs that are easier to prepare. Thanks to genomics and databases it is now relatively easy to find so many genes, in a variety of organisms, which can then be used or modified for transformation, or as bait to fish the gene from one's favorite crop by using consensus sequences derived from the databases. The databases even assist in designing the right bait to capture the genes. The time is becoming ripe to transform more species and to continue the domestication process of more species, at much lower costs than envisaged even half a decade ago. What is more, the gene constructs are often available, having been used in a few major species, the only species that industry (mistakenly) considers as profitable.[413] Even when the genes are not available, their orthologs are often known, or genome information is available to assist in finding the genes.

Presently, it is far more important to correctly formulate the questions to be asked; that is, to clearly define the needs of our underdomesticated crops and then pick from available tools and genes. It is really very important to understand the crop to be able to do this. We can learn from history and the field. The 2005 and 2006 crop seasons in the U.S. corn (maize) belt had severe drought, but the yield loss was not as severe as with similar droughts in the past. The maize had not been transformed for drought tolerance, but some of it was transformed for root-worm resistance. It has been suggested that it is the resistance to these root-pruning insects that saved the crop, allowing more roots to reach scarce water. Thus, this field experience tells us the unexpected; it may not be drought tolerance genes per se that maize (and surely other crops) need, it is resistance to the biotic stresses that are exacerbated by drought. Farmers who were reluctant to purchase insect-resistant transgenic maize because insect damage is sporadic will now be more amenable to it as insurance against drought, insects, and disease (as insects and their damage facilitate disease infection). We must learn to think more deeply about the needs; if this outcome could have been predicted, experimental proof of concept would have been proactive and would have led to using this gene earlier in maize. It could have led to countries like China to introduce Bt maize instead of Bt cotton, as maize is a far more important crop, and droughts become more severe every year. In other words, when we are told "to put drought tolerance in this crop" we now know to ask: "What are secondary implications and effects of drought?" If we can eliminate them, will we offset at least some of the losses due to drought?

Similar types of questions might be asked about all challenges we are given. Another example is: find genes that will allow a novel grain crop to dry more quickly. The right questioning will lead to asking: "if the crop were harvested later so that seeds would be dry, would the seeds be on the ground, due to crop shattering when the seeds are dry?" If that is the case, then one could find genes that prevent shattering. An in-depth survey will show that one size (gene) does not fit all cases. An antishattering gene construct preventing pod splitting, or silique opening from legumes or crucifers, is unlikely to have an effect on the abscission zone of the rachis where shattering occurs in cereal grain crops. Even in Graminae, many different genes encode shattering at different morphological locations in different species.[650]

Molecular biology continually becomes more sophisticated. The first generation of products is rather crude; the transgenes are under the control of constitutive promoters, and the genes are expressed at most times and places, including those where the gene product is not needed. This can lead to unwanted effects; for example, overlignification and cracking in young soybeans is due to too much product of the shikimate pathway when they are transgenically glyphosate resistant,[386] sterility in transgenic cotton with the same gene when the herbicide is used too late in the season.[1007] The overproduction of one product typically leads to underproduction of others from the same pathway, and such promoters can lead to yield drags. The increased sophistication in promoter technology can lead to having the gene turned on when needed (inducible promoters) and where needed (tissue-specific promoters), and the two types of promoters can be spliced to direct the gene to do both tissue and timing. Newer technologies will lead to the transgenes of choice being included or excluded from specific gene networks where they are or are not needed. Thus, the next generations of transformed crops will be better, especially if regulatory processes are modified to allow different uses of a gene, whose products have already been discerned to be safe (see section 4.1.1).

A cogent analogy—Transportation and human activities were totally changed by the mass-produced and inexpensive Model T Ford. This revolutionary first-generation, inexpensive automobile was not a 2008 Volvo. The revolutionary Model T Ford had exploding tires, poor brakes, transmissions that literally dropped out, engines that overheated and caught fire, axles that broke, steering that allowed it to easily flip; it was polluting and inefficient, and many an arm was shattered by the backlash received while starting it with a crank. It also lacked diversity; it was available only in black. If the precautionary principle had been invoked and the safety features of a 2008 Volvo, or even a 1950 Volvo had been required, our cities would still be polluted by horse manure. The first generation of transgenics has been far safer than the Model T, no one has died from them or even hurt by them, and they have had no deleterious environmental, ecological, or genetic effects, but the next generations are bound to be even better. Sophistication evolves due to competition and selection processes. Detractors tried to prevent the introduction of the Model T by arguing that they would scare horses, and so on; they enacted legislation meant to prevent the introduction of the Model T, but couched as "safety," for example, requiring that a flagman run in front of each car is it drove through a city. More moderate- and forward-thinking legislators produced traffic regulations and required driver licensing. No one considered having a totally risk-free automobile, but balanced the benefits with the risks. It would be wise to remember this analogy in the current debate.

3.1. Terminology

Not all concepts and terms will be explained in this chapter. The reader is referred to the excellent online glossary of genetic engineering terminology, which is available in three languages at www.fao.org/biotech/index_glossary.asp. A glossary had been contemplated for this book, but Wikipedia and the internet save paper and costs.

3.2. The Types of Functions to Be Engineered—Loss and Gain of Function

A major part of domestication was the arduous selection by prehistoric farmers of what we now know as recessive mutants, that is, loss of function. In most cases dormancy and shattering are dominant and uniform germination (lack of secondary dormancy) and nonshattering (retention of seeds till harvest) are recessive. Both are often encoded by more than one recessive gene,[415] rendering domestication even more onerous. Our forefathers must have been very astute when domesticating and breeding crops by eye, without knowledge of Mendel's peas. The task can be even more onerous when more than one recessive gene is involved, or the crop is a polyploid, where multiple copies of the gene may be active. Even patient and dedicated breeders find it hard to select for multiply-recessive functions. Thus, to obtain such traits in our underdomesticated crops we need a "loss of function" to remove an unwanted dominant gene function such as shattering or secondary dormancy, and achieve seed retention and uniform germination, without the necessity to select for recessive alleles. This breaches the glass ceiling by obtaining a new expression pattern using the crop's own genes.

In other cases, we want a conditional loss of function. For example, dwarfing has proven to be efficacious for increasing grain yield at the expense of straw. This can be achieved by mutating genes in the gibberellic acid biosynthesis pathway, as the hormone product of the pathway causes plants to grow tall. This does not work in all species, because the gibberellins are needed for other key functions. Here we need conditional suppression of gibberellin biosynthesis in the stem and in early stages of growth. Dwarfing is already recessive; nature is unlikely to have concomitant mutations in the promoter giving conditional loss of function of the pathway. That is why neither our forefathers nor the breeders have found such conditional mutants. Genetic engineering allows dominant constitutive or conditional suppression of traits such as shattering, dormancy, or too tall growth. Conditional suppression of a trait may well be both a loss of function (e.g., tall growth) as well as a gain of function (e.g., a stem-specific promoter for the gene preventing tall growth).

In many instances we need to gain functions by adding the structural genes that encode new functions, breaching the glass ceiling by bringing in genes from other species, or even genes that do not exist in the needed form in nature. In some cases, though, a gain in function is obtained by modifying controlling elements of an endogenous gene.

Plant genetic engineering is beyond the initial primitive days of modifying single genes. Whole pathways are being modified by adding genes for both gain and loss of functions to obtain the desired final product. This provides windows of opportunity to fully deal with the multitude of reasons that a crop with limited genetic base has reached a low glass ceiling; it can now be engineered with multitasking multiple genes to compensate for the low ceiling.

3.3. Where Are the Genes?

The hardest part of breaching the genetic glass ceiling is finding the genes to be used or even the (mainly recessive) genes to be suppressed. It is almost a rule of thumb that the more interesting the gene, the harder it is to find. Thus, the genes easiest to obtain are those encoding massive amounts of products (e.g., structural components of the cells, storage protein, or starch) or those involved in primary metabolism. To further domesticate a crop, only rarely will we want to tamper with primary metabolic pathways. We are far more likely to want to modify genes that are expressed in small amounts or that are turned on in very localized places (e.g., the genes encoding plant hormones, genes encoding the production of secondary metabolites, the genes involved in producing the excision zone leading to seed shatter). In second-ary metabolite production, the genes may be species specific. Still, more and more species are sequenced and the data almost immediately available in data-bases such as GenBank. Exceedingly complex, but user-friendly programs al-low rapid comparison of gene sequences, allowing the putative identification of genes.

Additionally analytical procedures such as matrix-assisted laser desorption ionization time-of-flight mass spectrometry (MALDI-TOF MS) allow one to easily obtain amino acid sequences from minute amounts of a partially pu-rified protein, assisting in finding the gene in a particular species. Each year more species are sequenced and many of the genes are putatively identified. Thus, it is likely that a close family relative of each crop will be sequenced in the near future. Analogy allows guessing what coding sequence might per-form the function, but see the caveats below.

Much of the donkey work for further domesticating crops will be in the biochemistry and molecular biology of finding the gene. Much of this can-not be done yet with kits, although once a putative gene is found, mundane (but expensive) kits and easy technologies are available to functionally verify the identity the gene.

Despite the caveats below, it is becoming exponentially easier to find the genes we need. It is now possible to isolate the few rows of cells that will be an excision zone in seed shatter by micromanipulation, isolate them during the ontogeny of the excision zone, and by comparison, ascertain which genes are turned on during excision, isolate these genes, antisense them, and see which singly or together can be suppressed to prevent seed shatter. The groups of techniques to do this were not available a few years ago.

3.3.1. Caveats on Gene Identification—Be Skeptical

It is far too typical in the scientific literature, and during seminars, to use the word "putative" once in referring to the possible function of a gene based on comparative sequence data alone, and then to forget that the word puta-tive was used. Ninety-nine percent homology between two gene sequences is not identity, yet at times researchers will find two genes with 50 percent ho-mology of sequences and first say that they may putatively have similar func-tions, and then assume that they do. Mother Nature is a scrounger. To make new genes from scratch would take, at present mutation rates, close to an eternity. Instead, Nature uses mutations that modify a preexisting gene that eventually can give it a totally different function. This is well borne out with genes of known functions: two cytochrome P_{450}s have 99 percent amino acid homologies in their active sites: one encodes a highly specific enzyme that modifies a very precise location on a particular sterol; the other is a rather general enzyme that participates in the oxidative degradation of a multitude of substrates. Similarly, a single amino acid substitution converted a carboxylesterase to an organophosphorus hydrolase.[785] Computer programs that assign putative function based on databases often misclassify the function of such genes, and naïve scientists unquestioningly believe the computer.

The same problem exists with finding genes that are turned on or off on chips based on expression levels of mRNA. The minimal detectable, statistically significant change is 70 percent up or down in the best of cases, although a doubling or a halving is usually considered significant. We know of many instances where much finer tuning modulates processes. For example, some herbicide resistances are correlated with a less than 10 percent increase in herbicide-degrading activity cytochrome P_{450}s,[868] others with a 20 percent increase in enzymes conferring oxidative stress tolerance.[1177]

What are the messages? More than one method may be required to find the genes of interest; the first method used may not be amenable to gene discovery. Carefully check the data of papers saying they may have the needed genes. The proofs might be there, but they may not. Glibness is an unfortunate trait in the wonderful tools that the bioinformaticists are developing. The best antidote for such glibness is a continual supply of skepticism during gene discovery.

3.3.2. Are There Worthwhile Genes for Everything?

There can be many reasons why a crop has lost its place in the farmers' fields, and not all problems are tractable by transformation with a simple gene or a few genes. Modern agriculture has been transformed by the Haber–Bosch process for generating nitrogen fertilizers by chemically fixing nitrogen from air in a process far more energetically efficient than the biochemical fixation by legumes. Thus, considering engineering legume nitrogen fixation genes into nonlegume crops is questionable from an energy consideration alone. The advent of inexpensive nitrogen fertilizer promoted the utilization of crop species responsive to such fertilizers, especially in places where nitrogen was a major limiting factor. Thus, most of the major crops now in cultivation (except soybeans) are highly responsive to nitrogen fertilizer. This is a trend that can hardly be reversed unless we wish to see more land under the plow. Some people would like to see poor-yielding crops that have been abandoned developed for the rural poor in the developing world, but this might keep them poor. Hopefully, it will be possible to find the genes that will increase nutrient use efficiency of abandoned or inefficient crops, although this may take a while.

3.3.3. Will the Crop Yield More without the Gene—To Grow or to Cope?

If the metabolic cost (in yield) to a crop from genes that assist in anything other than growth is greater than the external cost of its replacement, then such genes may be contraindicated. For example, if using genes for mobilization of bound soil phosphorus costs more in lowered yield than fertilizer, the farmer will use fertilizer. If the production of plant-produced phytoalexins to ward of insects and diseases costs more in yield than purchased fungicides and insecticides, the farmer will buy pesticides. The ecologists have recognized this as being due to two mutually exclusive evolutionary strategies of plants in general: "to grow or to defend,"[497] with defense being at the expense of growth. The enlightened gene jockey must understand this; to propose uneconomic solutions has little value in the agroeconomy. Economics change, so what may be uneconomic today may be economic in a few decades. For example, if organic agriculture (section 2.13) were to suddenly prefer engineered "natural" Bt toxins over synthetic insecticides or the few allowable "organic" insecticides, then the engineering of Bt into organic varieties of vegetable crops would rapidly ensue. If they would accept the herbicide bialophos, produced solely by bacteria (unengineered at that) in a fermenter as a herbicide, the engineering of resistance to bialaphos via the *bar* gene would be a reality.

3.4. Methodologies for Gene Isolation

3.4.1. Forward Genetic Profiling to Find the Genes

Genomic research has been revolutionized by the whole-genome sequencing of model plants, expressed sequence tags (EST) database development, and the rapid appearance of the data in public databases. As described earlier, the notation of gene function in these databases is at best tenuous. Thus, we still have a way to go from identifying interesting genes to ascertaining their function. Good biochemical data are required to trust the annotation of a database.

3.4.2. Reverse Genetic Profiling to Find the Needed Genes

The two approaches to reverse engineering are through biochemical methods or via molecular genetics. The choice of method depends on the gene function that one is pursuing.

3.4.2.1. Biochemical

If the product of the gene needing modification is known and can be isolated as a pure protein, MALDI-TOF MS can be used to microsequence peptides from the protein.[354,1162] For example, in a recent study proteins were extracted from eight *Arabidopsis* tissues and fractionated by two-dimensional gel electrophoresis. Six thousand protein spots were excised from the gels and analyzed by MALDI-TOF MS fingerprinting.[388] The proteins from nearly 3,000 spots were identified and found to be products of 663 different genes. Polymerase chain reaction (PCR) primers with likely nucleotide sequences can be generated to find likely gene sequences in the organism, at low stringency; after isolation and sequencing of the PCR products, high-stringency probes can be generated, and inverse PCR can be used to extend the gene sequence until the whole gene is isolated, or genomic libraries are probed to find the fragment containing the gene and sublibraries made. The putative gene fragments at different stages are cloned into yeast or bacteria to see whether the gene product is made, to verify that indeed the desired gene was isolated.

3.4.2.2. Molecular Reverse Genetic Methods

In *insertional mutagenesis,* mutations are generated by randomly inserting known DNA sequences (e.g., T-DNA or transposon insertion) that will randomly disrupt gene activity. Once the gene of choice is disrupted, it can be fished out by using inverse PCR with primers for the inserted DNA sequences. This allows one to sequence the gene that was knocked out.

If one finds a gene in the databases that may have the desired function, one can specifically knock it out by generating homologous gene replacement point mutations by TILLING (Target Induced Local Lesions in Genomes)[494] or by RNA interference (RNAi)/antisense (see section 3.6.1.1) to silence gene expression. This is often done as a verification of function, before either overexpressing the gene or using the gene to transform other species.

RNAi or antisense (see section 3.6.1.1) of a putative gene sequence can also be used to ascertain whether a gene with a putative function actually performs the stated function. If the function is suppressed, then this is the gene.

The techniques of insertional mutagenesis will not verify all putative gene functions. There can be false negatives, that is, cases were the function is not lost, yet the gene being tested has the desired function. This is because plants have a large degree of redundancy, with more than one gene performing the same desired functions. When genes are in "families" having a degree of homology, RNAi/antisense (see section 3.6.1.1) of a consensus DNA sequence found in all family members may give a positive response, where specific knockouts will not. This too has its drawbacks. When a consensus sequence to a large family is used, it is not clear which of the family members possess the actual function. Thus, it is unclear whether the correct gene is in hand for gain of function until it has been cloned into a system where this gain of function can be assessed.

3.4.3. mRNA Transcript Profiling to Find a Gene Where It Is Expressed

Profiling of mRNA expressed during specific developmental stages on DNA chips can be used to identify genes conferring traits of interest,[13,616,1021] as well as for other uses (Table 5). It may be used to find the needed trait in another species (e.g., disease resistance), or to find the gene that needs to be suppressed in the underdomesticated crop (e.g., genes for seed shatter). Transcript profile techniques can be qualitative or quantitative depending on detection characteristics. Differential display real-time PCR (RT-PCR) and cDNA-amplified fragment length polymorphism (AFLP) are used for qualitative analysis. They do show (unreliable) quantitative differences in transcript levels among samples but are best used for detecting the presence or absence of a transcript. Differential-display RT-PCR can be used to screen large numbers of transcripts by PCR with arbitrary primers, but because of the lack of sensitivity and reproducibility, it has largely been replaced by cDNA-AFLP. cDNA-AFLP is a modified genomic DNA-based AFLP technique where the use of selective PCR of adapter-ligated cDNA fragments improves reproducibility and specificity (Fig. 5). Quantitative transcriptprofiling frequently uses cDNA membrane macroarrays or cDNA—or oligonucleotide-

Table 5. Uses for DNA Microarrays

Purpose	Target sample	Multiplexed reactions	Demultiplexing probes on array
Expression profiling	mRNA or totRNA from relevant cell cultures or tissues	Amplification of all mRNAs via some combination of RT/PCR/IVT[a]	Single- or double-stranded DNA complementary to target transcripts
Pathogen detection and characterization	Genomic DNA from pathogens	Random-primed PCR, or PCR with selected primer pairs for certain target regions	Sequences complementary to preselected identification sites
Genotyping	Genomic DNA from organism	Ligation/extension for particular SNP[b] regions, and amplification	Sequences complementary to expected products
Resequencing	Genomic DNA	Amplification of selected regions	Sequences complementary to each sliding N-mer window along a baseline sequence and also to the three possible mutations at the central position
Find protein DNA interactions	Genomic DNA	Enrichment based on transcription factor binding	Sequences complementary to intergenic regions

Source: Modified from Stoughton.[1021]

[a] RT, reverse transcriptase; PCR, polymerase chain reaction; IVT, in vitro transcriptase.
[b] SNP, single-nucleotide polymorphism.

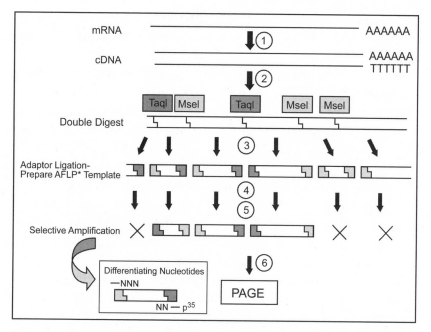

Figure 5. A cDNA-AFLP protocol. This procedure consists of the following steps: (1) the reverse transcription of mRNA using an oligo(dT) primer to produce cDNA; (2) digestion of double-stranded cDNA with a pair of restriction enzymes (in this case the enzymes MseI and TaqI); (3) ligation of adapters specific for the two restriction sites; (4) preamplification of fragments with primers specific to the two adapter sequences, but with a single nucleotide extension to reduce mismatching at the selective amplification stage; (5) selective amplification with adapter-specific primers with nucleotide extensions at their 3′ ends (two nucleotides for the TaqI primer, three nucleotides for the MseI primer); and (6) visualization of individual TaqI/MseI fragments on a polyacrylamide gel, as the TaqI primer is end labeled with 33P. See the paper by Vos et al.[1115] for hypotheses on the nature of the selective amplification. AFLP is a registered trademark of Keygene N.V. and the technology covered by their patents. *Source:* Redrawn from review by Donson et al.[291]

based microarrays. Macroarrays have denatured double-stranded DNA on nylon membranes.

In microarrays, tens of thousands of bacterial clones containing crop cDNAs are deposited on the chips and are hybridized with labeled cDNA generated from mRNA isolated from the tissue expressing the trait of interest. There are two types of microarray techniques: the cDNA microarray format and the oligonucleotide format. The choice depends on the types of probe sequences on the microarray chips and on the method of probe hybridization. The first microarrays had short oligonucleotides printed by ink-jet-type printers on glass and now use photolithographic synthesized arrays that have more than 5,000 times more spots per unit area than macroarrays. The Affymetrix GeneChip oligonucleotide format holds up to half a million 25-mer oligonucleotide probes photolithographically fixed on a small quartz wafer (www.affymetrix.com/technology/manufacturing/index.affx). It is necessary to first accumulate massive sequence information to attempt quantitative transcript analysis on a large scale. The process flow is schematically shown in Figure 6, with the functions in the center of the scheme performed by companies specialized in chip production and those surrounding the scheme performed by the researcher.

The last steps require the utilization of very sophisticated bioinformatics tools.[1021] Either total RNA or purified mRNA can be amplified from the plants. If the researcher is interested in nuclear traits, mRNA can be quickly purified by affinity chromatography on short columns that bind the poly(A) sequences that are on such nuclear-generated mRNAs. Chloroplast and mitochondrial DNA-encoded mRNA sequences lack this poly(A) tail. The mRNA is amplified by reverse transcriptase in vitro or by PCR, and labels fluorescing with different colors are attached to each batch of RNA from the different situations. The amplified material from different sources is hybridized with the cDNA on the chip (Fig. 7), such that they compete with each other for binding sites. The duration of hybridization, the stringency that determines the allowable mismatch of sequences, and the concentration of cDNA are all variables that must be determined by experience. Biases enter at all reaction steps and can be very sequence specific, such that the final fluorescent brightness of the paired probes is only vaguely relatable to abundance.[1021] Detection intensity for each probe on the membrane is digitally quantified showing the relative abundance of fluorescence from the mRNA transcripts between the material extracted from various tissues at various times or physiological states, which corrects for much of the bias.

Typically thousands of genes show differential expression of mRNA transcripts from the RNA pool. The bioinformatics gives a vague inkling of what the genes may be,[616] and the researcher has to guess which is (are) the needed gene(s). So far, despite widespread use, and the huge investments, it is not clear whether this method has proven itself in the field, that is, whether genes have been isolated on chips that have been put into plants that are useful in the field. Even if there have been, one should ask whether other techniques would have found them at a fraction of the cost. This consideration is important for the orphan crops being discussed in this book: can/should the outlay for chip technology be considered, if it is only to be used for gene dis-

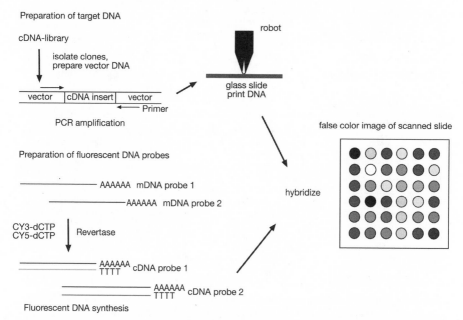

Preparation of target DNA

cDNA-library

isolate clones,
prepare vector DNA

| vector | cDNA insert | vector |

← Primer

PCR amplification

robot

glass slide
print DNA

hybridize

false color image of scanned slide

Preparation of fluorescent DNA probes

——————— AAAAAA mDNA probe 1

——————————— AAAAAA mDNA probe 2

CY3-dCTP
CY5-dCTP | Revertase

——————— AAAAAA cDNA probe 1
TTTT

——————— AAAAAA cDNA probe 2
TTTT

Fluorescent DNA synthesis

Figure 6. cDNA fragment-based microarray technology. Target DNAs are prepared by PCR amplifying inserts from cDNA clones. The amplicons are spotted onto microscope slides at fixed locations using high-speed arraying machines. Usually probes from independent samples are prepared and labeled with fluorescent nucleotides. Fluorescent tags with different excitation and emission optima can be used to label each probe differently. The probes are mixed together and hybridized to the same microarray. After washing off the unbound probe, the fluorescent DNA molecules hybridized to DNA fragments on the microarray are excited by light and the fluorescence signal associated with each spot on the microarray is read with an array scanner. The ratio of fluorescence emission at the optimal wavelengths for each probe reflects the ratio of the abundance of that sequence in the probes for each element of the microarray. The results are typically displayed as a false-color image. One color may indicate that the corresponding RNAs were more abundant in probe one than in probe two, another color may indicate that they were present at similar concentrations with both probes, and yet another color that the RNAs associated with these array elements were more abundant with probe two than with probe one. *Source:* Redrawn from review by Kuhn,[616] by permission of Oxford University Press.

covery? DNA microarrays have many other uses for plant functional genomics, in particular, in obtaining a basic understanding of coordinated metabolic pathways, signaling, and regulation.[13] It is sure to be of great use in the future, when further domestication will require more sophistication than we are presently proposing of a gene here and a gene there.

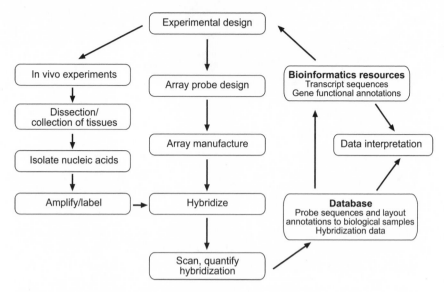

Figure 7. Process flow for microarray experiments. "Probe" here refers to the reporter sequence placed at a particular position on the microarray because it probes the sample for the presence of its reverse complement. Historically probe has referred instead to the biological sample. *Source:* Redrawn from review by Stoughton,[1021] by permission of Annual Reviews.

3.5. Shuffling to Improve on Nature's Genes

In many instances the gene found has the necessary properties, but is inefficient, even with highly expressive promoters. The lack of efficiency can be biochemically seen as a lack of substrate specificity, or a low affinity (high k_m) or poor activity (low V_{max}). The gene could be from a soil bacterium, and the resultant enzyme might not be able to withstand leaf temperatures, and so on. These problems could possibly be remedied by site-specific mutagenesis, but there is a good chance that this will not provide a sufficient level of diversity. Stemmer and colleagues[235,784,873] demonstrated the utility of a system in which DNA of a group of closely related genes from many sources is randomly fragmented with a DNase and the pieces are randomly reassembled by PCR into many new genes (Fig. 8). The vast majority of the recombinant proteins generated are inactive. This is immaterial if a good selection system exists; that is, a herbicide as a sole carbon source for recombinants containing the gene, or a recipient that is herbicide sensitive before the gene is introduced. It is then possi-

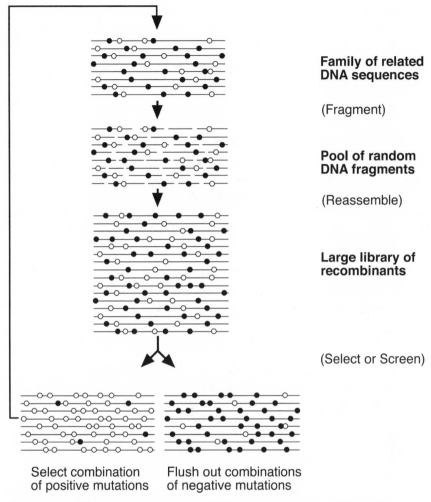

Family of related DNA sequences

(Fragment)

Pool of random DNA fragments

(Reassemble)

Large library of recombinants

(Select or Screen)

Select combination of positive mutations

Flush out combinations of negative mutations

Figure 8. Shuffling the same gene from different organisms to modify activity. Homologous recombination of pools of diverse but related individual sequences involves the partial unlinking of existing mutations by random fragmentation with DNase I. These random fragments are reassembled into full-length genes by PCR based on their DNA sequence complementarity, thus forming new combinations of mutations within a conserved framework sequence. When coupled with selection, this process allows rapid accumulation of useful mutations from multiple parental sequences, while at the same time flushing out detrimental mutations. The system requires selection that eliminates the vast majority of inadequate recombinations. Herbicides, disease toxins, temperature, and so on, are excellent selectors. *Source:* Redrawn from Stemmer[1011] by permission. Copyright 1994 National Academy of Sciences.

ble to select the few recombinants that have been fortuitously and randomly re-spliced in a meaningful manner and work better than their predecessors. DNA from the best recombinants is then again fragmented and reassembled in more cycles. In this manner enzyme activity is increased and substrate specificities of enzymes are optimized. Such changing of substrate affinity can have many uses in further domesticating plants, whether it is to obtain herbicide resistance or change from a toxic or allergenic product to a desired product, for molecular "pharming" (making pharmaceuticals in plants), etc.

Greater diversity of sequences (greater "sequence space") of a gene can be obtained by DNA shuffling than by mutagenesis (Fig. 9). This "DNA shuf-fling" or "sexual PCR" with reassembled sequences from many closely related genes should have great utility in producing highly active genes with the low-est possible metabolic load on the crop. This is shown well by bringing about of the evolution of an arsenate-degrading pathway.[236]

Shuffling has been used to modify the target of two herbicides, but not to alter the binding of the herbicides to obtain resistance. Phytoene desaturase (the target of norflurazon) and lycopene cyclase were both selected for by mo-

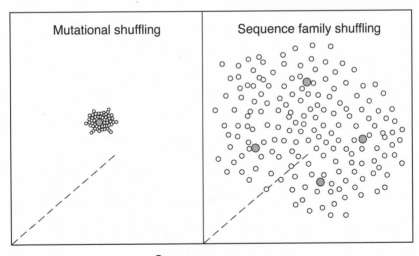

Sequence space

Figure 9. The relative breadth and efficiency of shuffling (shown in Fig. 8) is much greater than can be achieved by mutagenesis. Mutagenesis yields clones with a few point mutations that are typically 97–99% identical. The sequence diversity is vastly increased by shuffling the same gene from multiple sources, allowing more promising differences in function to be discovered. *Source:* From Crameri[235] by permission. Copyright 1998 Macmilllan Magazines Ltd.

lecular breeding and the products cloned into *Escherichia coli* to produce novel carotenoids, not to obtain herbicide-resistant crops.[953]

The herbicide glyphosate is effectively detoxified by various enzymes, but none is good enough by itself. One gene conferring detoxification, the *GOX* gene, is found commercially in glyphosate-resistant oilseed rape, where it augments the *EPSPS* target site resistance gene, which also by itself is almost good enough. A recently discovered microbial glyphosate *N*-acetyltransferase (GAT) activity was also insufficient to transgenically confer glyphosate resistance. Eleven iterations of DNA shuffling improved enzyme efficiency by nearly four orders of magnitude, fully sufficient to generate glyphosate-resistant crops (Fig. 10).[194] Shuffling was also used simultaneously to over-

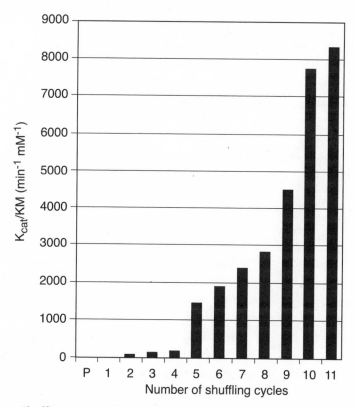

Figure 10. Shuffling enhanced the evolution of increased degradation of glyphosate by glyphosate-*N*-acetyltransferase (GAT). The total catalytic efficiency and specificity of new GAT enzymes continued to increase (as measured by k_{cat}/k_m ratios) in each shuffling iteration from the parent (P) level. *Source:* redrawn and modified from Castle et al.[194].

come the sensitivity to high leaf temperatures as well as increase the longevity of the GAT enzyme.[194] Plants with this gene may have fewer side-effect problems[412] than the plants on the market utilizing a target-site resistance by using an inefficient gene/protein.

3.6. Where/Why Genes Get Inserted

The genetic engineer or, possibly more accurately, the genetic surgeon has a variety of tools to provide the needed traits, as discussed below.

3.6.1. Gene Insertion

Gene insertion is the standard technique to provide both gain and, surprising as it may seem, partial loss of function. Gene insertion can be random, as in the old song "que sera sera, whatever will be will be," requiring many transformants to be generated, of which only a small proportion expressing the character in a manner that the researcher wants are retained for further screening. This process allows detractors of gene technology to correctly say "most transgenic plants are badly messed up," but they do not add that the badly messed up transformants do not make their way to farmers' fields. They are discarded early on, and only the best transformants are kept.

3.6.1.1. Targeted Gene Insertion

Whole genes can be targeted:

1. *to the cytoplasm,* where they will be expressed only in the cells or the plant (when systemic viruses are used as the vector), in such a manner that the genes do not become integrated into the plant genomes, and so are not passed on to further generations.

2. *to the nucleus,* where the genes will be randomly inserted into the nuclear genome. If one does this with a gene already found in the crop, or if multiple copies of a gene are inserted instead of a single copy, one can obtain loss of function or no function due to overexpression, or cosuppression where copies of the gene may suppress each other.

3. *to an organelle,* such as the plastids or mitochondria, where it will be randomly incorporated in the plastome (plastid genome) or chondriome.

4. *to the nucleus* at a specific site on a chromosome by the process "homologous recombination," that is, a homology to a specific DNA site that is encoded into the DNA to be inserted, and a "mistake" occurs during DNA replication such that the homologous material of the inserted DNA recombines instead of reading from the antiparallel DNA strand. This can be done in a few ways: into a specific natural DNA sequence (e.g., when it is desired that the gene of choice is inserted on a specific chromosome,[871] or into a receptor site, such as the Cre/Lox[392] or FLP-FRP[879] systems. There a receptor is first randomly engineered into the crop genome and the gene of choice is placed in a transformation "cassette" that performs targeted insertion, followed by precise deletion of DNA from transgenic plant chromosomes. The Cre-Lox recombination system is one of the best characterized and most widely used systems for these purposes. It is used primarily for the controlled excision of DNA fragments, in particular, selectable marker genes from the nuclear and chloroplast genomes, and for the targeted insertion of DNA into specific sites in the nuclear genome. Both systems have been used to engineer rice.[211,879]

Partial genes can also be inserted for various reasons including *gene modification* (also called chimeraplasty, or in humans, DNA or gene therapy)and RNAi/antisense techniques. Gene modification can be necessary to change an amino acid that cannot be changed by mutating a single nucleotide base, or a short stretch of amino acids. It is then possible to use homologous recombination (above) to change short stretches of DNA, including the sequences encoding the amino acid(s) to be changed. In humans, at present, it is considered ethical to do this only in somatic cells, for example, for correcting genetic diseases (e.g., cystic fibrosis where only lung mucosal cells are to be transformed). In plants one would want to do this in reproductive cells so that the modified trait would be inherited. Pioneering work with the haploid moss *Physcomitrella* for nearly a decade,[946] and more recently with the moss *Ceratadon*,[163] showed a more than 40 percent gene conversion with this technique. Some groups were initially successful in higher plants, albeit at rather low conversion frequencies.[104,1198] Chimeraplastic surgery is so far very tricky in plants, many groups have tried it unsuccessfully, including one group that reported failure in print.[928] Indeed, most groups with higher plants used the same gene, acetolactate synthase, conferring herbicide resistance. This gene can be mutated at many loci giving resistance at relatively high frequencies, and one group reported that the different gene sequences in the resistant plants they obtained had no relation to the sequence in the chimeraplast construct.[928] Two other groups reported success in rice and tobacco, with the DNA sequence of the product matching the expected sequence from the chimeraplast.[598,815] Higher transient frequencies were achieved in a regenerable wheat cell in vitro system, with the repair of point mutations more efficient than deletions.[290] How this will play out in permanent repair remains to be seen. A hopeful claim was made that chimeraplasty might also work with the chloroplast genome, based on correct chimeraplastic conversion of plastid DNA in a cell-free system.[596] Reports that plastid chimeraplasty works in cells, let alone cells that can be regenerated to plants, have yet to appear.

Chimeraplasty can also be used to totally obliterate the expression of a gene, for example, by encoding a sequence that will give a totally inactive gene. This could be used to suppress the dominant genes that cause shattering, dormancy, and height or those that are responsible the genetic enhancement of undesirable natural allergens or toxins in the crop. It could be used like RNAi/antisense technologies, which have proven successful in suppressing rapid overripening deterioration in tomatoes, and can be used to prevent starch formation in sweet corn, and so on. RNAi and antisense are transgenic techniques where recombinant DNA is found in the product. It can be argued that crop varieties obtained via chimeraplasty are not transgenic as the gene targeting does not involve the insertion of foreign DNA, just the use of a nucleic acid sequence as a highly specific mutagen; the plants are therefore not recombinant. Such plants should be governed the same way as those obtained through traditional mutational methods. There are no indications that this logic has been accepted by any regulatory authorities.

RNAi is based on posttranscriptional gene silencing mediated by double-stranded RNA. This is the most common way to silence a gene. RNAi is much simpler than excising a gene to completely delete it. An excellent recent book by Galun[370] describes the various ways it can be done, and the many uses. Here we just must know that it can be done and where it can be used. Basically, a section of the gene is either inserted into an "interference cassette" designed to produce hairpin RNA (hpRNA) with a double-stranded region. This interferes with mRNA production of the homologous gene.[1135] Such interference cassettes are now available as "kits," allowing easy insertion of the RNA piece to be used in a single step.[1152] In antisense, a section of the gene is inserted behind a promoter such that it will be expressed in the reverse order, and the construct is randomly transformed into plants. RNAi usually silences genes as if there was a null or deletion mutation. Depending on where the construct is inserted and which promoters are used, antisense transformants have the gene suppressed to different levels. This allows the researcher to pick transformants in the range of suppression required. For example, once it is known which is the major gene for shattering in a given underdomesticated crop, antisensing can be considered for its prevention. If the gene were totally suppressed, no threshing machinery could remove the seeds from the crop. Indeed, where hand harvesting and hand threshing are used, antisense transformants would be chosen that are easier to thresh than for varieties where harvesting machinery such as a combine is used, because when using such machinery the seed must be held more tightly. Thus, the variation achieved in different antisense transformants is desired. It is often not known in advance what level of gene suppression will give the desired result, so this variability of response is quite valuable even though it again allows the detractors of genetic engineering to state that "you cannot predict accurately what will happen when you transform plants." They seem to be saying that the generation of biodiversity is undesirable.

Overexpressive cosuppression can be obtained with partial gene sequences that match those in the plant are too highly expressed. As with RNAi/ antisense different transformants have different suppression levels, allowing the choice of the transformants that best suit the needs.

3.7. Transformation

Three major methods are used to obtain transgenic plants where the trait is stably inherited from generation to generation:

3.7.1. Agrobacterium-Mediated Transformation

The classic and, at one time, the easiest method of transformation is with *Agrobacterium tumefaciens*, the bacterial pathogen causing crown gall disease. During the normal ontogeny of disease it has a plasmid (Ti) that becomes integrated into the plant genome, and the DNA encodes overproduction of hormones that cause the gall (Fig. 11). The particular gall-causing genes have been removed from the plasmid used for transformation, that is, the plasmid has been "disarmed," and the virulence genes are replaced by the genes of choice; that is, the genes are "cloned" into the

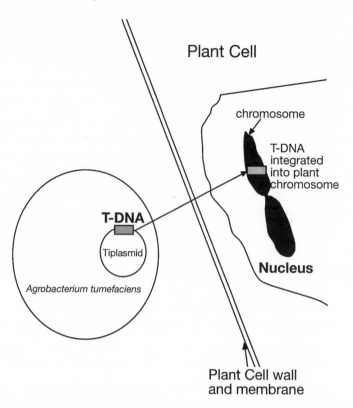

Figure 11. Transformation of plant cells by infection with *Agrobacterium tumefaciens.* After infection, the transgenes on the Ti plasmid insert themselves in plant chromosomes, and the *Agrobacterium* is "cured" from the plant culture with an antibiotic. *Source:* Redrawn from book by Stewart.[1019]

plasmid. Thus, whereas transformation into *Agrobacterium*-susceptible dicots is easy, building the gene construct in the Ti plasmid requires considerable expertise. The process had one major drawback: *Agrobacterium* naturally infects only dicot plants. This was overcome when it was found that dicot plants emit recognition chemicals that facilitate infection. Adding similar chemicals to *Agrobacterium* allows infection of various monocots. Thus, once the *Agrobacterium* containing the gene construct of choice is shown to work well in one species, it can be used to transform many other varieties of that species, as well as other species.

The *Agrobacterium* system still has a major drawback in this sophisticated era. The plasmid has only a limited size capacity for genes and must contain two standard genes before one starts: a plant (only) expressed gene that confers resistance to a chemical (antibiotic, antimetabolite) that allows one to kill off the *Agrobacterium* bacterial cells in a culture after the plant cells become infected; and another herbicide resistance or antibiotic resistance gene that allows one to kill all plant cells that are not transformed (a selectable marker). Room is left only for a few more genes. One cannot engineer a whole metabolic pathway into *Agrobacterium*. Building the constructs is not easy; highly specific and accurate spacers must be included between each gene inserted. *Agrobacterium*-mediated transformation is more likely than other procedures to insert a single copy of a gene, precluding overexpressive cosuppression, which often occurs when multiple copies get in. Typically callus-forming tissue, such as leaf discs or stem explants, is infected. After infection and "curing" (killing the bacteria with an antibiotic), shoots are regenerated from the callus, which are then rooted. It is a highly probable that cells that differentiate into gametes will also be transformed. *Agrobacterium* species are not the only bacteria that can be used to transform plants. Other bacteria can also perform this feat,[154] as discussed more fully in section 3.9.

3.7.2. Biolistics

The biolistic technique has become widely used. The naked DNA to be transformed into the plants, typically into meristematic tissue (embryos, young shoots, undifferentiated tissue cultures) is coated on microspheres of gold or tungsten and propelled into the plant tissues held in a vacuum chamber. The first equipment actually used gun powder and blank firearm shells to shoot the DNA into the tissue. This is now done with a burst of high-pressure inert gas (Fig. 12). The plants must be under partial pressure so that the noise of the explosive burst will not cause cell cavitation; sound waves as we remember from physics do not travel through a vacuum. If the ratio of DNA to metal beads to plant tissue is well calibrated, most of the transformants will contain a single copy of the inserted gene. If more DNA is added to obtain a higher frequency of transformation, multiple copies of the DNA can be inserted, suppressing gene function. One fascinating aspect is the success of cotransformation. It is not

Figure 12. Transformation with a helium-powered second-generation biolistic apparatus (*left*), and a first-generation gun-powder-operated gene gun (*right*). In both cases DNA containing the desired genes is coated on gold or tungsten microspheres that are propelled into meristematic tissue. Reduced pressure (vacuum) is used to prevent damage by shock waves produced by the noise of the explosive force. *Source:* Photos by Sagit Meir.

necessary to splice the genes of the selectable marker, and reporter gene (if used) with the various desirable genes into a single construct. Each DNA can be mixed with the others just before bombardment. The most interesting aspect is that most genes are inserted to one site in a transformant, that is, they organize as a single tandem sequence of genes, without a need to presplice them, and all are inherited together as a genetically linked block, which does segregate together in future generations. This block of genes is located at different chromosomal loci in each transformant,[660] the actual insertion site is still random. Selection for the selectable marker is used to eliminate untransformed cells, and shoots are regenerated.

3.7.3. Protoplast Transformation

Prior to the discovery that DNA could be shot through cell walls, a more subtle approach was used: remove the cell walls. Protoplasts are generated by removing cell walls with cellulolytic and pectinolytic enzymes, and kept in a high osmoticum so as not to burst. The DNA for transformation is added together with cell-membrane-breaching polymers such as polyethylene glycol along with calcium, to stiffen the protoplast membranes. After the DNA goes in, the protoplasts are allowed to regenerate cell walls, and often multiple media changes occur with different hormone and sugar mixes to generate calli, then shoots, and finally rooted plantlets. The selectable marker is used at an early stage to eliminate untransformed cells. Few people still use this technology for transforming plants, but it is commonly used with fungi.

3.8. Transient Transformation—Viruses and Endophytes as Vectors

Some viruses can be used to engineer expression of transgenic traits in somatic cells of plants (as described in section 4.8.10 and Chapter 8), but the traits are not inherited through true seed. Virulence genes, which cause crop yield reduction, are removed from the viruses, that is, they are "attenuated." The various genes of choice are then inserted into the viruses, and the viruses are used to systemically infect the crop. Such virus-vectored transformation can be used for single-generation transformation as well as with vegetatively propagated crops, although the technique has not been widely used. The technique would have the advantage that any variety of the species can be infected with the virus without the need for backcrossing, allowing greater crop biodiversity.

Endophytic bacteria and fungi, which normally appear in plant tissues, have been proposed as single-generation vectors for transgenes,[327] but the company developing this technology ceased functioning more than a decade ago. They were able to show efficacy, although there was a yield reduction with their first generation of material. One other group has been following up this type of research, so far reporting success only with reporter genes.[446]

3.9. Getting and Using the Needed Technologies

Much of the technology for creating transgenic plants, from promoters, genes, vectors, analytic techniques through to transformation techniques is tied up in patents. Although patents are important to reward the discoveries of astute inventors, problems can indeed occur in getting needed and novel products to market, especially in the developing world. Failure to receive a license on a single step could preclude a new variety from coming to market. Not all companies are willing to license patented technologies when they perceive little value from providing a license, especially when the crop is not one of the "big five" (the four major food crops and cotton). They often couch their unwillingness in terms of "product liability" should something go wrong, or other legalistic excuses, to cover themselves. The developers of golden rice had to contend with nearly thirty patented technologies, which they did after the technology was developed. If each licensor was to demand 2 percent royalties on sales, a typical number, it is obvious that a technology cannot be developed. To overcome this issue, various not-for-profit regional groups such as the African Agricultural Technology Foundation (AATF) have been set up to "acquire technologies from technology providers through royalty free licenses or agreements along with associated materials and know how for use on behalf of Africa's resource-poor farmers and establish partnerships to adapt agricultural technology to African circumstances as well as ensure compliance with all laws governing these technologies."[1] As good and useful as this is, it adds an additional cost to providing the technologies where needed.

There are cases where patents can legally be ignored; where there is no patent in the country where the crop is to be cultivated, there are no encumbrances. When inventors take out patents, they decide where patent rights are desired, and apply accordingly. If the inventor did not take out a patent in a country, there is no reason not to freely use the information in it, except one. The crop produced cannot be exported to any country where the inventor has a patent. That is moot for crops of subsistence farmers, where the crop gets no further than the mouths of farmers' families.

In most countries where patents exist, it is even illegal to perform basic research utilizing a patent, unless one possesses a license. Among basic researchers some abide by this legality and some ignore it. The reason that typically law-abiding basic researchers will allow themselves to ignore a patent is that the maximum punishment for patent infringement is triple damages on profits from commercial use. Triple zero equals zero. That is not the case with the biotechnology outlined in this book, where the goal is to get to the field. In such cases one must decide whether to obtain a license before or after development. As noted above, the inability to obtain a license on any one element in developing a transgenic crop can prevent a crop from getting to market. Very few in developing countries are willing to call the bluff of patent owners who refuse to license under reasonable conditions, and let the patent owner sue, and then fight the case in parallel in the court of world public opinion. If the crop is for subsistence farmers, or is providing food security, it should be possible to shame most patent owners into granting a license. Some countries do have laws allowing courts to force a license, under court-mandated terms.

Another solution, which can mutually benefit all the small players in developing new crop varieties is "open source biotechnology." Such technologies are being developed by a nonprofit independent organization CAMBIA under the auspices of international philanthropies.[178] CAMBIA develops and patents technologies for all the needs of plant biotechnology, from transformation technologies similar to *Agrobacterium*,[154] through vectors to genes, careful not to infringe on the patents of the giants. CAMBIA licenses the use of their technologies, with a royalty-free license, provided that any improvements on the technology be made known publicly and be license free. This is akin to open-source Linux computer software. Having the technologies as "open source" leads to what they call "collaborative invention" as all biotechnologists working with the open-source material further develop it for all, and innovations are quickly disseminated, instead of remaining proprietary knowledge within a company. This can be of distinct advantage to those trying to increase crop diversity, by further domesticating less-grown crops, especially those in the developing world where resources are thin.

3.10. Remember the Numbers Game—Listen to the Breeders

Transformation and regeneration are tedious tasks. The efficiencies are still low, and the time line is long. Still, it is imperative to generate large numbers of transformants. This is galling, especially to the academic gene jockey who wants to get on with a proof of concept, not develop a product. The ennui of transformation and regeneration often depresses the researcher by the time the first transformants are regenerated, such that the first sufficiently good transformant will be further characterized, without further efforts at transformation.

Let us start with what the breeders know. If they are doing mutation breeding they use very large numbers. Most mutations are lethal. Most of the surviving mutants are substandard in one way or another, and of the few remaining mutants, many that look good in a laboratory or greenhouse are poor in the field. With hybridization breeding it is almost the same. Most hybrids formed from crosses do not perform well. At early stages a large proportion of the better progeny are taken to the next level of testing. This continual use of many families is continued almost to the end of getting a new variety. Even then, many new varieties do not make it through extensive field trials and get released. Breeders use large numbers, with good reason.

So it is with transgenic regenerants and their progeny. First, there is the issue of stability. Plants have tricky ways to eliminate alien genes. In many species one can only be pretty sure that the transgene will not be eliminated from the genome until after the fifth generation. Thus, the claim of detractors of biotechnology that most transgenic lines are unstable, and that there are insertional mutations due to the randomness of present transformations is true in many instances.[635] What they do not say is that unstable lines are not released, but one may have to test many lines for a long time to find the stable lines for commercialization. They also do not provide a baseline comparing the effects of transformation with acceptable (to them) mutational breeding, or even the natural occurrences of mutations, gene deletions, rearrangements, and so on that occur naturally. Because transgenes are randomly inserted, an insertional effect can occur; the lethal insertion effects do not count, as they die young and we do not see them. Slightly deleterious effects cannot be seen in the greenhouse. Even if we have a plant with the desired property (e.g., disease or herbicide resistance), if there is a 5 percent yield loss, it is unlikely that it will get to market. It is hard to see a 15 percent yield reduction in the laboratory or greenhouse. Thus long-term field testing, typically three seasons, of many lines is imperative before release of the best. Added pieces of noncoding DNA or missing bits of DNA are all parts of the natural processes that occur that allow us to tell individuals apart, and unless shown to have consequences, can be considered inconsequential. Unfortunately, the detractors want proof that the insertional events are inconsequential, knowing full well that you cannot prove a negative.

Developing transgenics is like standard breeding: It is numbers and more numbers of transformants to obtain a product. In the end, the laboratory phase of developing a new variety is not the major expenditure, an issue those in basic research often forget.

CHAPTER FOUR

Biosafety Considerations with Further Domesticated Crops

Ascertaining the safety of any new product can be a contentious issue. Some individuals are intuitively against all current novelty despite their willingness to accept the novelty of a generation ago. The better regulatory systems weigh novelty against present practices, demanding scientific evidence of lesser risk than with current products or procedures. Where regulatory systems are meant to benefit the majority of the population, they utilize experts in science-based risk benefit analysis to perform a balanced evaluation of the data at hand. Scientists without such background can be wrong, especially when dealing with areas beyond their expertise: for example, Alfred Wallace, who in parallel with Darwin developed evolutionary theory, campaigned vehemently against smallpox vaccination, despite the obvious and documented advantages of the new procedure.

In our age, the transgenic technologies have incited similar, often illogical campaigns about the safety of the products, despite a level of regulatory scrutiny never seen before in dealing with a technology and its products. In the past, regulations were only enacted after a novel product was shown to have clear hazards. Part of this increased scrutiny on the products of biotechnology came from the responsible realization of the pioneering scientists involved, who a few decades ago called for a brief moratorium to develop a biosafety evaluation system, before the technology of gene transfer was used beyond microorganisms.[112] A wide array of often-conflicting regulatory systems has subsequently evolved, each system having very different philosophies, and too often not based on a consensus of expert scientists dealing in the area. Is it justified to use political/economic views and emotions to deal with risks?

Two risk issues need to be dealt with: (1) Is the product of biotechnology safe to be used or consumed? (2) Will the gene encoding the product introgress into other varieties of the crop or into related species in an unwanted manner, posing a danger to human safety or to ecosystems? Then it must be ascertained whether these risks are greater or less than for presently accepted and used technologies.

The regulatory frameworks dealing with this vary and will be discussed briefly and conceptually. Clearly we expect that the scientists using the tools of biotechnology should do so within the legal framework having jurisdiction, and more importantly, the scientists should prudently evaluate the biosafety of their products by using the best tools of science, independently of such frameworks.

The Canadian regulatory system evaluates all plants with novel traits, no matter how they were developed; transgenically or by breeding. They know of the cases where standard breeding gave rise to poisonous potatoes. Potato breeders know their crop, and saw to it that poisonous varieties were never released, although some with marginal toxicity such as cv. Lenape, the product of a cross with a wild potato were released and then taken off market when it was realized that this variety was just above the threshold.[828] This kind of problem is more likely to ensue from using whole-genome breeding than from excising only the needed genes from the wild and inserting them transgenically. To give a modern example, nontransgenic varieties of disease-resistant celery were developed in the United States[275,484] and are widely grown by organic farmers. They contain inordinately high levels of furanocoumarins (psoralens), known carcinogens, yet are under no regulatory scrutiny in the United States or Europe. These compounds act as "natural, organic" fungicides, replacing the registered fungicides normally used to prevent damping-off diseases. No one would dream of trying to receive regulatory approval of such mutagenic poisons such as psoralens as fungicides, knowing full well that there would be no approval. The organic seed companies know better than to try to register such varieties for cultivation in Canada, even if they are available elsewhere. The quirk in the system is that Canadian retailers can import the high-psoralen organic fresh celery with impunity from the United States.

The U.S. regulatory system deals with each transgenic case separately (but not with genetically bred crops, even when genes are brought in from a poisonous wild relative). The U.S. system is based on the traits and products in a multitiered system whereby some products are easier to register than others. Those where no novel proteins are expressed (i.e., results of gene suppression) and those where there is similarity to previously deemed safe products, or where less safe practices are replaced, all receive less scrutiny than plants producing novel products.

Europe has a system where the manner in which the gene was introduced and the promoters and selectable markers used are scrutinized and where the gene came from is far more important than what the transgene actually does. The system demands near absolute freedom of potential risk. Furthermore, the contemporary technologies or situations being replaced by the transgenic are not considered in practice. Thus, the use of Bt is not compared with the toxicological or ecological risks from insecticides, nor is the large drop in the fumonisin mycotoxins that occurs when Bt is used considered in the European regulatory deliberations.

It is hoped that by the time the products of transgenic domestication envisaged in this book are generated, that the regulatory structures around the world for their release will have been rationalized such that safe products will be released with greater ease, to the benefit of consumers and farmers. The problems with the current regulatory structures, along with intelligent, science-based suggestions for solutions, have recently been reviewed at length in a very thoughtful manner.[143]

Biosafety considerations should begin in the planning stages of a transgenic crop; the developers should not wait to start thinking about implications until after constructs are made and transformants are regenerated and tested. For example, a decade ago researchers then transforming oats were warned not to use herbicide resistance as a selectable marker, because of possibilities of gene flow to the biologically conspecific weedy *Avena* species,[408,977] some of the worst temperate grassy weeds in existence.[509] Nearly a decade later researchers were still using herbicide resistance as the selectable marker for oats.[824] They also did not deliberate on whether the salt tolerance encoding genes, their primary trait, might also provide a selective advantage to this pernicious weed, and they will have to consider containment and mitigation strategies. Thus, to have a product, they will have to start anew with new constructs designed to contain the genes within cultivated oats, or mitigate their flow to the weedy relative, and/or consider using a different selectable marker.

4.1. "Event"-Based Regulation, an Impediment to Increasing Crop Biodiversity

One of the many advantages of transformation technologies is that once a successful gene construct has been made, it can be used in different crops and different varieties of each crop, so that unique special traits in a variety need not be lost. This is the "science," but not the reality in many jurisdictions, especially the ones where gene technology has been used the longest. The reality is that the progeny of a single transformant (a single transformation "event" in legal terms) goes through the regulatory process, with all the necessary toxicology and allergenicity testing, and only it and its genetic progeny are registered. If the same construct is used to make transformants of other varieties of the crop, or is used in other species, the toxicology and allergenicity testing data do not hold for the new transformation events. The regulators seem to believe that the gene product of a given gene can be different in each transformant made with the gene. Thus, despite the ease and the rapidity that different varieties can be transformed, this is not done; the best that can be done is to backcross the registered event into other varieties. In many cases this cannot be performed easily or effectively because of the polygenic nature of quality-defining traits of many varieties, for example, basmati rice or long-staple cotton. For this reason only Bt short staple cotton has been available until the recent arduous backcrossing/selection from short-staple varieties in Egypt into their excellent long-staple varieties.[1058]

Even where an "event" can be backcrossed easily into most varieties of a species, there can be a severe drain on the genetic biodiversity of that crop that may not be readily apparent. After the multiple backcrossings into many varieties, there were claims that the genetic diversity of cotton[140] and soybeans[1001] had been regained. These are measures of average diversity, but does mean that there is full diversity, as the genes nearest the transgenic insert are closely linked to it and will not be replaced. Already more than 70 percent of the world's soybeans are transgenic (resistant to the herbicide glyphosate), all derived from a single transformation "event" by backcrossing. Soybean is a crop with needs for many highly local varieties based on climate, soils, and day and season lengths, yet it is a commodity crop with few considerations for quality. It proved easy to backcross this event into the multitudes of varieties for use where legal, and sometimes where not, as in Brazil where a third of the crop was transgenic for a few years before being legitimized by the politicians. What are the implications of this single event? Because soybean seeds are often saved by farmers, and are not hybrids derived from disparate parents, the gene for glyphosate resistance must be on both members of the pair of chromosomes (homozygous), or else part of the crop may not be resistant in the next generation. Thus, there is no genetic diversity near the transgene, providing some "linkage disequilibrium." Gene(s) linked to the transgene that can be deleterious under certain circumstances remain in all these transgenic soybeans. Can we discount this by saying that we have been growing these soybeans for a decade, and nothing has happened? Those with a short memory can indeed state this with surety that nothing will happen. Those with a longer memory remember that most of the world's hybrid maize had a similar "linkage disequilibrium" due to a gene for male sterility being on a bit of the mitochondrial genome for a few decades without problems.[1084] That is, it was without problems until one wet summer, when the world learned that the gene for male sterility happened to be linked to a gene for disease sensitivity, leading to severe crop losses, and a lesson learned.[1084] Learned then, but still remembered?

Event-based regulation makes it too costly to register different transformants, with the same gene located on different soybean chromosomes in different transformants. Each would have had to go through all the regulatory hoops, from the beginning. Is this good for the future of soybeans or any other major crop where all extant varieties derive from a single event? History says no.

There is a finite risk from having different varieties with the same cassette transgenically inserted on different chromosomes. They could mix by crossing, resulting in plants having multiple copies of the same cassette. Still, it is interesting how breeders have breeding plots with different lines adjacent to each other, without problem, as one percent crossover is not deemed a problem, and it is usually far less. Let us assume that two varieties are next to each other, each with the same cassette on a different chromosome, and they do cross. One percent at most will have double the number of copies, which will either be without an effect, or at worst there may be cosuppression. If one percent of the population will be less fit because it is the equivalent of non-transgenic due to cosuppression, it will not reproduce and individuals with the doubled cosuppressing copies will disappear. Thus, the risk exists but seems manageable.

The costs of performing toxicology and allergenicity testing, as well as environmental impact assessments are considerable, and justified when a novel gene encoding a new trait has been introduced into a crop for the first time. Can one explain how the same gene can have different allergenicity, toxicity, or environmental impact if it is located on a different chromosome in the same species? The only known risk is that the gene cassette may have inserted in a crop gene, disrupting it. This would be apparent to the breeder from the defective phenotype, and that "event" would be culled. Thus, event-based regulatory costs clearly present an impediment to enhancing diversity by engineering the same gene into "minor" crops, or underused crops, or specialty varieties, or landraces of the same crop. The economics may force untransformed crop species or varieties off the market, when the competing transgenic variety can be grown more cost-effectively, lessening crop diversity. An example of this is transgenic soybeans displacing conventional soybeans and other nontransgenic pulses.

Many developing countries already cultivate very few crops (e.g., monocultures of rice in southeast Asia, maize in large parts of Africa), and these are the areas where crop diversity is needed the most, for better, balanced nutrition and for food security. These same countries are presently formulating their biotech policies and regulations. Those developing the policies should understand the need for "preserving and enhancing crop biodiversity" and should avoid the mistaken direction of regulation based on "transformation events"; instead, they should base regulation on transgene function. Event-based regulation has delayed and precluded transforming many varieties, as well as transforming minor crop species with useful traits, which leads to less and less crop biodiversity.

No great scientific justification exists for going through the same whole regulatory process for the same gene construct (or the same gene with a different promoter), when transformed into a new variety or crop, whereas justification does exist for a limited regulatory package. Only the large multinationals gain from the status quo of transformation "event"-based regulation, as it keeps less well-financed competitors (the public sector and nascent biotech and seed companies in the developing world) off the market and keeps the multinationals' older material on the market. The world has taken on a hysterical paranoid approach based on the precautionary principle,[35] instead of the learned approach based on experience, termed "familiarity,"[677] which had been more prevalent. It may be harsh to use the term hysterical paranoia, but detractors have been crying "wolf" for decades now without any untoward and clearly documented negative effects from any of the transgenics released. There are few if any new technologies (or old ones) with such an excellent safety record. Immense resources have been invested in testing transgenics, and then for testing whether they are in foods, resources that would be used far better to test for prions, microorganisms, and toxins in foodstuffs. Familiarity had been used until paranoia prevailed. Familiarity allows one to use resources efficiently to study the unknown, instead of repeatedly asking the same questions of the already known.

Thus, there is a need for a multitiered, science-based system, which adequately deals with new transformation events with known genes to ensure safety, but not to start as if it was a total unknown. The same goes for stacking known genes. At present stacking is considered a new event, and the complete process starts anew. That precludes putting excellent known genes together in a package the farmer needs. Countries adopting a gene-based approach will be in the forefront, and their farmers will have a wider spectrum of appropriate varieties for each crop, and a wider variety of crops to sow and market, to the farmers' benefit and that of the national economy. Conversely, those following event-based regulatory systems will put their agriculture and economy into a disadvantageous position.

4.2. Food Safety

The basic issues of food safety are covered in an excellent, long review by Konig et al.[603] All the reviewed findings basically indicate that there is no evidence for an undo lack of safety with presently released transgenic crops. These findings will not be reviewed here. So many of the genetic traits needed to further the domestication of abandoned crops involve the downregulation of genes endogenous to the crop, and thus no new protein is made. Endogenous proteins may be made at different times or levels, or may be suppressed, which adds minimal risk to the product. The transgenic suppression of plant hormones to achieve shorter, higher yielding crops should need no more regulatory scrutiny than when this is done by breeding, yet there is more scrutiny than when this is done by repeatedly spraying hormones on plants to achieve the same ends. It is far harder and more time consuming to achieve dwarfing by breeding, as such traits are typically recessive. Transgenically, the same traits are dominant and without linkage effects. Many have called for the deregulation, or minimizing the regulation of such modifications, at least from the consumer safety side.[143] Indeed consumer safety can be enhanced by such technology. The further domestication of soybean by RNA interference (RNAi) to suppress the production of allergens that affect 30 percent of infants using nontransgenic soy milk formulas is a case in point.[496, 1158]

Non-plant proteins are often on or in our foodstuffs, without regulation, or regulation accepts them. Examples of this are the insecticidal *Bacillus thuringiensis* (Bt) whole bacterial sprays, so heavily used in organic agriculture, where all the other products of the bacteria remain with the Bt toxin as residues on the crop. If the single toxic Bt protein is in the crop because of plant expression, there is new scrutiny. Plant viral gene products are typically found in large quantities in foodstuffs; for example, much of the cabbages and other crucifers we eat contain cauliflower mosaic virus, yet the presence of infinitely smaller amounts of just its promoter raises the ire of some, when it is present transgenically in these vegetables.

So much of domestication has been to select/breed out alkaloids, poisons, and off flavors—compounds that provided resistance to insects and diseases. This required the subsequent use of fungicides and insecticides (subject to regulatory scrutiny). The pesticides are typically safer for the consumer than the poisons that were bred out. The pesticides are being replaced by transgenes encoding for new molecules, which should be demonstrated to be safer than both the endogenous compounds eliminated by domestication and the pesticides being currently used.

4.3. Gene Flow—Estimating and Overcoming Risks

4.3.1. Is All Gene Flow Bad?

Gene flow is a continuing process. It was going on before the advent of transgenics and will continue after transgenics. Gene flow is the source of so much of the biological diversity that everyone wishes to maintain. The paradox is that some who proclaim the loudest that biodiversity must be maintained also wish to stop gene flow, and they use politically laden terms such as preserving "genetic purity," which has the same connotation as preserving "racial purity." They have forgotten the terms "inbreeding depression" (which is what one can get from trying to maintain genetic purity) versus "hybrid vigor" (which one can get by mixing diverse gene pools). There has been gene flow from commercial varieties of crops to/from landraces growing nearby, only to the betterment, at times, of one party or the other. The farmer preserves the landrace, morphologically, tastewise, but actually (inadvertently) selects for individuals that have also picked up genes for disease or stress tolerance, or higher yields. This is especially apparent with the steadily improving maize landraces selected by Mexican farmers.[1160] The landraces of a century ago are not genetically identical with those two centuries ago or today, even if the farmers think they are. Those that desire no change are asking for extinction, based on the "red queen" principle that claims that a species must continue evolving to retain its ecological balance with other species that are also evolving.[1098]

Thus the cry to preserve the present landraces, and prevent them from exchanging genes with neighbors,[385] can only lead to the extinction of the landraces, not their preservation. Some transgenes, if they were allowed to flow to landraces, would clearly help their preservation and even expand their cultivation. If Bt genes would flow to a nearby landrace from a commercial hybrid, the farmer would happily select the landrace without larvae, as long as the crop looked like and tasted like the landrace. Thus, in discussions of gene flow, the question "so what?" must be a recurring motif. If there is a danger, it should be dealt with; if something good can come from it, let it be. The models used to show that gene flow to landraces can decrease diversity[385] ignore the farmer, who by selection keeps the landrace similar to but not identical with its forefathers, with the useful genes it picked up from strangers. The models of landrace extinction due to gene flow between wild species and related crops[539] would not come out to support the modelers' preconceived notions, if they would have incorporated the intelligence of the farmers.

In risk analysis it is first necessary to ascertain whether the trait will have a selective advantage (e.g., herbicide resistance) or be undesirable (a pharmaceutical trait) if the trait were to introgress in another variety of the crop, in a related weed, or a related wild or to ruderal species. The vast majority of our major crops (calculated as area planted) have no related wild or weedy species outside their centers of evolution (Table 6). This is not necessarily the

Table 6. *World's 25 Most Important Food Crops, Related Sexually Compatible Weeds, Weed Importance, and Weed Geographical Distribution*

Rank	Crop	Scientific name	World area planted[a] (M Ha)	World yield[a] (MT)	Related weeds: compatible (or not) with crop[b]	Weed importance[c]	Citation: compatible with crop	Geographical distribution
1	Wheat	*Triticum aestivum*	208	557	**T. aestivum**			Nepal[507]
		T. turgidum durum						
					Aegilops cylindrica		965, 1192	Turkey and USA[507]
					A. tauschii		1096	?Mediterranean; Iran[507]
					A. triuncialis		1096	?Mediterranean; Morocco and Turkey[507]
					A. ventricosa			?Mediterranean; Morocco[507]
2	Rice	*Oryza sativa*	151	585	**O. sativa**	Group C	201, 633	Worldwide; >50 countries[508]
		O. glaberrima			O. glaberrima		201, 633	West Africa
					O. barthii	Group C	201, 633	Sub-Saharan Africa; Nigeria[507]
					O. longistaminata*		201, 633	Sub-Saharan Africa
					O. rufipogon	Group C	201, 633	Continental and insular Asia to New Guinea and northern Australia; Latin America; Bangladesh[507]
					O. punctata	Group C	201, 633	Nigeria and Swaziland[507]
					O. officinalis*: not	Group C		

No.	Common name	Latin name			Taxon	Group	Ref.	Distribution
3	Maize	Zea mays	141	636	Z. mays ssp. mexicana		574	Mexico[574]
4	Soybean	Glycine max	84	190	G. soya		986	Northeast Asia: Korea, Taiwan, Japan, northeast China, Russia (Siberia); Japan[507]
5	Barley	Hordeum vulgare	55	139	H. vulgare			Argentina[507]
					H. spontaneum		471	Eastern Mediterranean to Iran and west central Asia; Iran and Jordan[507]
					H. murinum: not	Group C		Worldwide[508], 28 countries[507]
6	Sorghum	Sorghum bicolor	44	59	S. bicolor			Africa; USA[507]
					S. almum			Argentina, Australia, South Africa, and USA[507]
					S. halepense	Group A6	62	Worldwide[470], native southwest Asia and adjacent Africa; 51 countries[507]
					S. propinquum			Southeast Asia; Philippines[507]
7	Millet	Eleusine coracana	35	29	E. coracana ssp. africana		263	West Africa
					E. indica: not	Group A5	not[263]	Worldwide[509], 74 countries[507]
		* Pennisetum glaucum			P. sieberanum*		263, 689	West Africa, northern Namibia

(continued)

Table 6. *World's 25 Most Important Food Crops, Related Sexually Compatible Weeds, Weed Importance, and Weed Geographical Distribution (continued)*

Rank	Crop	Scientific name	World area planted[a] (M Ha)	World yield[a] (MT)	Related weeds: compatible (or not) with crop[b]	Weed importance[c]	Citation: compatible with crop	Geographical distribution
					P. purpureum: not	Group B	not[263]	Africa; Central America; Australia[509]; 21 countries[507]
8	Cotton	*Gossypium hirsutum G. barbadense*	32	57	*G. hirsutum,* feral			Mesoamerica and Caribbean USA[507]
					G. tomentosum: compatible?			
9	Beans, dry, green, and snap	*Phaseolus vulgaris*	28	26	*P. vulgaris*:* weed-crop-wild complex		102	Peru, Columbia
10	Groundnut (peanut)	*Arachis hypogaea*	26	37	*A. hypogaea*		n/a	Taiwan[507]
11	Rapeseed (canola)	*Brassica napus, B. rapa*	24	36	*B. napus*			Europe, Argentina, Australia, Canada, USA; 7 countries[507]
					B. juncea		361	Australia, Argentina, Canada, Fiji, Mexico, and USA[507]
					B. rapa (B. campestris)	Group C	438, 556, 1134	Worldwide (temperate climate); >50 countries[508]

No.	Common name	Crop species			Wild/related species	Group	References	Geographic distribution
		Hirschfeldia incana (*B. adpressa*)					639	Europe, Australia, south Africa, Argentina, USA; Argentina[508]
		Raphanus raphanistrum				Group C	223, 898, 1134	Worldwide (temperate climate); 65 countries[508]
		Sinapis arvensis (*B. kaber*)				Group C	638, 756	Worldwide (temperate climate); 52 countries[508]
12	Sunflower	*Helianthus annuus*	21	26	*H. annuus*		654	Mexico, South America, USA; 11 countries[507]
					H. petiolaris		900, 901	USA[507]
13	Sugarcane	*Saccharum officinarum*	20	1350	*S. officinarum*		539, 906	Taiwan[507]
					S. spontaneum	Group C	539, 906	Asia, Africa, Middle East, Mesoamerica; 33 countries[507]
14	Potato	*Solanum tuberosum*	19	311	*S. dulcamara:* not		not [711]	Belize, Canada, New Zealand, Turkey, and USA[507]
					S. nigrum: not	Group B	not 711	World-wide[509]; 68 countries[507]
15	Cassava	*Manihot esculenta*	17	188	*M. esculenta** *Manihot* spp.: all *M. reptans*		all 543, 778 779	? southwestern USA south to Argentina

(*continued*)

Table 6. World's 25 Most Important Food Crops, Related Sexually Compatible Weeds, Weed Importance, and Weed Geographical Distribution (continued)

Rank	Crop	Scientific name	World area planted[a] (M Ha)	World yield[a] (MT)	Related weeds: compatible (or not) with crop[b]	Weed importance[c]	Citation: compatible with crop	Geographical distribution
16	Oats	Avena sativa	13	26	*A. fatua*	Group A13	1057	Worldwide[509]; native to Europe, North America, Middle East and Central Asia; 56 countries[507]
					A. sterilis	Group A13	1057	Europe, North America, Middle East, and Central Asia[509]; 18 countries[507]
17	Oil palm	Elaeis guineensis	11	139	None			
18	Coffee	Coffea arabica C. canephora	11	7	None			
19	Coconut	Cocos nucifera	11	50	*C. nucifera*★; feral populations		475	
20	Chickpea	Cicer arietinum	10	7	None			
21	Sweet potato	Ipomoea batatas	10	137	*I. trifida*		276	Central and South America; Honduras and Mexico[507]
					I. aquatica: not	Group C	not 276	Africa, Asia; 32 countries[507]

	Common	Species	Area	Yield	Weed	Group C	Geography
					I. triloba: not	not 276	Asia, Caribbean, Meso- and South America, USA; 22 countries[507]
22	Cowpea	*Vigna unguiculata*	9	4	**V. unguiculata***	889	Niger, Nigeria (roadside weed)
23	Olive	*Olea europaea*	9	17	**O. europea***	1203	Mediterranean basin
24	Rye	*Secale cereale*	8	16	**S. cereale**		Argentina, Finland, Iran, Turkey, USA[507]
					S. montanum	1028, 1203	Mediterranean basin east through Turkey to Iraq, Iran; Turkey[507]
25	Grape	*Vitis vinifera*	7	62	*Vitis* spp.	all 819	
					V. aestivalis		USA[507]
					V. candicans		USA[507]
					V. hastata		Malaysia[507]
					V. rotundifolia		USA[507]
					V. rupestris		USA[507]
					V. tiliaefolia		USA[507]
					V. trifolia		Honduras[507]
					V. vulpina		India[507]
							USA[507]

Source: Warwick and Stewart,[1133] with permission.

[a] Area of production (million Ha) and world yield (million metric tons) for 2003 from the FAOSTAT website, http://faostat.fao.org.

[b] Species name in bold italic, listed as a weed in Holm.[507] *Bold italic, listed as a weed in Global Compendium of Weeds at website, www.hear.org/gcw/index.html.

[c] Holm Classification: Group A ranked 1–18 by Holm et al.[509]; Group B ranked 19–76 by Holm et al.[509]; Group C ranked as one of 104 worst additional weeds by Holm et al.[508] Unclassified weeds are not in the worst 180 weeds.

case of many of the crops to be domesticated, as described in the case histories below. Some of those are closely related to pernicious weeds. Still many of the domestication traits to be introduced will be neutral per se or unfit in wild or weedy species, just because they confer domestication properties (e.g., antishattering, antidormancy, and so on). In conventionally domesticated crops these traits are typically recessive (as described in section 2.1) and with transgenics they will be dominant. Although the dominance was bandied as an additional risk from gene flow,[481] a more in-depth analysis demonstrates that there seem to be no real cases where this can be perceived to be an issue.[416] One must first ascertain whether there is a risk that needs to be dealt with, and then discuss methods for containing or mitigating such risks (versus the "command" to abandon considering to generate transgenics if there is a risk). If there is no risk, one need not read beyond the first section.

Some domestication transgenes used to bring abandoned crops back into wider production may confer a selective advantage should they cross into related species, including their progenitors, or even other varieties of the crop.[318] This has led to popular press hysteria about "superweeds" evolving from transgenics, with claims and counterclaims by those with opposing agendas. Actually, continuing evolution of weeds and wild species, with either often picking up neutral or desirable traits from crops, has been going on since crop domestication began.[1139] Selective crop breeding has selected for traits that are appropriate for existence in a coddled agroecosystem. A related wild species living at the edges of ecosystems has two ways (not mutually exclusive) to become a weed: to evolve traits that help it adapt to the agroecosystem, or to introgress useful traits from the crop that help in such adaptations. Examples of the latter, where the proverbial "superweeds" evolved from introgression of (nontransgenic) traits from crops to their relatives are summarized in Table 7. There was little one could do to prevent sorghum from crossing with a relative giving rise to weedy johnsongrass (*Sorghum halepense*), or to keep foxtail millet (*Setaria italica*) from crossing with its weedy progenitor *Setaria viridis*, giving rise to giant green foxtail.

Because genomics studies of weeds have rarely been performed (the sequenced *Arabidopsis* is not a weed[412]), we know not whether weeds such as johnsongrass and giant green foxtail are rare events that got out of hand, or the result of multiple crossing events. If the former, vigilance would have prevented their spread, but not if they derive from continuously recurring crossing. It is only when an invasion is small that it can be easily eradicated.[742,1056]

Table 7. Natural "Superweeds"[a] that Evolved from Natural Gene Flow between
Domesticated Crops and their Progenitors and/or Related Weeds

| Natural "superweed"[a] | Derived from natural crosses between | | Comments | Reference |
	Crop	Wild/weedy/ progenitor		
Beta vulgaris	Sugar beet *B. vulgaris*	*B. vulgaris* ssp. *maritima*	Some weedy beets from crosses with progenitor, some feral	318, 1026
Setaria viridis var. *major*	Foxtail millet *Setaria italica*	*Setaria viridis* progenitor	Cross with	252
Sorghum halepense	*Sorghum bicolor*	*S. propinquum*	Cross with related species	312
Helianthus annuus	*H. annuus*	*H. annuus*	Intermediate between cultivated and progenitor	117
Oryza sativa	*O. sativa*	*O. rufipogon*	Cross with progenitor	640, 1090, 1103, 1104

[a]The term "superweed" is used tongue in cheek to denote hybrids between a crop and a related weed that is somewhat more adaptable or weedier than the initial weed type or the progenitor. Past experience has shown that there can be gene flow between crops and related progenitors or weeds, but this has not caused weedier progeny, which could change with novel transgenes that might provide a selective advantage.

The weedy beets of Europe seem to have evolved both from crosses between sugar beet and its progenitor, as well as by de-domestication due to back mutations to feral forms.[318,1026] It is not clear whether the conspecific weedy rice strains in cultivated of rice initially evolved by crosses with the wild or by back mutation to ferality, or both. There is ample evidence for continued bidirectional gene flow between the weedy/feral strains of rice and the crop.[640,1090,1103] Crops that are still not yet fully domesticated can be contaminated by less domesticated forms that are volunteer *cum* feral weeds. They exist and mix with the crop, considerably lowering the value of a crop. Examples of this are oilseed rape, where contemporary oilseed rape varieties can be diluted by older varieties or wild forms that are still high in erucic acid and glucosinolates.[277]

All this deleterious gene flow occurred before the advent of transgenics, and the hysterical fantasies about superweeds have made people expect that the situation can become worse when transgenics introduce traits that did not preexist in either the crop or its relative. Although the potential for problems exists, the solutions for the potential problems are far better with transgenics than nontransgenics. That is because gene technology can be used to both contain transgenes within the crop and to mitigate the establishment of transgenes, should they "leak" from the crop. These possibilities do not exist with traditionally bred crops. Had these transgenic techniques to limit gene flow been put into sorghum or millet, they could have delayed or prevented the evolution of johnsongrass or the giant foxtails, giving transgenics the upper hand at preventing unwanted movement of genes. The recently developed imidazolinone-resistant rice made by nontransgenic breeding allows control of feral rice, but the flow of the gene to feral rice forms is so fast that it is expected to be useful for only a few years.[376,1193] This would not be the same if transgenic fail-safes had been used. Similar molecular techniques can be used to inhibit crops from becoming volunteer weeds and prevent them from remaining as volunteer weeds that can further de-domesticate to feral forms. In rice, most of the gene flow would be within the field, to dedomesticating volunteer rice or feral rice, just the rice one wishes to control, and not to rice in adjacent fields, to which gene flow is almost inconsequential.[919]

Genes flow among related species that do not readily cross in a process coined "diagonal" gene transfer[412] to distinguish between vertical gene transfer in readily crossing species and horizontal gene transfer between totally unrelated species. For example, a nontransgenic DNA sequence typical of hexaploid wheat, found in modified form in some progenitors of wheat, was not found in more than 90 accessions of *Aegilops peregrina* (syn. *A. variabilis*) but was found in two geographically distinct populations of that species with more than 99 percent sequence identity to wheat.[1139] Wheat and *A. peregrina* do not have any homologous (identical) chromosomes that would allow vertical gene flow, but have homoeologous (similar) chromosomes allowing some diagonal gene flow. In agroecosystems, such inadvertent gene flow may be undesirable.

Most discussions so far have dealt with "containing" gene flow (preventing its movement) from agroecosystems to "natural" ecosystems,[22,318,411,412,482,545,818,1017] but there has been only a little discussion about preventing and mitigating dedomestication of crops to become volunteer weeds within the agroecosystem, or for mitigating gene flow after it has occurred.

4.3.2. Evaluating the Risk of Gene Flow

Much misinformation, disinformation, and widely inaccurately interpreted correct information has been promulgated about transgenic traits (cf. reference 905), especially by those with an antitechnology, antibiotechnology, and/or antipesticide bias.[1016] Conversely, those with potential commercial gains from sales of transgenic crops, and/or the increased sales of the pesticides to be used with them, portray transgenic crops as a risk free panacea to agriculture. The detractors often couch their agenda in political, moral, or environmental terms. Not all moral philosophers[593] or environmentalists[643] share these radical views, indeed they counter them with cogent arguments. We usually hear the radical cries of the pseudoenvironmentalists, while missing the more muted but cogent messages of environmentalists with realistic concerns. We are warned by the radicals that these crops can lead to the evolution of "superweeds" that will inherit the earth.[594] The rapid commercial release of such crops, often without broad-based scientific scrutiny about issues such as gene flow, leads to a certain degree of public skepticism about the needs, utility, risks, and values (beyond profit) associated with the use of transgenic crops. This skepticism continued into the scientific community when there were planned releases of herbicide-resistant wheat and rice, or pharmaceutical crops too close in genetics or geography to conventional varieties of the same crops. This was even evident in the overreactions within the biotechnology industry when mishaps occurred.[357] In risk analysis, there are issues that one needs to know, and there are data that would be nice to know. The line of demarcation between these is often debated and debatable, and the claim within the biotechnology industry that one only needs to know what regulators demand is scientifically untenable.

The severe pressures of the antitechnology groups on policymakers preclude much public-sector research in this area, which affects obtaining accurate information about the dangers from gene flow. These pressures also prevent domesticating/redomesticating crops where the agrochemical or seed industries perceive little profit. The situation is further complicated by well-meaning scientists who are drawn into the debates, but who lack the knowledge to balance the issues and then make scientifically untenable extrapolations from the known data. Frequent misunderstandings are caused by ecologists who normally work in natural habitats with a much slower pace of succession; crop fields are much more dynamic in their whole ecology, and many phenomena have to be seen in a short-term perspective (K. Ammann, pers. commun.). The agronomic needs for and benefits of transgenic herbicide-resistant crops are discussed in a well-balanced book with sections by detractors,[293] and more recent books cover the many other traits being inserted.[318,336,371,1019] Discussions of transgenic crops have often dealt with the purported environmental risks but have rarely dealt with the risks from a weed biology/science perspective, yet the major stated risk claimed by the detractors is of transgenic crops becoming volunteer weeds or introgressing with a wild relative rendering it weedier; the so-called superweeds. An attempt at such an assessment, based on weed science, containing a defined set of uniform criteria set in a decision tree format, was published for herbicide resistance traits alone.[422] The United Nations Industrial Development Organization (UNIDO) has expanded on that decision tree in a computerized internet version of such decision trees that can be a useful adjunct to risk analysis.[1085] Decision trees, by requiring discrete answers to sequential, stepped questions, lower the bias in arriving at conclusions vis-à-vis the risks deriving from a given hazard, but such decision trees should not be overinterpreted to be the ultimate decision mechanism.

4.3.2.1. Geography of Risk Assessment

Risk assessment must be performed on a local or regional basis, because the risks emanating from the gene flow of the same transgenic crop may vary from one agricultural ecosystem to another. Risks have been assessed on a case-by-case basis for transgenic crops presently released in all major law-abiding jurisdictions. The scientific criteria used in various countries have not been uniform. The lack of uniformity is so great that most countries have a double standard that delineates between transgenic and nontransgenic crops. Only Canada seems to have overcome the double standard and deals with all novel traits. Indeed, just before dealing with oilseed rape, their first transgenic crop, they delineated criteria "to evaluate plants with novel traits,"[47] and then specifically evaluated oilseed rape in the context of these criteria.[46] In a series of documents they further evaluated various herbicide-resistant oilseed rapes.[48,49,51] One should realize that the decision process was based on their perception of the risks to regional agricultural ecosystems in western Canada and on the scientific knowledge of the time, but not on the risks to other regions (including the eastern provinces) that may be importing the same crops, where they could be risky. The risks to other agroecosystems can be lesser or greater, even within Canada, such that questions have been asked about how to deal with such risks when a commodity moves from one geographical area where there is no risk, to another where there is.[409] In the case of oilseed rape, it is claimed that there are no weedy relatives in western Canada, but there are weedy relatives in the eastern provinces, and just south of the border into the United States from the western provinces and in many importing countries. The importing countries typically have quarantine measures that prevent bringing in tiny amounts of seed for sowing, but not mega amounts for processing, even though seeds fall off of vehicles.

On a more international scale the Organization for Economic Co-operation and Development (OECD) and UNIDO have been developing a series of "consensus documents" on the biology of various crops (with regard also to related weeds) so that there can be a common starting point to evaluate each cropping situation. The documents on oilseed rape,[52] potato,[54] and soybean[53] have been released.

4.3.3. Unimportance of Gene Source for Risk Analysis

We discuss where transgenic crops have value, where there is a hazard and their use might be contraindicated, what the implications of their use are on gene flow, as well as what precautions and monitoring are needed. How a gene got there is not as important as what the gene does to the crop, and how and whether it will move to weeds, or whether the crop will become a weed. There is little difference between mutagenesis-derived sulfonylurea or imidazolinone-resistant soybean, maize, and oilseed rape, and other crops with the same gene transgenically introduced. The most poisonous plant or virulent pathogen will have more than 99 percent of genes that are identical to innocuous organisms. Scientifically, it is important what a gene does, not where it came from. European Union (EU) regulations consider the source, based on a perceived demand of the populace who typically ask about a suitor "does he come from a good family." This is quite different from the North American who says "I don't give a damn about his parents, what can he do?" No wonder the regulatory structures are so different when the cultures that they come from are so different.

4.3.4. Generalizing from Hazards to Risks

Because of genetic variability of crops and weeds, and the variability of the traits being introduced and their effects on the crop and related species, one cannot make easy generalizations about the risks of gene flow. The first question to ask is: Is there somewhere it can flow to? The second question is: If so, where? Each case of predicting the risk of and from introgression must be evaluated based on its merits, often after performing basic biological, genetic, and epidemiological studies. Among the issues that must be considered:

1. What is the benefit to agricultural food and fiber production of having a transgenic trait in a certain crop?
2. What are the risks from and implications of having that trait pass into another variety or into a related weedy or wild species?
3. What are the risks from and implications of having the transgenic crop becoming a volunteer weed in agricultural ecosystems, in ruderal or more pristine ecosystems?

The final decision in any given jurisdiction is ultimately a balance (to use a positive term) or a compromise (to use a less positive term) between science, economics, local benefits, local values, local interests, pro and con interest and pressure groups, as well as local politics. Politicians often use "science" as a cover for clearly political decisions.[864] Still, there is good reason that the criteria for risk assessment of transgenic crops should be uniform, using universal criteria and processes of examination. This need for uniform risk assessment procedures takes on greater importance in dealing with international trade. Should one country be forced by international trade agreements to import live seeds of a commodity crop for processing, when the same seeds would be too risky to sow, yet might escape in the importing country?[409] The onus is on the importing country to demonstrate that it is not erecting illegitimate, protectionist, and artificial trade barriers. Indeed, to prevent trade wars, political expedience and compromise has led to allowing importation of commodities initially claimed to have untenable scientific risks.[864] "Science" has been used for bargaining purposes, and even more so since the precautionary approach was transformed by the EU, without giving any reason, into a precautionary principle.[35] If two countries use identical scientific risk assessment criteria for their agroecosystems, and it comes out that the risks are much greater in the potential importing country than to the exporting country, the importing country could have the beginning of a case. Even then, this case would not be foolproof; the importer would have to show that the benefits of importing are greater than the potential costs of mitigating procedures (i.e., eradication of volunteer or introgressed weeds). Indeed one could envisage an involvement of the insurance industry in risk assessment should there be a requirement that importers or exporters insure themselves against such risks. In this particular issue, the importation of seed for processing to food and feed, one of the best mitigation procedures has hardly been discussed: the use of nuclear irradiation to sterilize the seed before shipment. This would render the seed nongerminable, while killing grain pests and mycotoxin-producing fungi. At present only spices are commonly irradiated.

4.4. The Risks Associated with Transgenic Crops

The many hazards posed by transgenic crops are discussed below vis-à-vis their risk potential.

4.4.1. The Risks from Antibiotic Resistant Selectable Markers

One purported risk that is often mentioned has turned out to be a red herring scientifically, but not yet politically. This is the purported risk that the antibiotic resistances used as selectable markers will spread horizontally to soil microorganisms. It had been posited that this horizontal spread poses a great risk: that such genes will then move to human pathogens, rendering them antibiotic resistant. Besides horizontal gene transfer from crops to bacteria being a dubious and unproven possibility, despite efforts to obtain such information, and besides the well-demonstrated fact that that humans' overuse of antibiotics supports the rapid evolution of antibiotic resistance, there are good biological reasons to negate the scenario. One must remember that humans did not invent antibiotics, soil microorganisms did, to ward off competing soil microorganisms. They did this long before animals appeared on the planet. In doing so, the organisms that produced the antibiotics had to invent resistance, to keep them from intoxicating themselves. Some of the neighbors they were inhibiting also evolved antibiotic resistance.

Biophilosophy aside, is there supporting evidence that antibiotic resistance gene flow does not pose a risk? German soil microbiologist Kornelia Smalla wished to prove that there might be a danger of antibiotic resistance coming from transgenic crops and to do so started a large-scale baseline study to ascertain the frequency of antibiotic resistant microorganisms in soils where transgenic plants had never been planted. The risk of relatively easy horizontal gene flow among microorganisms is well documented, so it was important to know whether resistant organisms were present before the release of transgenic plants with antibiotic-resistant selectable markers in European soils. She and her group collected small soil samples throughout Europe. In every soil sample they collected they found microorganisms that are resistant to the antibiotics being presently used as selectable markers, except for the thin layer of soil above permafrost in polar regions.[666,994,995] The findings of the ubiquity of antibiotic resistance in soils is no surprise to the biophilosopher. The antibiotic resistant genes were originally isolated by the genetic engineers from soil microorganisms. A logical application of the precautionary approach would be to have a moratorium on the cultivation of transgenic crops where antibiotic resistance is the selectable marker only on soil above permafrost, until it can be proven safe, and allow it elsewhere. The present EU regulations will not allow the cultivation of transgenic crops with selectable marker genes that have resistance to any medicinally used antibiotic after 2008, a "minor" perversion of the precautionary approach, as it ignores the clear scientific evidence.

4.4.2. The Risks of Transgenic Crops Becoming Volunteer Weeds

Many crops in the world are, at the same time, noxious volunteer weeds in agriculture. In fact all crops that have propagules, be it vegetative or generative, that survive the off-season (winter in temperate climates and the dry season in many tropical climates) can cause a weed problem. Not all do, because it is the growing system that determines whether their weedy potential will be fully expressed. Weediness is a term that is of major importance in risk analysis. Weediness is a very complex concept. It describes the degree to which a species has the potential to manifest itself as a weed, in other words the degree to which a species may become a perceived problem to/by humans. The weediness of a species is determined by a combination of many factors.[85, 1155] The risk analysis of all these factors is important and can be judged by using local agricultural experience: does the species involved always cause economic damage if it is not actively controlled? In that case it is a major weed (or major volunteer). If economic damage does not occur as a rule in the absence of active control, but only on occasion, the species is a minor weed (or minor volunteer).

For example, more than 20 percent of oilseed rape may shatter when harvest conditions are poor (Chapter 16). This represents ten times more seed than is normally used in planting. Oilseed rape seeds may persist and appear in following crops. The control of volunteer rape in cereals is comparatively easy with phenoxy-type herbicides and incurs no cost, because the same herbicides are used to control other weeds. The control of volunteer oilseed rape in other crops may be difficult. It is practically impossible to control one rape variety in another because their response to herbicides is identical, unless the new one is a transgenic variety with a resistance different from the previous one. Separating varieties may be very important however, because different varieties may have different chemically desirable/undesirable contents (glucosinolates and erucic acid in rape on the negative side, more healthy oils on the positive side). Yet the transgenic varieties released in Canada have already intermingled, showing that gene flow was greater than had been predicted,[452] even though the problems that ensued have been deemed to be minimal,[451] so far.

Similarly, volunteer potatoes infest many subsequent crops as a competitive weed and are difficult to control. Volunteer potatoes also disable rotational systems set up to prevent the carryover of soilborne potato diseases.

Conversely, the introduction of transgenic herbicide resistant crop varieties obviously provides possibilities for management of existing volunteer weed populations of the same crop. However, in several cases this may not be a blessing, but a curse in disguise, because there is in general little or no reason to believe that the resistant varieties do not engender their own appearance as volunteers, becoming problems in other crops where the new herbicide had been used. For example, if one were to develop a transgenic 2,4-dichlorophenoxyacetic acid (2,4-D) or dicamba herbicide-resistant oilseed rape, it would be a most problematic volunteer weed in wheat, where either of these chemically related phenoxy herbicides is the key herbicide. Newly built-in modes of action might help for some time, but one should be well aware of the fact that weed (and volunteer) control very rarely reaches 100 percent. This could allow resistance gene stacking caused by each new transgenic crop crossing with its feral predecessor in volunteer populations—a real hazard.

Sexually propagated transgenic crops are obviously at a greater hazard in the field than the vegetatively propagated ones, because no natural gene stacking can take place within the volunteer or feral population in the vegetatively propagated crops. Moreover, only sexually active outcrossing crops are potentially able to rapidly spread their genes to relatives, which increases the hazard from their use. Crops not expressing weedy traits are unlikely to behave as weeds after the introduction of transgenes, because it takes more than single traits to make a plant that is weedy in other crops (see section 2.2). For example, bolting (prematurely flowering) beets are a problem in beets, but are not weedy in most other agronomic crops. A pleiotropic effect of astounding character would be needed to achieve weediness after introduction of only one transgene, especially transgenes that are designed for further domestication, as described herein. Domestication is typically (but not always) the opposite of weediness.

Volunteer weeds are not a problem unique to the introduction of transgenic crops; such problems are exacerbated by the introduction of each new selective herbicide or many other advantageous traits into cropping systems. Indeed, herbicide-resistant transgenic crops can even mitigate the problem with some crops that have a high volunteer potential. Biotechnology can allow the introduction of genes into the at-risk crop that confer resistance to herbicides that are rarely used in other crops. This allows the control of volunteer weeds with other herbicides whose resistance should not be introduced into the at-risk crop. More herbicide resistance genes are needed to attain this goal, yet we have seen a contraction in the number of transgenic herbicide resistances on the market because of poor sales of all but glyphosate-resistant transgenic crops.

Though it is not likely that agriculture will perish under a load of biotech-derived highly resistant volunteer weeds, it is conceivable that some cases may prove to be serious in an agronomic sense. For example, the high efficacy of the herbicides, the high dependency of agriculture on herbicides, and the expected very large-scale use of too few herbicides with herbicide-resistant transgenic crops together will create the stage for a worldwide selection pool for resistance. This is beginning to happen with the glyphosate-resistant crops, where many weeds are evolving resistance under the strong selection pressure due to the widespread adoption of glyphosate-resistant crops and the repeated use within seasons.[483] Still, herbicides are not the only tool to manage weeds.

4.4.3. The Risks of Transgenic Crops Becoming Weeds Outside Agriculture

Some weedy characters do pose threats to relatives residing beyond agriculture in ruderal and wild ecosystems. Disease, herbivore, and abiotic stress resistances, enhanced nutrient uptake, and so on, may confer increased fitness in the wild, depending on how spotty, sporadic, and intense the stresses.[448] The extent to which such increased fitness will enhance the proportion of a given species in the wild due to such traits is still mainly an open question.[448] Volunteer weeds can evolve into feral populations just outside agriculture. These populations are often not subject to more than incidental control and could therefore promote the genesis of feral crop types that derived from dedomesticating volunteers, and then migrating back to arable land bearing, from the plant's point of view, a most useful set of resistance genes 'for it to compete in the agricultural environment.

Crops themselves can evolve into weeds (cf. references 86, 135) and weeds continue to evolve mimicries of crops.[92, 401] This does not imply that transgenic traits per se will make a wild species or a weed weedier or less controllable. This depends on the transgenic trait, that is, with transgenic herbicide resistance on the available alternative management methods, with disease resistance on the incidence of the disease and its resistance in the wild, and so on. Although many traits differentiate between the few hundred major agricultural weed species and the (at least) hundreds of thousands of plant species, it is doubtful that any one trait can turn a wild species into a weedy one. A trait such as herbicide resistance is far less likely to confer any advantage to a wild plant outside the agroecosystem where the herbicide is used, compared with pathogen or insect resistance, or altered response to abiotic stress factors, if the wild relative does not have those traits already, as many do.[955] Many such resistances were bred out during domestication yet remain in the wild relatives.

World mobility and trade have moved thousands of wild species to new habitats, where most have remained at low density for many generations. Only roughly one percent of these naturalized (the value-loaded synonyms are alien, imported, or invasive) plant species evolved into pests of agricultural or natural ecosystems.[1070,1155] Some view this differently, viewing even the movement of an imported species, even when it establishes at low density, as being undesirable. Using the "any density" criterion, 15 percent of the ornamental or landscaping species introduced to Europe from the Americas now grow outside of human plantings.[609] In the end both groups come to the same numbers, based on a "tens-rule"; about 10 percent of introduced species spread, and of about 10 percent of those that spread, 10 percent will eventually cause problems as pests.[609]

The reasons for naturalized species becoming pests are seldom understood and depend heavily on local conditions and the history of a given place related to its biodiversity. Tropical islands with their relatively young and unbalanced flora and regions with old floras that have not been subjected to invasions for millions of years are definitely more susceptible to having naturalized species become weedy. Data are available to show that the higher the biodiversity, the more stable the plant communities.[1148] An approach to dealing with this issue would be to perform long-term ecological research and to monitor both naturalized plant species and transgenic crops after they are introduced. The conundrum is that long-term research must be performed on a large scale to be meaningful, that is, the equivalent of full-scale introduction. Active monitoring, with specialists going out to search for problems, is probably not cost effective, as the problems often take very long to evolve, and the initial problem population will be in an unpredictable place. Spread from one focus is at least as common as concurrent evolution in a number of foci. Thus "reactive" monitoring, with education of the public to be on the lookout for such problems and to quickly report them, with a rapid response plan ready for eradication of populations that get out of hand, seems more practical and cost effective than active monitoring.

4.5. Inherent Biological Factors Governing the Risk of Gene Flow to Related Species

Below are some questions one must ask in assessing the consequences of gene flow.

4.5.1. Transfer Studies

Some crops are botanically identical with neighboring weeds. Botanically identical is defined as the ability of a crop and wild species to cross and have fertile offspring without more than minor incompatibility barriers. Taxonomists have often given botanically identical (conspecific) crops and wild species different species names, because they can be morphologically distinguished, even though they meet the botanical criteria for identity, for example, *Oryza sativa* (domestic and feral rice) = *O. rufipogon* (perennial wild rice) = *O. nivara* (annual wild rice)[1103]; *Setaria italica* (foxtail millet) = *S. viridis* (green foxtail)[252]; *Avena sativa* (domestic oats) = *A. fatua* (wild oats) = *A. sterilis* (wild oats),[251] etc. Apart from the easy gene transfer by cross-pollination from crops to their wild relatives (often progenitors) belonging to the same botanical species, it is necessary to consider the "diagonal" transfer to closely related but separate species where compatibility barriers exist but can occasionally be overcome in nature. Still, the breeders long ago found that not all crops will cross with relatives, not even with their progenitors in the field, requiring lab tricks, such as embryo rescue to facilitate gene introgression, that usually do not work.

Far more apocryphal reports of crop gene introgression into weedy and wild species are available than validated reports, as this was an area that did not interest breeders who made observations that they never published. There are even fewer data from controlled field experiments. The apocryphal reports include observations that traits such as green flag leaf bred into cultivated barley, oats, sorghum, or rice soon appear in their closely related ruderal and weedy species in the same areas. It never seemed important enough to substantiate the incidence of such traits. This could be done by analyzing herbarium specimens of the wild species, before and after introducing the new traits, if herbaria remain extant. Herbaria have been used to ascertain that hybrids have been formed between wheat and many of its relatives,[38] but they do not tell us for sure whether the hybrids were infertile, or if there could have been backcrossing to the wild. Thus there is specifically a need for analyzing both the risks and implications of introgression in these cases with complimentary field experiments.[37,353]

Most early experiments showing that traits can be transferred from crops to wild relatives resorted to model or artificial systems; hand pollination after emasculation of the weed, male sterility or self-incompatibility in the weed, massive amounts of crop pollen, and/or embryo rescue of the mostly sterile rare progeny. The fear of transgenic crops precludes performing gene flow experiments in the field, especially in Europe where the fears are greatest, and the demand for more information is most intense.

The older epidemiological/apocryphal reports are actually more relevant to risk analysis than many of the artificial laboratory experiments; the older results indicate that such transfers occasionally can occur in the field, the time until predominance, and the competitive advantage (if any) of such introgressions. In rare cases natural gene flow from (nontransgenic) crops to weeds has produced weedier weeds, especially in *Sorghum* spp.[312] (Table 6).

Most studies rating risks of movement do not differentiate between the reports of field transfer and the studies showing it could occur by hand pollination. In the rush to obtain such data, erroneous information has managed to be published in reputable journals. For example, there was a report of 30 percent hybridization from transgenic pollen up to distances of 1 km, more than 70 percent when in proximity.[993] Seventy percent is double the theoretical maximum, as calculated by a group that reanalyzed the data.[233] The polymerase chain reactions (PCRs) used to obtain this fantastic hybridization rate were not controlled by assaying nontransgenic plants, nor were any other biochemical or molecular methods used to verify the PCR data, yet this artifactual study is likely to be quoted as fact. Similarly, another case (later renounced in the same journal) of the misuse of PCR is likely to be quoted as fact, claiming that transgenes have moved to Mexican maize varieties.[877] Other studies with similar material could find no supportive evidence.[826] Few studies dare to comparatively estimate how long it will take to have traits (such as Bt for insect resistance, or herbicide resistance) introgress and predominate in wild populations versus how long it would take the same insect or weed resistances to evolve by natural selection (if they were not there already), versus the expected commercial lifetime of the trait in agriculture. Recently, a classic case involved a worry that a transgenic disease resistance might move from carrots to conspecific weedy carrots = Queen Anne's lace (both *Daucus carota*). It was then observed that the weedy carrots were naturally resistant to the disease, and gene flow could provide them with no advantage.[955] This would not be the case for carrots engineered for herbicide resistance, to withstand parasitic *Orobanche* species.[75] In the latter case there is a risk that must be compared with the devastation caused by the parasite. To paraphrase Canadian rulings concerning oilseed rape, the advantage of being able to get the control of the weed is greater than the possible problems of gene flow, as long as adequate means are available to control the wild carrot. Other jurisdictions might disagree with this pragmatic, practical approach.

4.5.2. Weediness

Would the newly transgenic crop really be weedy once the selector providing an advantage is no longer present? There is an analogy and lesson from the wild *Brachypodium distachyon* living on gravelly sand stone outcroppings that evolved triazine resistance along roadside shoulders, where crushed sandstone was used as substrate.[428] When triazine usage stopped, the weed species reverted to being the minor ruderal and wild species it had been before it evolved resistance.[418] If the wild type was never weedy, why expect the transgenic biotype to be more fit? Too often one forgets that most ruderal species lack weedy properties, which is why they are ruderal. A transgenic trait is unlikely to tip the balance. It is not the single trait that differentiates between weeds and ruderal species, it is the difference in response to two ecologically different environments.[418]

A change in growing system may well mean an opportunity for a ruderal weed to become a real weed as with *Bromus sterilis* in the United Kingdom when British wheat farming went to minimum tillage.[704] Indeed, if the herbicide is used in the herbicide-resistant transgenic crop, then susceptible weeds growing in its midst will not introgress the resistant genes; dead weeds do not readily have sex, although a few weeds might remain in a sprayed field or near it. The genes could introgress into nearby unsprayed weeds. Seed set from oilseed rape on emasculated plants was measured 1.5 km from pollen source.[1063, 1064] Can the progeny compete or survive in feral populations without selector? Without emasculation, resistant pollen fertilized 24 percent of conspecific plants in the immediate vicinity but <0.017% just ten meters away. Others have found similar results with this crop,[367] as well as with rice.[919] Escape to more benign environments allowing cross-breeding may also occur in time, via the seedbank. If transgenic crop seeds carry over as volunteers to other crops where related species are serious weeds, gene exchange might happen through this route, and long-distance dispersal becomes unnecessary.

4.5.3. Fitness

Does the transgene provide traits that can increase fitness when the selector is not present (e.g., when the herbicide is not used with herbicide-resistant crops, the disease or insect pressure with those resistant crops, or the mineral deficiency with mineral-mobilizing crops, and so on)? An unequivocal "no" is hard to provide; a transgene might conceivably have other unexpected pleiotropic traits. Herbicide resistance potentially confers less feral fitness advantage than disease or insect resistance, or resistance against adverse abiotic conditions such as drought or cold. This is because most herbicide-resistant genes are intrinsically unfit,[113] and hybrids outside of agricultural and ruderal ecosystems are not likely to be confronted by the herbicide. As long as the weed can be controlled by other means, the unpredictable is unlikely to lead to Frankenstein-like superweeds. Continual monitoring for newly resistant weeds is always called for, not just for those evolving from introgression with resistant crops. It is easier to eliminate nascent resistant foci than huge areas after spread.[742,1056] It becomes exponentially harder to deal with larger infestations, requiring a readiness for fast responsiveness to problems. Hybrids bearing insect or disease resistance can confer a fitness advantage or disadvantage in crosses with the wild, depending on the pest pressure, see reference 448 for an excellent and critical review. Few have addressed the quantitative aspects of fitness advantages from transgenes; assuming there is an advantage, how much of an imbalance will it cause in the wild? Will the recipient species predominate, or will its proportion in the population increase by a statistically significant increment that has little ecological significance?

A lack of knowledge about the unfitness of Bt resistance and the number of genes involved led to erroneously modeled predictions based on assumptions of a single target for Bt, resistance that would be dominant, and the expected relatively neutral fitness of the mutants, all allowing rapid evolution of resistance.[416] The modeled results set into motion a whole scheme of Bt resistance management with refuges, which cost compliant farmers quite a bit, and profited noncompliant farmers with much higher yields at lower production costs. The level of noncompliance was high enough, though, to discredit the models, which should have been based on at least two targets, and individuals resistant to the levels of Bt in transgenic crops being highly unfit.[416,1039] This should teach us that we really need to know quite a bit about modes of action and unfitness genes so that we can allay false fears or preemptively deal with real problems.

4.5.4. Opportunity for Outcrossing

Are the wild interbreeding species near the crops? If they are on different continents or far-removed ecosystems, what is the problem? Conversely, engineering glufosinate resistance into domestic *Avena*[824, 1003] in an area where interbreeding wild oats is a problem was inexcusable when other selectable markers could have been used.[408] Because small-scale separation cannot be legally enforced in practice, it would be wise to organize separation of cultivation on a regional scale, along the lines practiced in standard quarantine listings used by the regional plant protection organizations such as (North American Plant Protection Organization (NAPPO) and European and Mediterranean Plant Protection Organization (EPPO).

The outcome of outcrossing in forest trees will take decades to measure. Many years will pass before transgenic forest trees will have a chance to outcross with wild relatives, as many years will pass before they reach sexual maturity, and many more years will pass until the progeny reach maturity. With transgenic herbicide resistance there is little risk of gene flow increasing the invasiveness of forest trees, as herbicides are used only during early forest establishment. A rare result of introgression with a wild species will have but little chance to use its introgressed trait. Other transgenic traits should have greater risks, should they become introgressed into the wild.

4.5.5. Coinciding Flowering Time

Do the weed and crop have overlapping flowering times? Otherwise, mating is complicated. Note that flowering-time differences may vary within different areas of the range of a species, and that in some exceptional years overlap may occur where it normally does not exist.

4.5.6. Sexual Compatibility

Three important factors control compatibility.

1. Is the weed self-incompatible, preferring foreign pollen, enhancing chance meetings? Predominantly self-pollinated species accept alien pollen (and thus genes) more slowly than outcrossing species.
2. Is the resistant pollen more, or less competitive than conspecific pollen? Pollen competition in some cases is exceedingly strong, and unfitness from resistance or from pollen being of another species statistically delays gene transfer.[202,762,1013] Pollen longevity and decrease of viability with time are also very important, especially when competing with wild-type pollen. Pollen may be able to fertilize hand-emasculated flowers a kilometer away, but is it far too weak to compete with native pollen by the time it has traveled that distance?
3. How easily can interspecific barriers be overcome? Being in the same genus often engenders worry, which is not borne out in practice, when there is a good literature search. It was considered to be "highly likely" that genes will move between cultivated barleys and *Hordeum glaucum*,[985] while earlier attempts to transfer paraquat resistance from *H. glaucum* to cultivated barleys had proved futile[529]. Likewise, it was considered likely that genes from potato will cross into Old World weedy

Solanum species.[985] The reciprocal was possible by protoplast fusion,[124,425] but there have been no successes in sexually crossing potato with anything but Andean *Solanum* spp.[310] The weedy European *Solanum* species had many traits breeders had wished to convey on potatoes. Their repeated futility was the reason for the rare publication of the negative data.[310] Because the breeders had tried all the lab tricks to keep hybrids alive, there should be little worry that natural hybrids might make it in the field. Tomato has recently been placed in the genus *Solanum*, but it too will not hybridize with Old World weed *Solanum* species.

4.6. Agronomic Factors Governing the Risk of Gene Flow

Many factors must be taken into account and understood in translating the hazards to actual risks.[422,1085]

4.6.1. Cropping Systems

The cropping system is a major determinant in selection pressure because it determines the niche available for a weed; and thus for a transgene to become established. How often is the weed in question a weed in other crops where the same selector appears? The selector is the raison d'etre for using the trait, that is, the herbicide, with herbicide resistance, mineral deficiency when mineral mobilization is used, pest pressure when insect/disease/herbicide/stress resistance is used. Will the rotational herbicides in the rotational crops control resistant weeds? The cropping system should not only be judged on a farm scale where specialized farmers may grow a crop continuously in monoculture, be it minor or major, but also on a regional scale. In other words, how large is the selection pool for successful transfer of the gene?

4.6.2. Selector Trait and Selector Action

Continuous selection with the same selector will enhance the possibilities of establishing introgressed transgenes or evolving resistance of a transgenically conferred resistance. One must completely understand the modes of action of the genes and selectors. If the Bt insecticide toxin gene in crop A confers resistance to Lepidopteran (moth) insects and a different Bt is in crop B conferring resistance to Coleopterans (beetles), and the crops are rotated, it is not as if the same mode of action is continually used or selection pressure applied. Resistance to the photosynthesis-inhibiting herbicides diuron and atrazine can be achieved by mutating the *psbA* gene, but each at mutually exclusive domains on the gene product such that a rotation would be as if two different genes and selectors were present. Clearly there is a need to know and understand a considerable amount to make meaningful risk analyses.

4.6.3. Magnitude and Frequency of Trait/Selector Use

Where a small chance exists of successful hybridization and establishment of a resistant gene in the wild, or a chance of the transgenic crop becoming a volunteer, the ensuing weed problem can be minimal to large, depending on the use patterns of the selector. The problem intensifies if the crop is cultivated in monoculture, or if the same selector is used on different transgenic crops in the same area. In other words, one must consider the agricultural system concerned as a whole. The problems diminish where there is much more variety, both in selectors, especially when coupled with the added use of other measures and crop rotation. The repetitive situation selects for the establishment of escaped transgenes; the highly variable cropping system may be an effective safeguard against their becoming established in the population.

4.7. Systems for Assessing the Risk of Gene Flow

Scientists should only assign the risks. It is up to a much larger group to decide what level of risk of gene flow is acceptable, and to balance the risk with the perceived advantages from the transgenic crops, and then compare the risks of the transgenic with the risks accrued from current cropping procedures in that crop. The larger group making these decisions must also understand scientific talk. It is scientifically untenable to declare that something is totally risk free. This does not mean, as some would like to interpret, that scientists are unwilling to be categorical and interpret a finite risk to be a considerable risk. One cannot guarantee that an airplane will not crash into a cave and kill you, but you usually do not worry about it. Scientists will not say that it cannot happen, because it can. Various systems have been used to assess the level of risk.

4.7.1. Decision Trees

A decision tree is a system of keys and operates as a ranking system to aid decision making. The final considerations not only include biological and agricultural factors dealt with here on gene flow, but also human health, economical, environmental, ethical, and political factors. In the first key the hazards imposed by the weediness and dispersal of the gene are ranked in relation to the inherent biological characteristics of the crop and the related weed species. If the hazard is found to be very low or no reasonable hazard can be conceived, no further examination of risk is needed, otherwise further keys assist in evaluating the risk level. The final level that is acceptable is less a scientific decision and more of a decision for the community through its elected politicians and their appointed regulators. The reader is referred to the published[422] or the expanded internet version of decision trees for transgene flow risk analysis.[1085]

4.7.2. Other Risk Analysis Schemes

Tabular systems have been constructed that consider many of the same parameters and codify risk levels on local or regional levels for crops and their indigenous relatives, where the authors define the level of risk for the reader.[37, 353]

Much can be gained by using the right transgenes in the right way for the right purposes in the right crops. Conversely, much can be lost by injudiciously using the right or wrong genes in the wrong crop. Methods that safeguard transgene technologies are urgently needed to improve the much desired sustainability of the agricultural systems of the world. Transgenic crops are but a tool in the building of sustainable systems in the future. Therefore we need them, and at the same time, we must avoid misuse that may endanger the needed transgenic solutions. The human mind is like a parachute, it is best used when open. We need an open mind to weigh where transgenic crops can be beneficial to food production and where they might be detrimental. We must present regulators with scientific tools to weigh these issues, as too often regulators are swayed by the unsubstantiated emotions of activists on one side, or by the quick-profit, short-term economic interests of industry on the other, to state the extremes. Like all-powerful tools, certain transgenic crops have important long-term roles in food production, others are contraindicated. The same transgenic crop may have important uses in one agroecosystem and be foolish in another.

4.8. Containing Gene Flow

When a decision tree or other deliberations arrive at an unacceptable risk of gene flow, solutions are needed instead of surrender. The solutions must keep transgenes from becoming established outside the crop. Two general approaches deal with risks when they are perceived: (1) contain the transgenes in the novel variety so that flow is precluded; (2) mitigate gene flow effects if there are inevitable "leaks" in the containment system. Mitigational systems should also prevent volunteer weeds from establishing/reaching maturity so that they cannot evolve into feral problems. Containment and mitigation of gene flow are discussed below in the general context of bidirectional containment of genes to and from the crop, as well as mitigation.

Several molecular mechanisms have been suggested for containing genes, especially transgenes, within the crop (i.e., to prevent inflow from and/or outflow to related species), or to mitigate the effects of transgene flow once it has occurred.[246,411,412,818,1017]

4.8.1. Containment by Targeting Genes to a Cytoplasmic Genome

The most widely discussed containment possibility is to integrate the transgene of choice in the plastid or mitochondrial genomes.[580,682] The opportunity of gene outflow is limited due to the predominantly maternal inheritance of these genomes in many but not all species. This presently arduous technology of transforming genes into chloroplasts, which so far is limited to a few crops, does not preclude the wild or weedy relative from pollinating the crop, giving rise to the same F_1 hybrid that would have been obtained if the crop had been pollinated by the weedy/wild species, but bearing the plastomic or mitochondrial trait. Then, if the wild or weedy species acted as the recurrent pollen parent, the plastid or mitochondrial trait could be fixed by backcrossing the wild/weedy relative. Even though the hybrids into the crop and into the wild may be the same, the likelihood of a crop-wild hybrid of any of the major crops discussed surviving in the wild is minimal but finite, especially over long durations. Thus, the problems ensue when the crop and its relative inhabit the same ecosystem in proximity.

The claim of strict maternal inheritance of plastome-encoded traits in many species[127,247,682] has not been substantiated; large enough scale experiments were not performed by the proponents to demonstrate that the frequency is below a low number, for example, less than the frequency of typical mutations. Tobacco[76] and other species[250] have between a 1 in 1,000 and 1 in 10,000 frequency of pollen transfer of plastid inherited traits in the laboratory. Thus, maternal inheritance of plastomic traits is not absolute, as had been thought; pollen is not devoid of plastomic DNA. Pollen transmission of plastome-encoded traits can only be easily detected using large samples together with nuclear and plastomic selectable genetic markers. A large-scale field experiment set out using a *Setaria italica* (foxtail millet) with chloroplast-inherited atrazine resistance, with the resistant plants also bearing a nuclear dominant red-leaf-base marker. These were allowed to naturally cross in the field with five different male-sterile, yellow- or green-leafed herbicide-susceptible lines. It was easy to discriminate which plants were the results of cross-pollination due to the nuclear markers, and from within this group ascertain how many had the plastome-inherited trait. Chloroplast-inherited resistance was pollen transmitted at a frequency of 3 in 10,000 in more than 780,000 hybrid offspring.[1126] At this transmission frequency, the probability of herbicide resistance movement via plastomic gene flow is orders of magnitude greater than by spontaneous dominant nuclear genome mutations. Thus, chloroplast transformation is probably unacceptable for preventing transgene outflow, unless stacked with additional mechanisms.

As noted above, plastomic inheritance will not at all impede gene inflow, and it is more likely that a hybrid with a wild species can establish in a crop field than in the wild. Maliga[682] discounts the relevance of the findings with tobacco and *Setaria* as being due to plastids from interspecific (closely related) cytoplasmic substitution, where barriers to pollen transmission can break down.[583] *S. viridis*, the wild progenitor of *S. italica,* is basically conspecific with it,[252] so this argument does not hold. There are two other problems with this denigration of the relevance of pollen movement of plastome-encoded genes: (1) it is just such interspecific movement that could be a problem between crops and related species; (2) Maliga[682] ignores the other previously published cases of intraspecific transmission of plastomic traits by pollen at about the same frequency among biotypes of the same species.[250]

4.8.2. Nuclear Male Sterility

Much male sterility is cytoplasmic, inherited on the chondriome (mitochondrial genome); chondriome engineering is yet unknown. A simple fail-safe mechanism may be possible with nuclear sterility in hybrid crops. If a dominant transgene of choice is placed in the male sterile line in close linkage with the male sterility gene, there will be little possibility of introgression to wild or weedy relatives in crop-production areas. Care will have to be taken in the seed-production areas when the male sterile line is restored. Such areas must be kept free of related weeds, a typical precaution in seed production, in general, practiced well before the advent of transgenics. This fail-safe mechanism will be easier to bring into practice when better methods are developed for position-specific transformation, and the position of a major nuclear male sterility gene is known so that it can be closely linked to preclude segregation from one another in future generations.

Present transformation technologies give rise to random insertion on chromosomes, with just the beginning of homologous recombination to desired targets (section 3.6.1.1). There would be little value in having the herbicide resistance trait segregate from male sterility.

Instead of using indigenous genes, if one is engineering male sterility[1153] into the crop by one of the newer technologies for nuclear male sterility, then another primary gene of choice could be coupled in tandem with the male sterility gene, to lower the risk of gene flow of the second primary gene.

4.8.3. Male Sterility with Transplastomic Traits

A novel additional combination that considerably lowers the risk of plastome gene outflow within a field (but not gene influx from related strains or species) can come from utilizing male sterility with transplastomic traits.[1126] Introducing plastome-inherited traits into varieties with complete male sterility would vastly reduce the risk of transgene flow, except in the small isolated areas required for line maintenance. Such a double-fail-safe containment method might be considered sufficient where there are highly stringent requirements for preventing gene outflow to other varieties (e.g., to organically cultivated ones), or where pharmaceutical or industrial traits are engineered into a species. Plastome-encoded transgenes for nonselectable traits (e.g., for pharmaceutical production) have been transformed into the chloroplasts together with a trait such as tentoxin or atrazine resistance as a selectable plastome marker.[248] When they are coupled with the recently reported transplastomic engineered male sterility[929] to further reduce outcrossing risk, plastome transformation can possibly meet the initial expectations. It will still not overcome having a wild or weedy species being a recurrent pollen parent, with selection for the plastid traits, a risk that must be further evaluated.

4.8.4. Genetic Use Restriction Technologies (GURT) a.k.a. "Terminator"

Other molecular approaches suggested for crop transgene containment include seed sterility utilizing the genetic use restriction technologies (GURT) (or the "terminator genes," as they are referred to in the popular press).[237,817,818] In these proposed systems, the transgenes of choice are inserted behind a chemically induced promoter that causes the inactivation or physical excision of the genes of choice in the flowers. The inducer was to be turned on in the seed stage, that is, before sale to the farmer. The plants would grow normally after induction, expressing the transgene throughout vegetative growth, but neither the seed nor the pollen carry the transgenic trait to further generations. In the excision version of GURT technologies, the pollen and seed are viable (in contrast to press reports), but they do not carry the proprietary or patented transgene. This does not prevent farmers from saving seed (as intimated in the press), but the saved seed does not bear the transgenic trait. The GURT systems were originally developed to protect intellectual property, not to prevent gene flow. Not enough experimentation with these systems has been documented to know whether the transgenes would be excised from all offspring after induction, that is, whether the induction is incomplete. This is immaterial for the first proposed use of the technology, to require purchase of seed every generation; for this 90 percent effectiveness is sufficient. Near 100 percent efficiency is required to prevent transgene flow, and the leakage rate is unknown at present. Theoretically, if the inducible controlling (terminating) element of the transgene is silenced by mutation, expression would occur, a potential defect in principle and possibly in practice. The frequency of loss of such controlling elements is yet unclear, as no large-scale field trials have tested this.

About half a percent of the crop area sown is planted with seed for future planting, that is, seed that is not "terminator" induced. This area would have to be contained by other means to prevent transgene flow, and if the transgenes do escape, there is no way to "terminate" them once out.

4.8.5. Chemically Induced Promoters

If a transgene encoding the desired trait is placed behind a strong, chemically induced promoter, the desired trait will be expressed when the chemical inducer is used. Such a promoter system was patented for use with a glyphosate-resistant 5-enolpyruvylshikimate-3-phosphate (EPSP) synthase gene.[546,547] The chemical inducer can be treated together with glyphosate, as glyphosate kills slowly, and inducers supply products within hours. This strategy allows the control of the transgenic crop as a volunteer weed the following season. The herbicide can be used without the inducer as a treatment just before planting the rotational crop, or in the naturally resistant crop to control the volunteer weed. If the crop were to introgress the gene into a weed, the weed could be controlled by the herbicide (without inducer). If the herbicide-resistant gene were to introgress into a wild species that does not inhabit agroecosystems where herbicides are not used, it would be of little value and would probably have enough of a fitness penalty so as not to establish. Similarly, other transgenes of choice would have no benefit (selective advantage) to wild or weedy relatives if the transgenes are not turned on by the inducible promoter.

A system that turns on transgenes such as this may be preferable to one that turns them off, such as "GURT." Theoretically, if the GURT gene is silenced, a possibility for introgression exists. If the inducer gene is silenced, then those individuals possessing the mutant are killed and the germ line is nontransferable. Still, there is the remote possibility of an inducible promoter mutating to a constitutive promoter.

Unfortunately no inexpensive, foolproof, inducible promoter system is available for plants. The copper- and the alcohol-inducible promoters that have been developed do not work in the field; enough copper is in most soils to trip the former, and environmental conditions are often sufficient to cause enough alcohol be naturally present in a crop to trigger the latter.[1204] There is also an estrogen-inducible promoter that can be used,[240] but the inducer is expensive as well as being a hormone. The plant hormone auxin can also be used with a new inducible promoter,[196] but auxins can have side effects on the plants. Antibiotic-inducible promoters are known; but widespread use of antibiotics in the field would not be allowed, wisely so. For certain resistance traits, it might be advisable to use pathogen-[1200] or wound-inducible[149] promoters, but these will not be fail-safe mechanisms, as the same disease or insect that induces the gene in the crop will do so in the related species.

4.8.6. Recoverable Block of Function (RBF)

Various technologies have been developed by using the *barnase/barstar* gene system. *Barnase* encodes a potent ribonuclease, which, when expressed, kills a cell because it chews up the whole protein-manufacturing system. The action of *barnase* is held in check by *barstar*, a strong repressor gene that prevents *barnase* action. It is only when *barstar* is not present that *barnase* can exert its lethal effect. In a strategy called "recoverable block of function (RBF)"[622,623] to prevent transgene flow, *barnase* is inserted in a large synthetic intron inserted in the midst of the gene of choice in such a way that transcription of the two genes is in opposite directions. Both genes are thus genetically linked, in what the developers refer to as a "blocking construct," where both are inherited together. This is in the same manner as proposed for transgenic mitigation (see section 4.9). To prevent this lethality, they propose a "recovering construct" containing *barstar*, but under an inducible promoter. They demonstrate the efficacy of this with a heat shock promoter. Both constructs are inserted (randomly) on different chromosomes. In crosses with a wild/weedy species, or another variety, all F_1 progeny of the parents of homozygous progeny will live if *barstar* is induced, all will die if not.

If the crop is a heterozygous hybrid containing *barnase* and *barstar* half of the progeny will die and half will have uninduced *barstar*. This would not be an effective failsafe if it were not for the inducibility of *barstar*. The transgene cannot be expressed in the progeny containing *barnase* and *barstar* without the promoter action. As heat shock might occur in the field, they also developed the same system using a tetracycline inducible promoter for *barstar*[622]. The RBF system seems like a complicated way to use an inducible promoter, when a simple one has already been proposed (see section 4.8.5).

4.8.7. Repressible Seed Lethal Technologies

A presently impractical technology has been proposed to use a "repressible seed-lethal system."[949] The seed-lethal trait and its repressor must be inserted simultaneously at the same locus on homologous chromosomes in the hybrid the farmer sows to prevent recombination (crossing over). Such site-specific transformation technologies are not yet workable in plants. The hemizygote transgenic seed-lethal parent of the hybrid cannot reproduce by itself, because its seeds are not viable. If the hybrid could be made, half the progeny would not carry the seed-lethal trait (or the trait of interest linked to it) and they would have to be culled, which would not be easy without a marker gene. A containment technology should leave no viable volunteers with the transgene, but this complex technology would kill only a quarter of the progeny and half would be like the hybrid parents and a quarter would contain just the repressor. Thus, the repressor can cross from the volunteers to related weeds, and so can the trait of choice linked with the lethal trait, and viable hybrid weeds could form. The death of a quarter of the seeds in all future weed generations is inconsequential to most weedy and wild species that copiously produce seed, as long as the transgenic trait provides some selective advantage.

4.8.8. Trans-Splicing to Prevent Movement

A system has been proposed and partly demonstrated[1027] that was designed for the generation of transgenic hybrids, where only part of the segregating F_2 generation would bear the transgenic trait. Enzyme splicing in *trans* was demonstrated using the *DnaE* intein, which reconstituted to a functional DnaE protein. The gene for herbicide resistance, *ALSII* was fused in frame to *DnaE* intein segments capable of promoting protein splicing in *trans* and was expressed as two unlinked fragments. Cotransformation with the two plasmids led to production of a functional enzyme by protein splicing in *trans*, which then conferred herbicide resistance.[1027] If each plasmid integrates into a different chromosome, introgressing into a readily crossing weed will give 25% of the weeds resistance, which is hardly fail-safe. If one of the genes is on a nuclear chromosome, and the other in the plastome, the rate of introgression will be half that of a whole gene being on the plastome. The rate of introgression will be near zero if one half of the gene is on the plastome and the other half on the chondriome, but chondriome engineering is still close to science fiction.

4.8.9. A Real Chaperon to Prevent Promiscuous Transgene Flow from Wheat to Its Wild Relatives

There is ample evidence that inserting transgenes to allopolyploid crop chromosomes that are homoeologous to related weed/wild species would not preclude transfer,[1139] yet wheat needs transgenes. Thus a specific method was conceived for wheat, based on the presence of a specific gene in wheat, and the possibility of gene insertion in specific chromosomal locations.[1140] It has long been known that wheat bears a *Ph1* gene located on chromosome 5B, which specifically prevents the promiscuous pairing of homoeologous chromosomes, preventing recombination among the three genomes of wheat and with related species.[812,902,962] Reduction in homoeologous pairing may also be due to the rapid elimination of many sequences from homoeologous chromosomes of progenitor species after allopolyploidization,[338,833] which increases the differences between the homoeologous chromosomes.

When wheat with its *Ph*1 gene hybridizes with a wild polyploid species there is little or no homoeologous pairing in the F_1, as long as *Ph*1 is watchfully present as a chaperon. Because an unpaired chromosome has a probability of only 0.25 of being present in the gametes (due to random segregation in the two meiotic divisions),[752] about 75 percent of the progeny resulting from the backcross of F_1 to the wild parent will lack wheat chromosome 5B and, consequently, *Ph*1. In these plants missing in *Ph*1, single wild and wheat chromosomes will promiscuously pair homoeologously and recombine with abandon, transferring genes with impunity.

*Ph*1 is located in the middle of the long arm of chromosome 5B (5BL),[536] about 1 centimorgan (cM) from the centromere,[963] and has been characterized molecularly.[433] Using the novel systems of targeted introgression (also called, confusingly to geneticists, homologous recombination),[646,871,890,967] it should be possible to insert the transgenes of choice in proximity to *Ph*1 on chromosome 5BL.[1140] Thus, the transgene of choice will remain genetically linked with ever-watchful *Ph*1 and will segregate with it in the hybrid and backcrosses, and thus not introgress the chromosomes of wild/weedy relatives.[1140] During backcrosses with the wild/weedy species, the excess wheat chromosomes are selectively eliminated due to lagging during anaphase.[1130] Chromosome 5BL with the *Ph*1 gene and the linked transgene will be retained in a small proportion of offspring only as long as the transgene confers a selective advantage, for example, when the transgene is for disease or herbicide resistance, and the particular disease or herbicide is present or used.[1140] In seasons where other herbicides are used or the disease is not active, the selective disadvantage will eliminate 5BL and its linked transgene. Such a solution, especially if coupled with other solutions, for example, mitigating genes (section 4.9), could considerably lower the risk of gene flow between wheat and its relatives. *Ph*1 was recently localized to a 2.5-Mb region of chromosome 5BL containing a structure consisting of a segment of subtelomeric heterochromatin that inserted into a cluster of *cdc*2-related genes, genes that affect chromosome condensation.[433] The correlation of the presence of this structure with *Ph*1 activity makes the structure a good candidate for the *Ph*1 locus.[433] When *Ph*1 is fully isolated and described, it might be possible to use it directly as a mitigator gene, no longer necessitating targeted insertion, randomly inserting it linked to the gene of choice.

An alternative way to curtail the movement of a transgene from wheat into wild relatives is by inserting the transgene in tandem with a suicide gene on any chromosome arm other than 5BL, and inserting a gene encoding a suppressor of the suicide gene product on chromosome arm 5BL adjacent to *Ph1*.[1140] The linkage between the suicide-suppressor gene and *Ph1* on 5BL will prevent the transfer of the suppressor to a wild chromosome and, consequently, the establishment of this gene in the wild population. The suicide gene can encode any heterologous protein that is toxic to plants that possess it.[1140] *Barnase* is such a gene (section 4.8.6), and another suicide gene encodes a ribosome-inhibitor protein (RIP) that destroys ribosomes.[531] The suicide-suppressor gene can be any gene that encodes a heterologous protein that inactivates either the suicide gene or the toxic protein encoded by the suicide gene, for example, *barstar*. Any backcross progeny that have the tandem transgene-suicide gene, but are without the suppressor of the suicide gene, will die. Chromosome arm 5BL will be eliminated during the continuous backcrossing to the wild parent because it cannot recombine with homoeologous material because of the presence of *Ph1*.[1140]

This concept might be less efficient at preventing gene flow if the recipient wild/weedy species contains a gene that suppresses the *Ph1* gene. There is evidence that several diploid relatives of wheat, for example, *Aegilops speltoides, Amplyopyrum muticum,*[903] and several diploid and tetraploid *Agropyron* species have such a suppressor.[209] Thus, this might necessitate ascertaining that transgenic wheat will not transfer genes to the indigenous wild/weedy diploid relatives in each locality where transgenic wheat is to be cultivated to ascertain the value of this chaperoning system. Clearly either mechanism using the *Ph1* chaperon can be effective in preventing gene flow from wheat to *Aegilops cylindrica* in the parts of the world where it is the sole problem weed known to stably introgress genes from wheat.[1140]

4.8.10. Transient Transgenics

Endophytes are fungi and bacteria that normally grow inside plant tissues, often with a symbiotic function. Attempts had been made to use endophytes to carry useful genes into plants (e.g., Bt genes) by pressure infiltrating the endophytes into seeds.[326,1066] The advantage of the concept was that it was not variety specific and that the endophytes would not be transmitted via seed to the next generation. The technology as developed caused a yield reduction, probably due to an overload of the endophytes. The same concept could be used to carry other genes conferring useful traits, if potent highly expressed genes are used with unobtrusive sparsely growing endophytes, or if disarmed viruses are used as vectors.

It would be conceivable to insert certain useful traits on systemic RNA viruses or in endomycorrhizae that are expressed in the plant but are not carried through meiosis into reproductive cells. One could introduce the transgene of choice into a disarmed pathogen such as a non-disease-producing strain of *Clavibacter* that also dwarfs the infected crop.[188] It is necessary to transfect the crop every generation, but the transgenes would not spread sexually.

The possibility that such a procedure might work was borne out in many cases with dicots showing that they express encoded genes, for example, reference 607. It was possible to infect *Arabidopsis* with Tobacco Etch virus carrying the *bar* gene; the plants were resistant to the herbicide glufosinate.[1147] Cucurbits were artificially infected with an attenuated Zucchini Yellow Mosaic potyvirus containing the same transgene and the plants were herbicide resistant in the field.[975] Wheat streak mosaic virus carrying an NPTII antibiotic resistance gene was used to infect various grains.[218] *NPTII* was expressed (immunologically) but it was not shown that the plants were antibiotic resistant. The virus carrying the genes was expressed in the roots following leaf infection, though not in all tissues.[218] All of the viral transfections described were performed on leaves.

The ability of viruses has recently been greatly enhanced by a reconstruction of the viruses, using only synthetic replicons attached to the gene of choice[692] and letting the plant perform some of the functions typically performed by the virus. The replicons were not systemic, because they do not contain genes for virus coat protein that allows such movement. The replicons cannot be scratched into leaves like viruses, and instead are introduced into *Agrobacterium*, which is vacuum infiltrated into the crop leaves.[692] This obviates the need for systemic movement, as most leaf cells are infected by the *Agrobacterium*. Still, only a few cells produced recombinant protein, but they did so after introducing silent nucleotide substitutions and addition of multiple introns.[692] Up to a millionfold increase in expression compared with wild-type virus is claimed, a level that makes the plant quite sick because 80 percent of the soluble protein produced is recombinant. With lower expression levels, the authors claim the ability to produce large quantities of recombinant protein by this "magnifection" process.

Hopefully, such infection procedures could be used to introduce useful genes into crop leaves (see Chapter 8) (or seeds using the pressure infiltration systems[188]). The transgene could not move to other species, and the following generations of crop seed would not contain the transgene. Additionally, the transgene DNA would not be found in crop seed, an advantage in today's scientifically irrational market place.

Considerable technological obstacles to efficient infection will have to be determined. The safety issues about the mode of disarming the virus must be considered, and ensure that indeed there is a total lack of gene introgression from the virus to the plant chromosomes, and near total nontransmission of the virus through ovules or pollen, must be ensured. Still, there are many crops, especially those with related, introgressing weeds (e.g., sorghum, barley, rice, sunflowers), where such a technology could be worthwhile in safely solving problems without the fear of gene flow. An alternative to seed inoculation is described in Chapter 8, where it is proposed to infect standing forage crops by modified commercial sandblasting equipment.

If the trait conferred by the virus has a strong selective advantage, and the crop can be vegetatively propagated, or has that potential, there could be a limitation to this technology due to vegetative spread of the transgene.

4.9. Mitigating Transgene Flow

None of the preceding containment mechanisms (Fig. 13) is absolute, but the risk could be reduced by stacking a combination of containment mechanisms, compounding the infrequency of gene introgression. Still, even at very low frequencies of gene transfer, once gene transfer occurs, the new bearer of the transgene could disperse throughout the population if the transgene confers just a small fitness advantage. Thus it is necessary also to utilize technologies that will prevent the establishment and/or spread of transgenes in the population (i.e., mitigation strategies).

If a transgene confers even a small fitness disadvantage, the transgenic crop volunteers and their own or hybrid progeny should only be able to exist as a very small proportion of the population. Therefore, it should be possible to mitigate volunteer establishment and gene flow by lowering the fitness of transgene recipients below the fitness of competitors, so that the volunteer or hybrid offspring will not reproduce. A concept of "transgenic mitigation" (TM) was proposed in which mitigator genes are linked or fused to the de-

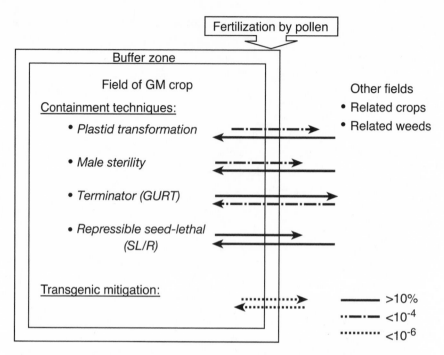

Figure 13. Containment systems allow gene flow in one or more directions, either from crop to related species, or vice versa, and are leaky in the direction expected to work. The incomplete effectiveness of such systems indicates a need for stacking containment systems to bidirectionally contain gene flow, to compound the safety factor, and to mitigate the effects of genes that leaked from containment. The seed production fields of the GURT technology will allow gene outflow. Gene flow in both directions will be lost when the GURT is turned on. *Source:* Reproduced from Gressel and Al-Ahmad,[424] with permission of Routledge/Taylor & Francis Group, LLC,

sired primary transgene.[411] Thus, a transgene with a desired trait is directly linked to a transgene that decreases fitness in volunteers (Fig. 14). Transgenic mitigation could also be used as a stand-alone procedure with nontransgenic crops to reduce the fitness advantage of hybrids and their rare progeny, and thus substantially reduce the risk of persistence of feral hybrids.

Figure 14. (opposite page) Transgenic mitigation to prevent hybrid establishment of desirable transgenes to form superweeds (S) by coupling the transgenes in tandem with genes that are neutral or positive for the crops, but render volunteers or hybrids unfit to compete outside of cultivation forming superwimps (W). TM, transgenic mitigation; GC, gene of choice. *Source:* Reproduced from Gressel and Al-Ahmad,[424] with permission of Routledge/Taylor & Francis Group, LLC, copyright 2005.

A. Endo-Feral

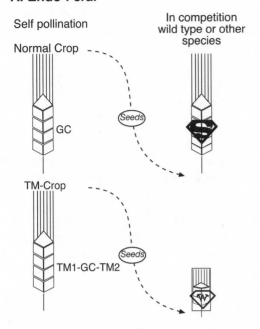

Self pollination

In competition
wild type or other
species

Normal Crop

GC

Seeds

TM-Crop

TM1-GC-TM2

Seeds

B. Exo-Feral

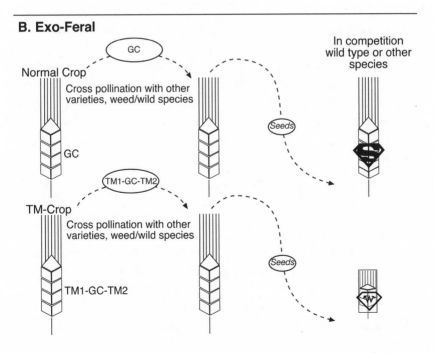

GC

Normal Crop

In competition
wild type or other
species

Cross pollination with other
varieties, weed/wild species

GC

Seeds

TM1-GC-TM2

TM-Crop

Cross pollination with other
varieties, weed/wild species

TM1-GC-TM2

Seeds

This TM approach is based on the premises that: (1) tandem constructs act as tightly linked genes, and their segregation from each other is exceedingly rare; (2) the gain of function dominant or semidominant TM traits chosen are neutral or favorable to crops, but deleterious to volunteer progeny and their hybrids because of a negative selection pressure; and (3) individuals bearing even mildly harmful TM traits will be kept at very low frequencies in volunteer/hybrid populations. The strong competition with their own wild type or with other species should eliminate even marginally unfit individuals and prevent them from persisting in the field.[411]

Thus, it was predicted that the primary gene of agronomic advantage being engineered into a crop will not persist in future generations if it is flanked by one or more TM gene(s), such as genes encoding dwarfing, strong apical dominance to prevent tillering (in grains) or multiheading (in crops such as sunflowers), determinate growth, nonbolting (premature flowering) genes, uniform seed ripening, nonshattering, antisecondary dormancy, and so on. When such a TM gene (or genes) is in a tandem construct with the transgene of choice, the overall effect would be deleterious to the volunteer progeny and to hybrids. Indeed a TM gene such as antishattering will lower the number of initial volunteers. (There is typically a small amount of shattering due to imperfect harvesting equipment, which usually leaves some seed behind.) Because the TM genes will reduce the competitive ability of the rare hybrids, the hybrids and their progeny should not be able to compete and persist in easily measurable or biologically significant frequencies in agroecosystems.[411, 412]

Once TM genes are isolated, the actual cost of splicing the TM constructs is minimal, compared with the total time and effort in producing a transgenic crop. The cost is even inconsequential in systems where cotransformation allows introducing genes into the same site such that the tandem construct is made by the plant.

4.9.1. Demonstration of Transgenic Mitigation in Tobacco and Oilseed Rape

Tobacco was used as a model plant to test the TM concept: a tandem construct was made containing an $ahas^R$ (acetohydroxy acid synthase) gene for herbicide resistance as the primary desirable gene of choice, and the dwarfing Δgai (gibberellic acid-insensitive) mutant gene as a TM mitigator.[22] Dwarfing would be disadvantageous to the rare weeds introgressing the TM construct, as they could no longer compete with other crops or with fellow weeds, but it is desirable in many crops, preventing lodging and producing less straw with more yield. The dwarf and herbicide-resistant TM heterozygous tobacco plants (simulating a TM introgressed hybrid) were more productive than the wild type when cultivated alone (without herbicide). They formed many more flowers than the wild type when cultivated by themselves, which is an indication of a higher harvest index (Fig. 15). Conversely, the TM transgenics were weak competitors and highly unfit when cocultivated with the wild type in experiments simulating ecological competition (Figs. 15 and 16). The inability to achieve flowering on the transgenically mitigated plants in the competitive situation (Fig. 15) led to a zero reproductive fitness of the transgenically mitigated plants grown in a equal mixture with the wild type at the spacing used. This is representative of weed spacing in the field. The highest vegetative fitness was less than 30 percent of the wild type.

From the data above it is clear that transgenic mitigation should be advantageous to a crop growing alone, but disasterous to a crop-weed hybrid living in a competitive environment. If a rare pollen grain bearing tandem

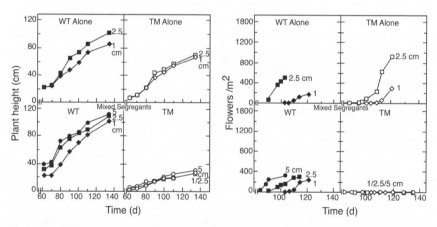

Figure 15. TM (transgenic mitigation) suppresses growth and flowering in competition with wild type (WT). A TM dwarfing gene, in tandem with a herbicide resistance gene, was transformed into tobacco plants (open symbols) that competed with the wild type (closed symbols) (*right panels*). They grew normally when cultivated separately without herbicide (*left panels*). The wild-type and transgenic hemizygous semidwarf/herbicide-resistant plants were planted at 1, 2.5, and 5 cm from themselves, or each other, in soil. See reference 22 for further details. *Source:* Reproduced from Gressel and Al-Ahmad,[424] with permission of Routledge/Taylor & Francis Group, LLC, copyright 2005.

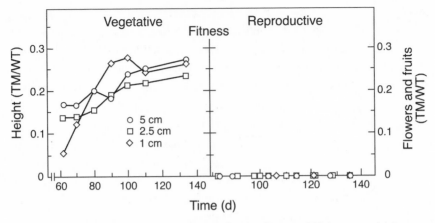

Figure 16. Suppressed vegetative and reproductive fitness of TM transgenics in competition with wild-type tobacco. The points represent the calculated ratio of data for TM to wild-type plants in Figure 15. *Source:* Reproduced from Gressel and Al-Ahmad,[424] with permission of Routledge/Taylor & Francis Group, LLC, copyright 2005.

transgenic traits bypasses containment, it must compete with multitudes of wild-type pollen to produce a hybrid. Its rare progeny must then compete with more fit wild-type cohorts during self-thinning when hundreds of plants are cruelly thinned to leave a single plant that establishes, replacing a parent. Even a small degree of unfitness encoded in the TM construct would bring about the elimination of the vast majority of progeny in all future generations, as long as the primary gene provides no selective advantage that counterbalances the unfitness of the linked TM gene.

The same construct was inserted into oilseed rape and the progeny selfed,[21] as well as crossed with its weedy relative *Brassica rapa* (syn. *B. campestris*)[19]. The hybrids were competed with the respective wild types. The results are basically the same as with tobacco; the progeny of the weedy individuals introgressing into the transgene in a TM construct were unfit to compete with their weedy cohorts or the crop. Should some seeds of such hybrids fall in an area where there is little competition with weeds, the dwarf transgenically mitigated individuals may further reproduce but are not a threat to crop production. The rare hybrid offspring from escaped pollen bearing transgenic mitigator genes would not pose a dire threat, especially to wild species outside fields, as the amount of pollen reaching the pristine wild environment would only be at a minuscule fraction of the pollen from the wild type.

It is probably wiser to flank the gene of choice with two mitigator genes (e.g., dwarfing and nonshattering), so that seeds from the few surviving dwarf plants are harvested and discarded. The use of two flanking mitigator genes will mathematically compound (the yet unknown) infrequency of mutation to loss of function. Large-scale field studies will be needed with crop/weed pairs to continue to evaluate the positive implications of transgenic mitigation to risk mitigation.

4.9.2. Risk that Introgression of TM Traits Will Affect Wild Relatives of the Crop

A widely acclaimed (in news releases) recent model claims that "demographic swamping" by crop transgenes would cause "migrational meltdown" of wild species related to the crop, especially if the introgressed genes confer unfitness.[481] This proposition that recurrent gene flow from crops, even unfit gene flow, could affect wild relatives deserves some discussion, as it flies in the face of Darwinian concepts of survival of the fittest.[416] If correct, it would reduce the utility of transgenic mitigation.

Data for conventional crops already belie this possibility that recurrent gene flow from transgenic crops with less fit genes will cause wild populations to shrink. Major domesticated crops are not fit to compete in wild ecosystems, so their normal genes should confer a modicum of unfitness. Crop wild hybrids continually form at a low frequency, yet no evidence has been published that demographic swamping has occurred from recurrent gene flow from conventional crops to wild species in natural ecosystems or that concrete situations exist where the model may apply. Haygood et al.[481] supply no data or examples to support their model simulations. Indeed, considerable evidence has appeared that many crops exist near their wild or weedy progenitors, without causing the extinction of the progenitors, despite continuous gene flow of crop genes, which are unfit naturally in the wild.[417] For example, some grass biotypes have lived for 2,000 years on Roman mine tailings in Wales, having evolved heavy metal resistance, and are unfit compared with the wild type on normal soil. Their pollen has blown to their sensitive cohorts centimeters away without the swamping[144] despite what the models predict.[481] Maize and teosinte grow in proximity, and F_1 hybrids form, but the teosinte has not been "swamped" by massive maize cultivation in proximity,[282] as would be predicted by this model. At the worst, hybrid swarms often appear at the boundary between crop and wild species, but they remain contained.

There are other flaws in their model that are based on questionable premises and assumptions, not borne out by plant biology.[416] Three problematic issues that seem to invalidate their model for the vast majority of conceivable crop/wild species systems are discussed below:

1. To get the level of swamping that they discuss,[481] the wild relative and the crop would have to live in the same ecosystem. There is typically geographic separation between agroecosystems and wild ecosystems, with pollen flow decreasing exponentially with distance, usually to a low asymptote due to wind currents or insects not fully following simple physics. Far more wild pollen should always be produced in the wild ecosystems, so hybridization events in the wild from crop pollen will be rare, even from the masses of pollen occurring within the agroecosystem. Thus their basic assumption of crop pollen swamping wild-type pollen in the wild is probably invalid. Indeed, even when they assume an enormous 10 percent of hybridizations in the wild each generation coming from crop pollen, according to their model it will take about twenty generations of recurrent pollination for the unfit crop allele to become fixed in just half of the population (Fig. 17), and fifty generations for an unfit gene to asymptotically reach 80 percent of the population. The model is also contradicted by experiments[20] where a replacement series was used and a transgenic crop bearing an unfitness gene was competitively intermingled with the wild type and progeny counted. When nine times more unfit individuals swamped fit individuals, the result was less than 90 percent unfit progeny, and simple extrapolation demonstrates that every year with recurrent selection the unfit genes will gradually disappear (Fig. 16).

2. Haywood et al.[481] assume that the crop and the theoretical wild species growing in its midst will have synchronous flowering, no self-fertilization, and no genetic or other barriers to cross-fertilization; indeed, this negates the definition of speciation. It is exceedingly rare for crop pollen to fertilize another species without any genetic barrier in the wild relative. They do not suggest any cases where this might happen, but in reviewing the literature on interbreeding wild relatives of crops, one sees that it might only occur with conspecific wild sunflowers, which might fit this criterion, but even in this case genomic deterrents

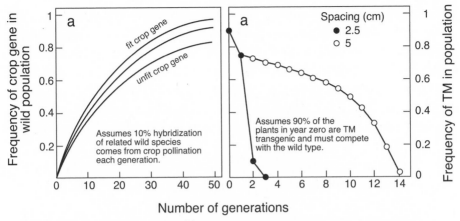

Figure 17. Will unfit transgenic hybrid plants establish? Two views compared: (A) Modeling of gene introgression under recurrent gene flow from crops to wild relatives by Haygood et al.[481] If validated, it would preclude the use of transgenic mitigation as a fail-safe mechanism to prevent transgene establishment in volunteer weeds and related weeds in agroecosystems. Note that the model starts with 10 percent introgression per year. Modified from Haygood et al.[481] (B) Elimination of the transgenically mitigated plants from the population under competition with wild type, starting with a more stringent 90 percent of the population as 2:1 hemizygous to homozygous TM transgenic segregating plants in season zero. The data points are based on flowers formed per season per unit area of the TM and the wild-type plants grown together in a replacement series at close spacing. The data for first season were used as the starting frequency for the next season, using the data in Fig. 7a of Al-Ahmad et al.[20] to obtain the interpolated data, continuing in this manner for season after season. *Source:* Reproduced from Al-Ahmad et al.,[20] by permission of Springer Science and Business Media.

to introgression appear (reviewed by Stewart et al.[1017]). Weedy sunflowers, growing in or near domestic sunflowers have often introgressed some crop genes, but truly wild sunflowers growing on a native prairie are far less affected by a crop. Positive transgenes from crops might eventually establish in prairie populations but unfit ones would surely be competed out.

There are fertilization barriers of different chromosome numbers, nonhomology, and so on, that limit fertilization of wild relatives by crops of oilseed rape and wheat, so they are outside the model stipulations.

3. Their model[481] assumes animal-type replacement rates where just a few progeny per mating are typical, allowing lower fitness to indeed become fixed. Most wild relatives of crops produce copious amounts of seed to replace parents. Hundreds to thousands typically germinate in the area occupied by a parent, and the process of self-thinning is ferociously competitive, eliminating less fit individuals. Self-thinning, except by sperm during fertilization, is far less an important factor in animals than plants because of the low progeny number and because most animals are "perennial" and most wild relatives of crops reproduce a single generation in their adult life.

Their conclusion that "the most striking implication of this model is the possibility of thresholds and hysteresis, such that a small increase in (unfit gene) immigration can lead to fixation of a disfavored crop allele"[481] flies in the face of evolutionary evidence, and decades of classic and contemporary field data showing that only near-neutral genes exist in pockets of the evolutionary landscape of plants, and blatantly unfit plant genes are not known to exist in such pockets unless all the fit genes are somehow removed. The evolutionary gene pockets of teosinte[282] clearly demonstrate that their predictions do not hold in this case, where they should, if correct. Just as endogenous unfavored gene mutations exist in the wild at a frequency lower than the mutation rate, crop transgenes that have a fitness penalty will exist in the wild at a rate lower than the immigration rate. As discussed above, the immigration rate to the wild is very low. When a model contradicts reams of data, more likely than not, the model is invalid.

Haywood et al.[481] further contend that their model would work if the crop were heterozygous for the unfit gene (and many hybrid crops have the transgene in a single parent and are thus hemizygous). The data in Fig. 17 clearly show that when even 90 percent of the starting population contains a hemizygous unfitness gene, these plants cannot compete with their nontransgenic sibs, let alone the wild type. Part of the problem may be that they[481] (p. 1880) "assume (that) the number of plants surviving to maturity does not vary from one generation to the next," a questionable assumption for unfit phenotypes when they must compete with fit cohorts and other species.

Where might their model have some validity? The model might be valid for a few weeds (not wild species) related to crops in agroecosystems or ruderal ecosystems, but not in wild ecosystems. When flowering weeds are at a low density in an agricultural ecosystem (and perhaps close by in ruderal systems) the model might be predictive, but would it be so bad to see a weed go extinct? Because weeds are (inadvertently) man-made domesticated species,[1133] should not people also have the right to eliminate them? The nature of weeds is such that they do not go extinct, as much as the farmer would desire. It is far more likely that such evolutionarily threatened weeds would evolve exclusionary mechanisms that would block evolutionarily threatening gene flow, for example, they could evolve a shift from outcrossing to predominant self-fertilization that would protect them from crop pollen bearing unfit genes.

In summary, to quote Nobel laureate Manfred Eigen: "A hypothesis has two possibilities, it can be right or wrong; a model has a third alternative, it can be right, but irrelevant." The model of Haygood et al.[481] may be right for certain animal systems but irrelevant for the vast majority of plant systems. Their peculiar assumptions demonstrate that Eigen forgot the fourth alternative applicable to a model—it can be wrong and irrelevant.

4.9.3. Traits and Genes for Mitigation

Some possible traits discussed above for TM constructs just exist as named genes that are inherited, others are also mapped to positions on various chromosomes, and a few are actually characterized as sequenced genes. Thus, not all TM traits have known sequenced genes that are immediately available for insertion in tandem constructs. Still, there can be many different ways to confer a TM trait, and thus, more than one gene might be available.

4.9.3.1. Secondary Dormancy

Unfortunately, *Arabidopsis,* the typical source for genes, has already been sufficiently domesticated that it is unlike related weeds; the laboratory strains no longer have strong secondary dormancy.[1094] A mutant that is insensitive to abscissic acid and lacks secondary dormancy was found in a wild, undomesticated *Arabidopsis* strain.[1009] Such a gene might be useful.

4.9.3.2 Shattering

Physiologically, one way to avoid premature seed shattering is to have uniform ripening. Early maturing seeds of oilseed rape on indeterminate, continuously flowering varieties typically shatter. Determinacy, with its single uniform flush of flowering is one method to prevent shattering, but this often shortens the season, reducing yield. The hormonology of the abscission zone controls whether shattering will occur and it is possible that if cytokinins are overproduced, then shattering will be delayed. The cytokinin pathway is well documented and there are genes that could be put in constructs for cytokinin overproduction.[619,708]

A *SHATTERPROOF* gene has been isolated from *Arabidopsis* that prevents seed shatter by preventing seed dehiscence[652] by delaying valve opening on the silique. This may be the ideal gene for closely related oilseed rape. Many other genes control flowering, including *TERMINAL FOWER1* or (*TFL1* and *TLP2*) from *Arabidopsis*, which has orthologs among fruit trees,[324] and tomato where the ortholog is called *SELF PRUNING*.[190] A single amino acid transversion of *TERMINAL FLOWER1* converts this repressor into *FLOWERING* Locus (*FT*), an activator of flowering.[461]

None of these genes may necessarily be appropriate for grass crops where the mode of shattering is quite different from *Arabidopsis* or for legumes where the pods open, releasing seeds. Indeed the grasses have multiple pathways for seed shattering, relating to the different mechanisms used even in the same species to shatter.[650] As shattering is typically dominant, and the nonshattering of domesticated species is recessive, our ancestors had to select for many recessive traits to obtain what we now cultivate. In the wheat, *Aegilops* complex shattering can be due to[337]: spike disarticulation (breakage), where the rachis is brittle at the base, and the whole spike shatters as a unit, and occurs only in *Aegilops* species at present; barrel type (B) spikelet disarticulation, where shattering occurs at the lower side of the junction between the rachis and the spikelet base. This only occurs in wheats with the D genome and is controlled by the dominant *Br2* gene on the long arm of wheat chromosome 3D; wedge-shaped (W) spikelet disarticulation where the rachis breaks at the upper side of the attachment with the spikelet base. This is controlled by the dominant *Br1* gene on the short arm of wheat chromosome 3A.

The first cultivated wheats were selected for nonshattering at these loci, but they had tough glumes (encoded by *Tg*) and were hulled (controlled by gene *q*, a transcription factor controlling many domestication traits). Recessive mutants for free threshing (naked) wheat had to be selected to obtain the wheat used now.[337] When wheat dedomesticated and became feral, as it had in Tibet,[1029] different accessions had different shattering types.

Rice has a recently cloned shattering gene *qSH-1*,[604] which has a high degree of homology with barley gene *JuBel2*, which does not seem to control shattering in that species,[650] and it seems that maize-shattering genes are not orthologous to any of these.[650] Because multiple genetic pathways seem to control shattering,[650] it is unlikely that there will be a "one gene fits all" single antishattering mitigation gene for all the grass species cultivated as crops. Instead, special antishattering mitigators will have to be found for each crop-weed combination, based on the shattering mechanism in the related weed.

4.9.3.3. Dwarfing

Many of the genes controlling dwarfism seem to have an unknown function. Still many other genes are known that control height.

— *Gibberellins.* Preventing the biosyntheses of gibberellins reduces height,[1136] which is the basis of the chemical dwarfing agents used commercially on wheat. The three enzymes and genes controlling various steps in gibberellin biosyntheses are known and cloned.[485, 491, 632, 1167]

— *Arabidopsis* mutations bearing mutations in any one of these genes is dwarfed, and the dwarfing is reversible by gibberellin treatment. Overexpression of a gene coding for *ent*-kaurene synthase, causing cosuppression also mimicked the mutant phenotype. Additionally, a defective GA receptor gene has recently been isolated that confers gibberellin insensitivity when transformed into grains (ΔGAI) by competing with the native receptor, thereby inducing dwarfing.[843] Despite dwarfing oilseed rape at the early rosette stage, plants transformed with the ΔGAI gene do have flowers on long stalks.[21] This is despite a requirement for gibberellin activity to bolt. Presumably either less ΔGAI is produced in the stalk or more native receptor, or a different gibberellin receptor is produced when it is time to flower.

— *Bolting.* Some processes such as flower stalk bolting are controlled by specific gibberellins; in radish GA_1 and GA_4 are responsible.[792] This might suggest that oilseed rape has different receptors in the stalk controlling bolting. It may be necessary to characterize the genes coding for the monooxygenases and dioxygenases that are responsible for these later steps of GA biosynthesis.[485] Some of these genes have been isolated.[619] Another antibolting gene has recently been reported in Chinese cabbage, *BrpFLC*, which encodes a MADS-domain transcription factor.[651] Its level was higher in varieties that required longer vernalization, suggesting that its transgenic modulation could be utilized to prevent premature bolting in climatic conditions where this is a problem. It is still unclear which genes this transcription factor controls.

— *Brassinosteroids.* This group of hormones also causes elongation of stems in many plant species, and their absence results in dwarf plants. A 22 d-hydroxylase cytochrome P_{450} controls a series of these steps in brassinosteroid biosynthesis,[217] and plants lacking the enzyme are dwarfed.[80] Plants are also dwarfed when they produce normal levels of these growth regulators but are mutated in the *bri1* gene coding for the receptor.[796] Additionally, suppressive overexpression of a sterol C24-methyltransferase in the pathway also causes dwarfing.[948]

4.9.3.4. Shade Avoidance

Various forms of the pigment phytochrome interact to detect whether a plant is being shaded.[268, 997, 1071] It is advantageous for a crop plant to grow taller when shaded by a weed, but not so when shaded by cohorts, as less grain is produced on the latter, taller stalks. The engineering of suppressive overexpression constructs of one of these phytochromes led to plants that did not elongate in response to shading.[909]

4.9.3.5. Activatable Genes for Susceptibility to Herbicides and Other Toxicants

At least one gene and the chemical pair is already available: a bacterial P_{450} that activates an experimental sulfonylurea proherbicide.[800] It has been used under a tapetum-specific promoter to prevent pollen formation but could be used under a general promoter that would allow the use of the proherbicide to cull crop-weed hybrids, as well as volunteer crop weeds.

Other such proherbicides with exogenous activating genes could also be envisaged for use as mitigating fail-safes should the primary transgene escape.

4.9.4. Special Cases Were Transgenic Mitigation Is Needed

We have described above some general antiweediness genes that can be used to engender a modicum of unfitness to volunteer offspring and their hybrids with other cultivars and species. There are some special cases where other genes can be envisaged for use to design an unfitness on volunteers or on feral forms coming from the crop. The TM genes are typically still neutral or positive to the crop but give unfit offspring.

4.9.4.1. Mitigation for Biennial and Annual "Root" Crops

Mitigating genes should easily prevent both the premature and volunteer flowering in sugar beets, carrots, onions, celery, radishes, and other biennial or two phase crops, where the vegetative material is marketed and premature flowering (bolting) is detrimental. This could easily be effected by preventing gibberellic acid biosynthesis,[485] either in a TM construct and/or by permanent mutation of the kaurene oxidase gene by using a chimeraplastic gene conversion system,[1198] a system that as yet is hard to use in plants (see section 3.6.1.1). Suppression of kaurene oxidase, a key enzyme in gibberellic acid biosynthesis, would require the use of gibberellic acid to "force" flowering for seed production. There should be a concomitant biosafety requirement that seed production areas be far removed from areas where weedy or other feral or wild relatives grow to prevent pollen transfer.

Delaying of bolting and flowering by using other transgenes has been demonstrated. Curtis et al.[239] engineered a fragment of the *GIGANTEA* gene, the gene encoding a protein that is part of the photoperiod recognition system, into radish using an antisense approach. Bolting was considerably delayed, and thus seed production could come about without reversal mechanisms if seed producers waited long enough. Li et al.[651] isolated a *Flowering Locus C* gene that controls bolting in Chinese cabbage and demonstrated that the higher the transcript level, the less bolting there is.

If, despite all isolation distances, a TM construct or a mutant in a seed production area introgresses with a wild species, the progeny will be biennial or too delayed, that is, the transgenic hybrid would be noncompetitive with cohorts that reproduce in a single year and do not need to overwinter.

Other transgenes can be considered for mitigating the risks of introgression with root crops, such as genes promoting partitioning to roots, which would be advantageous to cultivated root crops, but detrimental to feral forms.

4.9.4.2. Rendering Crops Obligatively Vegetatively Propagated

Some vegetatively propagated crops, such as potato and elite tissue culture propagated forestry material, also flower. In forestry, this is especially problematic as the long-term implications of gene movement may take longer than human lifetimes to measure. The introgression of traits from these species to wild populations has been discussed.[412,659] Some landscaping trees such as decorative plantings of olives create an urban problem of allergies from pollen.[383] Such ornamental trees could be vegetatively propagated if there were a way to prevent allergy-causing pollen clouds and messy fruits (chapter 23). The possibility also exists that preventing allocation of resources to sex will increase the growth of the vegetative tissue where vegetative propagation is possible, which could be advantageous in many ornamentals and in forestry. Thus, a TM trait that prevents pollen formation or fruit set could be coupled to herbicide resistance or other primary traits. An ideal gene for doing this is *barnase* under the T29 tapetum-specific promoter.[691] The ribonuclease is only produced in the tapetum tissue in the flower and prevents pollen formation with no other ill effects. There is a good chance that the shelf life of many flower species (e.g., roses and carnations) could be enhanced as well by preventing pollen production; fertilization starts the process of floral degeneration and fruit set (chapter 22). Additionally, a flower-specific promoter from poplar coupled to a cytotoxin gene caused flower ablation,[992] requiring vegetative propagation of the trees.

If one has an important crop in which transgenics are exceedingly worthwhile, yet the risks of gene flow are too great, one could envisage using a pollen sterility system coupled with flower drop, as described above. The crop could then be propagated by artificial seed, for example, artificially encased somatic embryos produced in mechanized tissue culture systems.

4.9.5. Special Issues in Forestry

Forestry needs herbicide resistance, along with insect and pathogen resistance, which are primary traits, all of which pose theoretical benefit to wild or weedy species. Herbicide resistance is a transiently needed trait that would be used as a selectable marker for other genes and then in the field. Much of the risk analysis will be based on the other traits. There also has been considerable worry about imported forest species cross-pollinating the locals, a process that began well before transgenics.[609] Indeed, many poplar trees in nature are now hybrids. Commercial forestry increasingly is using vegetative propagation, so pollen control via *barnase-T29* (section 4.9.5) could prevent the pollen flow of primary transgenes. Although it will be possible to ascertain whether a construct with *barnase-T29* prevents pollen production in poplar, it will take decades to ascertain this in other tree species having much longer juvenile periods before they reach sexual maturity and flower.

The vegetative propagation of these trees is tedious, but rewarding for the other characters. As overproduction of gibberellins through multiple copies is a dominant trait, the rare, related poplar pollinated and setting seed will surely not succeed in establishment in the wild. This thus represents a TM trait that could be easily coupled with any herbicide or pest resistance trait.

4.9.6. TM Genes for Crop-Produced Pharmaceuticals and Industrial Products

Pharmaceuticals and industrial proteins, especially enzymes and antibodies, can be produced inexpensively in plants, without the need for animal tissue culture cells grown in a medium of expensive serum albumin that is all too easily contaminated with pathogenic mycoplasms, prions, and viruses. Although compelling economic and biosafety considerations propel the production of pharmaceuticals in crops, there are equally good reasons to exclude the pharmaceutical and industrial transgenes from introgressing other varieties of the crop, or related species, or to remain in viable volunteers in the field.

The containment of pharmaceutical transgenes has been physical, as evidenced by recent human error that reportedly allowed temporary volunteer escape of "Prodigene" maize containing such genes. The biological containment strategies described above may be preferable to a dependence on physical containment by fallible humans, and the transgenic mitigation strategies should work as well. Pharmaceutical transgenes in maize are expressed in embryo tissues, and a potential tandem mitigating gene could be any RNAi-type suppression of genes that affect the endosperm, for example, the various "shrunken seed" loci, especially those where sugar transformation to starch is inhibited.[219] Such shrunken seeds, with their high-sugar content, are somewhat harder to store than normal maize, but are extremely unfit in the field and are unlikely to overwinter and produce volunteers. Hybrids with other varieties would have shrunken seeds that would be culled during seed cleaning. Their volunteers would be unfit and also could not overwinter as volunteers in the field. Because of the triploid nature of corn endosperm where two-thirds of alleles are derived from pollen, it is important that expression of pharmaceutical encoding genes be only in the diploid embryo. Should such pollen fertilize a few seeds in adjacent fields, they will be shrunken and will sort out during processing. Thus, such a technology would mitigate against both outflow and influx of pollen.

4.9.7. Mitigation in Species Used for Phytoremediation

Plants have been used to correct human error over the ages. The few species capable of revegetating Roman lead and zinc mine tailings in Wales[1000] taught us that a limited number of species can withstand toxicants: some by exclusion, and others that can withstand toxic wastes after they have been taken up. Plants with the latter type mechanism are of interest for phytoremediation. Ideally, one might consider that it is best to use the species that naturally take up particular toxic wastes, but these are often slow growing (e.g., mosses, lichens, or the *Thlaspi* species that take up heavy metals)[611] and may have a potential to be weedy. If the desired wild species do not exist locally, there may be a reticence or legal issues about introducing them into the ecosystem, toxic as it may be, because of fear that the plants or their genes may spread to other areas. Two types of multicut species are usually considered for phytoremediation, with the cut material burned to extract the heavy metals or to oxidize the organic wastes: herbaceous species such as *Brassica juncea* and *Spartina* spp. (cord grasses), which are most efficient at dealing with surface wastes, and trees such as *Populus* spp. for dealing with deeper wastes.[850] Thus, heavy metal tolerance has been brought into *B. juncea* (Indian mustard) from *Thlaspi* by protoplast fusion (along with many other genes).[297] *B. juncea* wild type had been used commercially, because it grows rapidly and is easy to cultivate as a crop, but especially because of its inherent ability to take up heavy metals. This ability has been enhanced by mutant selection (in tissue culture) for heavy metal resistance.[959] It was better yet to transgenically transfer genes leading to enhanced glutathione content[109,1201] to make the necessary phytochelatins that complex the heavy metals. A single cropping of *B. juncea* does not clean up a toxic site; many growth cycles are required, with multiple harvests and natural reseeding. *B. juncea*, even more than its close relative *B. napus* (oilseed rape), is not fully domesticated, and the multiple cycles of cropping would allow the possibility of selecting for ferality. Thus, mitigation seems necessary to prevent volunteers from becoming feral. The issues with species such as poplars are discussed in detail with issues of forestry. One gene that might specifically fulfill the need for a mitigator gene is overexpression of a cytokinin oxidase,[123] which reduces the levels of isopentenyl- and zeatin-type cytokinins. This in turn leads to phenotypes with far-reduced shoot systems (unfitness to compete) but with faster growing, more extensive root systems,[1144] all the better for extracting toxic wastes. Genes that delay or prevent flowering may also be useful with the *Brassica* species, allowing multiple cuts of larger vegetative plants and preventing gene flow. Such genes are discussed in the general mitigation genes.

4.10. Concluding Remarks

Systems exist that can theoretically preclude a crop from hybridizing with the same or related species, whether by containing gene flow or by preventing the establishment of hybrid plants in the field by mitigation. Thus, if a risk is discerned, it should not preclude developing transgenic crops; it should stimulate the imagination to devise and test systems to deal with the potential problems.

CHAPTER FIVE

Introduction to Case Studies

Where the Ceiling Needs to Be Breached

There are volumes and series on potential new crops and on crops that have been abandoned.[79,456,457,469,472,538,696,781,849,853,1143] A representative grouping of crops is chosen to illustrate the diversity of the problems and the diversity of the solutions. No "unified theory of plant domestication" exists, or single way to redomesticate or further domesticate a crop. In too many cases the reason for lack of cultivating a crop is "low yield" or other poorly definable criteria, where the imagination and/or a trip to GenBank does not provide a solution. This too will change with time. If the reader has a favorite crop not covered in the following case studies, look for the nearest analogous case and see the way it has been covered, or just follow the principles of discussing and defining the problems and seeking the solutions, interpolate, extrapolate, imagine, and improve. This discussion is mainly dedicated to crops for which it should be simple with a few known or easy-to-isolate genes to continue the effort of domestication. At some point in the future the author will be charged with being far too conservative in outlook.

Each case study in the following chapters was chosen for unique reasons: the type of crop; adaptability to specific growing conditions; specific needs, special issues; different types of genes or techniques to be used. The abandoned, orphaned, and minor crops are the crops most likely to have reached a low genetic ceiling. Some major crops are also discussed, because major crops also need to have their biodiversity expanded. Despite having reached a high ceiling, the ceiling is there and the food biosecurity demands necessitate breaching even high ceilings. This has already been demonstrated with the few transgenic crops that are on the market and had been cultivated cumulatively on a billion acres by 2005. Clearly farmers discovered the need for transgenic herbicide and insect resistance in soybean, cotton, oilseed rape, and maize that had not been met by breeding, otherwise the farmers would not be growing these crops on such vast areas.

The list of case histories discussed is not exhaustive. If many crops have the same problem, only one is discussed and the reader is expected to extrapolate conceptually, if not necessarily specifically. Although it was considered better to use crops as examples and to let the reader extrapolate, before specific crop examples are dealt with, four pervasive and general crop constraints are discussed in the next four chapters.

Parts of the case histories, where it is suggested how the ceiling may be breached, are in many ways "science fiction," as it has not been done, and the genes needed may not yet be available yet in the local library (e.g., GenBank online). Suggestions are made regarding what one might look for and where one might find the genes, and it is hoped that most readers will have a more intuitive or innovative imagination than the author. Some of the crops discussed have not even been seen by the author, nor is the author a crop scientist with in-depth knowledge of any of the crops. Still, problems were discovered with some crops during the literature search that were known to the medical community, but not to any crop scientists that were canvassed who are experts with those crops.

5.1. Continuing the Green Revolution—Why Has Dwarfing Stopped?

On being questioned by Pooh about the existence of the East Pole, Christopher Robin replied, "I guess there is one, but no one ever talks about it." The green revolution was a singular success in enhancing yields by the simple expedient of reducing the height of wheat and rice. This not only increased the harvest index (grain weight divided by total stalk weight), but also vastly decreased lodging. Lodging was more of a problem in the old tall varieties when fertilizer was applied, so fertilizer could not be productively supplied to them. The green revolution varieties flourished better when supplied with nitrogen. The breeding of such dwarfed varieties was not easy. The recessive dwarf mutants had to be found and then crossed and backcrossed into extant varieties. There were yield-reducing linkages between the dwarfing genes and other genes on the same chromosomes that had to be eliminated by waiting and selecting for random meiotic crossover events. The problems were so great that this incredibly valuable third world development has not been adopted in parts of Europe, where they still prefer to lace their wheat with growth-stunting hormones, instead of doing this genetically. The precise mutant genes responsible for the green revolution varieties have subsequently been identified. What took more than a decade by complicated breeding can now be quickly performed by antisense or RNA interference (RNAi) technologies in a much shorter time, without linkage problems. Different levels of dwarfing that are found among separate transformants could provide the breeder and farmer with considerable leeway. Such dwarfing genes could be stacked with other useful genes and/or used as mitigator genes as part of strategies to deal with gene flow (section 3.3.2). Other tall crops produce large amounts of stalk material. Why have they not been dwarfed to increase the harvest index and reduce wastage in the field? Has it been done but, "like the East Pole, no one ever talks about it"? This is surprising, as recent studies where oilseed rape was dwarfed by using a truncated form of the gibberellic acid receptor (Δgai) showed a more than 30 percent yield increase[21] while testing a single gene construct and a few transformants, not as breeders would do. In a much more laborious study, a dwarfing gene was found in wild *Brassica rapa*.[758.] It was crossed into oilseed rape (*B. napus*), embryos were rescued and backcrossed to oilseed rape, but there was no yield increase because of a linkage to an insect susceptibility that will have to be crossed away,[757] illustrating the utility of the transgenic approach versus interspecific crossing.

Perhaps this issue of dwarfing should be readdressed with taller crops such as maize, rye, and sorghum, with some of the crops being domesticated (e.g., tef, the amaranths, and quinoa). Interesting excuses have been offered on why these crops should not be dwarfed, for example, the farmers use the straw for feeding and construction. Feeding the added grain from dwarfed varieties would more than compensate for the inefficiently fed straw. The farmers who use straw for construction do so because they are too poor to purchase construction materials they would prefer. Greater yields would supply this income. Temperate maize hybrids are somewhat shorter than those cultivated half a century ago; presently, "corn is (no longer) as high as an elephant's eye." Truly dwarf maize might have some disadvantages and has long been targeted by breeders. The known dwarf mutants are also severely affected in floral development.[465] This could be due to linkage effects, so the reason exists to target dwarfing transgenically by antisensing. The use of different tissue and development stage-specific promoters could allow normal elongation where needed, and dwarfing when elongation is not needed. Many tropical maize and sorghum varieties are much taller than common present-day varieties, and one could question the validity of making so much photosynthate into so much stem tissue instead of grain.

An easy way to ascertain the potential value of dwarfing is to treat the favorite tall varieties of any crop with plant growth regulators that suppress the biosynthesis or effects of gibberellic acid, the plant growth hormone responsible for stem elongation. It was just such experiments that gave the clue that dwarfing oilseed rape might well enhance yields.[21]

5.2. Genomics and QTLs (Quantitative Trait Loci) versus Transgenics

In a very thoughtful paper, Naylor et al.[781] describe what they believe are the major needs for further developing orphan crops. Much of their discussion is an extrapolation of the use of genomics and various analytical tools that have proven useful for breeding for quantitative traits in major crops. Very little genomic information is available about many of the crops described in the case histories (Fig. 18). More often than not, the reason crops have become abandoned or are orphans is that they have major gene trait problems that cannot be tweaked by breeding using selection. The problems are the result of missing genes or the need to suppress expression of endogenous genes without affecting the rest of the genotype, and not in the constellation of native genes being expressed. Marker-assisted QTL breeding deals with the constellation of genes, allows the use of fewer crosses, and gets the breeder back to near the original genotype with fewer backcrosses, but it requires large amounts of genomic data.

Some crops are major crops because they are not missing key genes; thus, great effort can be invested in quantitative breeding to improve them by small

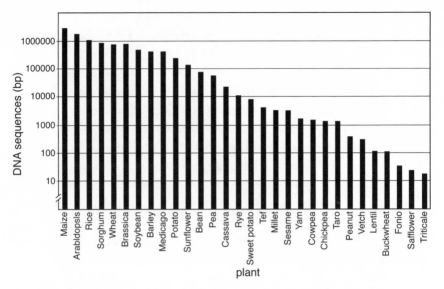

Figure 18. Orphan crops have few sequenced genes; the numbers of sequences deposited in GenBank for selected crop species as of early 2006. Some species are more equal than others.

increments. Many abandoned crops were orphaned, or remain in genetic slum ghettos because they miss key genes. Only after the needed missing genetic information is added can one really deal with modulating the quantitative traits in such species. It is possibly misdirected to initially go into QTL marker-assisted breeding in such crops, as seems to be the direction proposed,[781] until the missing key genes are added. This was borne out in a long study to further the domestication of maize for Africa by marker-assisted breeding for natural stem borer resistance. After many years, the project was "re-evaluated,"[129] and Bt has been quickly and somewhat successfully engineered into maize to fill the same purpose.[760] It is perhaps expedient that this push toward marker-assisted breeding has taken place; there are no regulatory constraints to this use of molecular biology and thus work can begin immediately. This expedience can prove to be wasteful, if the key constraints to a crop have not been dealt with, as the final effort may be to use the markers to breed for traits that do not exist within the genome.

An analysis of the case studies below will show that the selectors and the breeders have neared a dead end in many species, and it is only exogenous genes that will return such discards to the pantheon of widely cultivated species.

5.3. Are All Underdomesticated Crops in Immediate Need of Transgenics?

While transgenics ultimately offer the possibility to introduce genes from a huge number of sources, not all underdomesticated crops are in immediately perceived need; their constraints can either be addressed more easily by conventional breeding, have little to do with missing a single trait, or have more to do with market forces than biology. A few examples are provided below based on our reviews of the available literature.

5.3.1. Bambara Groundnut (Vigna subterranea)

Bambara groundnut is an indigenous African crop grown by some subsistence farmers throughout sub-Saharan Africa. Its culture and needs alone were the subject of a published 173-page symposium volume.[490] It is a nutritious crop that does not seem to be missing major biotic or abiotic traits. In many places its yields are enhanced by rhizobium inoculation, and despite the considerable genetic variability available, little breeding has been done. Its main problem is that it seems to have fallen out of favor of the farmers and the consumers. It is rarely available in markets. Will a large-scale breeding effort, raising yields and lowering price, bring it back in to favor? Would growing it on a large scale, and not as an intercrop as at present, reveal constraints only amenable to transgenics? Time will tell.

5.3.2. Mashua (Tropaeolum tuberosum)

Mashua is an interesting South American starchy tuber crop that also has been intensively analyzed for further development.[403] Pests and diseases were not constraints. The main limitation appeared to be the low acceptance by consumers, far below that of other parallel starchy tubers. The flavor of some mashua lines is so strong that it deters potential consumers, even those accustomed to spicy foods and eager to try "novel" foods. Depressing specific flavor traits is amenable to gene suppression, but this is not the only problem with mashua. The shelf life of mashua is short because the tubers lose water rapidly in storage, and the tubers deteriorate within weeks after harvest, yielding a product with little or no market value. Some cultivars do not tolerate storage at all; the tubers of such cultivars sprout or rapidly rot almost immediately after harvest. Still, there is considerable genetic variation among varieties, so selection and breeding might easily overcome the problems, which do not really relate to the crop but to market acceptance. In summary, the main limitations for expanded use of mashua at present relate essentially to demand, marketing, and postharvest aspects, not to production or to issues easily solved transgenically.

5.3.3. African Yambean—A Case Where There Might Be a Solution

African yambean is unknown to consumers. A group of west African agricultural scientists was asked at a recent meeting to describe crops that they remembered but are no longer used because of intractable problems. Hands flew up, and memories of the African yambean (*Sphenostylis stenocarpa*) as a very tasty bean were recounted with gusto from their youth. They stated that it has a hard seed coat, which required that it had to be cooked overnight to soften it, something the now urban consumers, as well as many rural consumers are unwilling to do, in part, because of fuel availability, as well as a lack of willingness to watch the pot. They thought that the crop was poorly researched, yet knew that it was resistant to insect attack and assumed that the hard seed coat is the factor conferring insect resistance. It is worthwhile to ascertain if this assumption is correct because, if no relationship exists between seed coat and resistance, the seed coat could be "thinned" by breeding or by RNAi technology of a major coat component, and the genes that are not in the seed coat conferring insect resistance could be isolated and cloned for use with other legumes.

Our African colleagues stated emphatically that if this bean could be cooked in normal time, it would be a gourmet hit, and be highly nutritious and storable. A search of databases indicated that more has been published by Africans than is known to the African scientific community, with the most recent review by Potter,[863] which is the source of the unreferenced older information cited below. The African yambean should not be confused with the New World tropical legume also confusingly called yambean (*Pachyrhizus* spp.).[1038] The African yambean is one of seven species in the genus *Sphenostylis*, and its tubers are also edible. The tubers are also responsible for the perennial aspect of the species, as the shoots die back during dry seasons, but it is mainly grown as an annual, from seeds, for seeds. Some of the other species of this genus are collected in the wild, but this is the only one to be truly domesticated. It has significant differences from its own wild forms, including pod and seed size and, as expected, delayed shattering of the pods. The wild material causes sore throats or headaches after being eaten, but the tubers of the domesticated types do not.[863]

An early survey (cited in Ene-Obong and Obizoba[320]) revealed that indeed in the areas where the yambean was eaten, it was cooked overnight. The hard seed coat (hull) is about 10 percent of the weight of the seed. Yambean is indeed is highly insect resistant because of factors that are mainly in the tissues underlying the seed coat. There are early reports of cyanide and oxalate. There are heat-inactivatable trypsin inhibitors[823] as well as insecticidal lectins.[821] A major lectin was isolated with the idea that its gene might be useful for transforming other legumes, but it did not affect the legume pod borer (*Maruca vitrata*), a moth, but it was effective against weevils[673] and Hemiptera.[813] The crop is also high in tannins and phytate, all of which are reduced during soaking treatments.[33] The raw stored product causes diarrhea, poor protein utilization and growth in rats,[12] another reason why it is easy to store in conditions of subsistence agriculture. High doses are lethal to rats, lower doses induce drowsiness and can extend phenobarbital-induced sleep,[69] which may explain the folkloristic use of infusions of African yambean to calm violently drunk humans.

This nitrogen-fixing species is easy to cultivate on poor soils, but the yields on good soils are not as high as soybeans. Unlike most African and introduced legumes, it is not a host to the parasitic witchweed *Striga gesnerioides*, but has the additional advantage that it stimulates *Striga* seeds to germinate.[114] Interesting genes clearly can be found in this crop.

The antinutritional properties of African yambean were localized mainly in the cotyledons, and not in the hard seed coat, so indeed the seed coat may not have major responsibility for resistance to insects. But is there really a need for the transgenic approach? Two groups claim that two to three hours of cooking are enough to get a soft bean, devoid of the heat-labile antinutritional compounds,[12,320] which is less than for cowpeas. Six to twelve hours of presoaking in water can reduce the cooking time by a third,[320] and soaking in brine reduces it by two-thirds.[12] Why don't consumers know this information that was generated a decade ago? Or do they, but the scientific experiments were uncontrolled? The researchers measured softness of the cooked yambeans, but do not report having done the critical organoleptic taste panel testing to see whether the product tasted the same, or whether there were aftereffects. Neither of the authors dealing with short cooking time cites an earlier report that twelve to fourteen hours of cooking are needed to inactivate factors reported to cause diarrhea.[70]

We are not in a position to have the answers, but an excellent crop, with great attributes for Africa, is not being cultivated. If the product is unsafe, there is a place for biochemists and pharmacologists to find the compounds requiring the long cooking, so that the genetic engineers can find the genes to be suppressed. If indeed a good, safe product can be obtained by presoaking and short cooking, the engineering needed is with advertising hype, packaging for the nascent African supermarkets targeting the middle class with "new, short-cooking African yambeans," with soaking and cooking instructions on the package. If the product is also targeted to Northern Hemisphere gourmet markets, the packaging in Africa could also include the magic words "for export." If the price is right, the product and "new" technology will spread to the farmers as the market is realized. The gene jockeys' present job with African yambean will be to find the anti-*Striga* and anti-insect genes in this crop to transfer them to other crops.

5.3.4. Upgrading Niger as an Oilseed?

Choosing species to discuss is not easy, if one wishes only to provide examples of the types of problems one wants to deal with. Thus, edible-oil quality is dealt with in one species (Chapter 11). One can state the problem: Very few oilseeds are grown in Africa and most of the edible cooking oil is imported; is there not an appropriate species that can be further domesticated? A perusal of the literature shows that a superficially appropriate crop may exist; there is a native African oilseed crop: niger (*Guizotia abyssinica*), domesticated in Ethiopia (where it is called noog or noug), which might be appropriate. This composite crop should not be confused with another oilseed *Brassica niger*. Niger is also grown in India, where many scientists have studied it.

The oil of this species has problems to address. It is highly polyunsaturated, with a content of 77 percent linoleic acid,[27] which leads to its being highly unstable due to a potential to be easily oxidized.[886] Perhaps this rapid rancidity leads to the seeds quickly losing germination potential.[270] It is an obligate outcrosser, which complicates a breeding effort. The residue (meal) after oil extraction has a high quantity of trypsin inhibitor, and the protein is low in threonine, leucine, and isoleucine,[120] so it is not the best for feeding to animals. Even in optimal growing conditions in the United States, the yield potential hovers at a very low half ton per hectare.[563] Niger is susceptible to diseases, so one wonders why there is much interest in it. Much is still to be done with breeding, for example, selecting for plants with a determinate growth period to shorten the time to harvest.[7] As it is, much of the production of niger is marketed as a pretty, colored birdseed, indicating its low economic esteem. Thus, even though the species can be transformed,[769] I decided to pick oil palm as the example of an oil species needing further domestication. Oil palm is an extremely high-yielding crop and represents over a quarter of the edible oil produced in the world. It has the converse problem: an exceedingly high content of unhealthy saturated fatty acids (Chapter 11). If all the other problems of niger can be addressed by breeding or biotech, the issue of modifying the polyunsaturated fatty acid content can be easily addressed, as outlined in Chapter 11, by antisensing or "RNAi'ing" a key desaturase, which would give a healthy oleic acid oil.

5.3.5. *Domesticating Arabidopsis to Detect Explosives*

Detractors of genetic engineering often refer to the products as "Frankenfoods," alluding to Frankenstein-type monsters. Only one product of engineering, widely sown, should be given that title. Too much of the world's agricultural and other lands have been sown, and continue to be sown with the most horrendous species engineered by man: *Dynamita horrida* A. Nobel (landmines). It is claimed that more than 100 million landmines are buried and active in planet earth today. Another 100 million are stockpiled and ten million are produced annually. More than a million people are said to have been killed or maimed by landmines since 1975. About 26,000 people are killed or injured each year. Another significant problem is that large areas of agricultural land (in Cambodia, estimated 40%; in Angola, estimated 90%) are unused because of land mines, with severe socioeconomic consequences for the population/countries.[61] Agriculture is a precarious enough occupation without landmines, yet there are no good ways to remove them, and their cost in life and limb is far greater to the civilian population than to the military that plants them.

Do genetic engineers have an answer? Here we have a situation where the answer is perhaps "no" and that is because the projects seem not to be thought out prior to beginning the research. "We'll cross that bridge when we get there" syndrome may have been a part of the project plan. Thus, we discuss this serious need for landmine detection, because it can be forgotten that a good project needs more than excellent genetic engineering. One must know how an engineered plant will be cultivated.

A few groups have studied whether plants can be used to detect and/or biodegrade explosive material. The results have been fascinating from a biochemical point of view. A bacterial 2,4,6-trinitrotoluene (TNT) nitrate reductase has been transformed into various plants, and their roots can degrade buried TNT in the soil.[460,896]

Whereas the plants above might be useful for phytoremediation of soil near munitions factories, it is hard to envisage their roots penetrating into buried landmines and chewing up the explosives. A Danish company[61] has been working for several years using technologies they developed[137,716,717] for a biodetection system in *Arabidopsis*. The plants will change color from green to red in the presence of NO_2 leaks from landmines, which result from the slow bacterial degradation of the TNT. They envisage that their *Arabidopsis* may significantly speed the removal of landmines in cultivatable areas to permit the subsequent use of cleared areas for agriculture to maximize socioeconomic benefits. They have addressed one biosafety issue, claiming that because *Arabidopsis* is an obligate self-pollinating plant, it will not spread its transgenes to other *Arabidopsis* plants in the environment. Furthermore, they state that "male-sterility can be introduced into the genetically engineered plants in order to eliminate the risk of spreading pollen." They have also introduced genes that will not allow the seeds to germinate nor set seeds unless a specific growth hormone is added to the plants.[61]

They seem not to have addressed another issue of biosafety—how can this be used in the field? First, this method relies on seepage of NO_2 from leaking mines, it has the potential to miss the more recent landmine types that are specially sealed to obscure such detection.[474] Second, how can one ensure that puny *Arabidopsis* can compete with the indigenous vegetation? To establish any crop one must have a good, weed-free seedbed. So, first one will have to plow and disc the landmine-infested field (or run a spray rig across it with herbicide, statistically slightly safer for players of Russian roulette), then drill (sow) in the *Arabidopsis*. These indicator plants are hopefully herbicide resistant, so other species will not overgrow them, and then one must spray herbicide to prevent the weeds. After germination one would have to get down on hands and knees and crawl through the field to see whether the *Arabidopsis* changed color. *Arabidopsis* does not grow well at above 27°C so we might have to wait until there is global cooling to use this technology in the tropics, where many landmines are located.

A century ago, the famous engineer Izambard Kingdom Brunel who designed many important, still-standing bridges, stood under each one when the first overloaded trains were sent across to make sure the bridge met specifications. For the sake of the health of the genetic engineers who designed this *Arabidopsis*, it is hoped that they will not try to meet that type of challenge in a minefield with their product.

Three biological methods could allow biodetection of landmines. Two would require putting their very complex construct into other plants. One of the plant scenarios might even work (with the caveats stated above about limits to what could be detected).

1. Engineer the construct into a mythical superweed that can be seeded from the air; the superweed will magically germinate without preparing a seedbed. It will then crowd out all other species and turn red when growing over mines. There are some biosafety and environmental issues with this. Among others, fields covered with superweeds would be of little use to the farmers, even when all the plants are green, signaling no more landmines.

2. Engineer the construct into a mixture of broad spectrum, disarmed viruses, infect standing vegetation in the field, possibly by a low-flying helicopter using sandblasting (Chapter 8), such that grass and legume species in the field will be transiently transformed. This still leaves the issue of viewing the reporter plants. One group has developed technology to remotely view the green fluorescence when the green fluorescent protein is used as a reporter gene.[1018] It could be used instead of the anthocyanin production in the present NO_2 detection system, and the fluorescence of the green fluorescence could be remotely detected by laser-induced fluorescence imaging.[1018]

3. Use other more mobile organisms than plants to detect the landmines. African giant pouched rats as well as dogs have been trained to detect landmines and are probably light-footed enough not to set them off.[474] One advantage of using animals is that they can be trained to smell more than one kind of explosive, whereas the *Arabidopsis* technology is based on NO_2 alone.

This seems to be a case where there has been a disconnect between molecular biologists, agronomists, and ecologists, demonstrating why such disconnects are to the detriment of getting good and needed products to market. Other disconnects are described in the case studies, accentuating the need of collaboration.

CHAPTER SIX

Evil Weevils or Us

Who Gets to Eat the Grain?

The day-to-day existence of the nomadic hunter-gatherer before crop domestication was fraught with starvation, as there were few ways to store food during seasons when game was sparse and there was not much to gather. Domestication of cereal grains and legume pulses and members of not too many other families (e.g., buckwheat, safflower) changed that. Despite the seasonal nature of harvest, dependent on warmth (in temperate areas) and rainy seasons in the subtropics and tropics, the domesticated seeds could be stored. But this storage was fraught with problems. The stored grains (the term grains will be used for all saved seeds from cereal grains, pulses, etc.) were infested with insects, especially weevils (mainly *Sitophilus* and *Bruchus* spp.) but also grain moths (mainly *Tribolium* spp.), who found the stored seeds a happy home. Even grain in prehistoric archaeological sites from the Egyptian pyramids to northwest Europe in Roman times was found to have weevil infestations.[166] Post-Biblical Jewish literature forbade eating of grain infested with grain insects,[734] perhaps due to an understanding that the insects are vectors of mycotoxin-producing fungi (Chapter 7). Clearly, the weevils were camp followers, traveling inside grain as grain spread from the Middle East, as

Sitophilus weevils are flightless. Ancient Roman writers suggested sprinkling the remnants from pressing olives, soot, or the juice of *Sedum* on the stored grain to deter insects,[249] which would have added an oppressive flavor to the grain.

Farmers considered the larvae in the grain to have arisen from the grain. In one area in contemporary Nepal, "most farmers believe that stored product pests emerge spontaneously from the grain. Pest growth is initiated and triggered by grain moths whose respiratory heat creates weevils."[444] We were all taught about the experiments to prove that maggots do not arise by spontaneous generation from meat, but it is less well known that Van Leeuwenhoek spent four months studying the life cycle and sexual apparati of grain weevils as part of the effort to discredit such ideas of spontaneous generation.[927] Until this day farmers in Nepal perceive insects as a "mistake in God's creation"[444] and accept their damage with too much equanimity.

These insects lay their eggs in or on the grain, and the developing larvae eat out the insides. Hungry adults eat the grain outright. A farmer storing grain for the seasons without food, or worse, for extended periods of drought as with the Biblical seven bad years, may find little to eat in the stored sacks. The weevil parts are allergenic and can induce asthma attacks.[495,664,924] These allergy problems are typical among grain mill workers and bakers,[495,541] suggesting that there is a significant level of ground weevil parts in commercial flour. Additionally, the insects are often vectors for disease pathogens that attack the grain (Chapter 7). Rodents can also cause great losses, but the rats and even the mice are too big for biotech to handle, at present.

The drier the seed going into storage, and the drier the storage conditions, the less damage to the grain. Even though these insects are exceedingly drought tolerant, they do require some water for life. It is easier said than done, especially early in domestication when people stored their grain where they lived, in dank caves or leaky huts. It was difficult to adequately dry the seed of underdomesticated crops that had been harvested while still damp. Such crops had to be harvested before they were fully mature; if the seeds were really dry, they shattered and fell to the ground before harvest. This is another reason why the lack of shattering has been and is such a key point in early domestication, and is still critical for some crops. Warm, humid climates, and poor, leaky roofing facilitate the lifestyle of the grain insects.

6.1. The Magnitude of the Problem

The grain weevils (Curculionidae) are well known as major primary pests of stored cereal grains and have spread throughout most of the world.[141] The losses can be of more than one type: actual weight loss, which is due both to eating as well as opening the grain to more drying; quality loss, which is a function of the percent grain bored and the amount of insect parts and excrement in the grain; nutritional loss due to insects heading preferentially to the high-protein, vitamin-containing germ; loss in seed viability in grain stored for planting. Much of this sums up as commercial loss and can be greater than all the other losses when grain becomes no longer marketable.[142] Grain insects cause massive crop losses in sub-Saharan Africa, especially the larger grain borer, which causes a loss of 40 to 60 percent in storage (Table 8). For example, after eight months in storage, damage to untreated grain in Zimbabwe was 76 percent, but it was 36 percent, 17 percent, and 10 percent, respectively, for grain treated with the insecticides malathion, pirimiphos-

Table 8. *Constraints to Food Production in Africa due to Grain Weevils*

Country	Area planted ('000) Ha	Yield (kg/ha)	Estimated yield loss (%)	Yield loss ('000 T)
Large grain borer in maize				
Botswana	83	112	19–27	1.8–2.5
Cameroon	350	2,429	31	263
Congo	1,463	799	18–28	210–327
Ghana	713	1,315	20.0	188
Kenya	1,500	1,800	23–41	621–1107
Malawi	1,446	1,099	14–18	222–286
Mozambique	275	896	23–39	57–96
Sierra Leone	10	928	40	3.6
Tanzania	1,457	1,795	34	889
Uganda	652	1,801	50	2,114
Burundi	155	1,087	29–47	49–79
Pod borer *Maruca vitrata* in cowpeas (countries where data available)				
Malawi	79	683	30–88	16–47
Senegal	146	323	12–70	6–33
Niger	3,000	117	23–60	80–210
Tanzania	147	320	18–83	8–39
Kenya	150	484	32–71	24–51

Source: Compiled by J. Ochanda. From reference 431, by permission.

methyl, and methacrifos.[391] These insecticides are only appropriate for grain destined to planting and not for human consumption. The grain weevils are able to establish themselves on whole, undamaged grains of maize, sorghum, rice, and wheat as long as the grain is not exceptionally dry. *Sitophilus zea-mais* is the dominant species on maize, while *Sitophilus oryzae* is dominant on wheat but also attacks legumes. The bostrichid beetle *Prostephanus trun-cates* (the larger grain borer) is a highly destructive primary pest of maize, especially maize stored on the cob. This insect has spread from its previously more limited indigenous range in Central America[705] and also attacks wheat, rice, legumes, both cocoa and coffee beans, as well as stored cassava, yams, and sweet potatoes.[142] Six genera of bruchid beetles attack legumes. The eggs are typically laid on the green pods with the larvae boring into the seeds, and infested seed often causes pods to shatter. Quarantine seems to be unable to keep the grain pests from spreading, even to Australia; can you catch each arriving tourist wearing a necklace made of pretty seeds?

6.2. Conventional Technologies

6.2.1. Physical and Chemical

Artificial drying is used in the developed world, and the grain is then stored in hermetically sealed, dry storage facilities with low oxygen, and/or with fumigants such as phosphine, and in some cases nuclear irradiation to kill all living forms (including the seeds). Lower irradiation doses are used for sterilization of the beetles[25] to provide grain without reproducing weevils. What can the subsistence farmer in the humid tropics or subtropical climates do to prevent losing 75 percent of the grain stored in their homes? It is indeed a pitiful sight to see farmers sorting through grain to find kernels to grind for food or use for planting. The infested grain is fed to animals or used to produce beer, often with dire consequences to the consumer. (Those northerners who may find these uses for infested grain "gross" should not be told what kinds of apples are used to prepare cider for human consumption.)

Sieving was used in the past to separate insect parts from grain and flour, but could not remove the infested grains still containing larvae and pupae for continued storage. Now equipment is becoming available that separates infested from whole kernels by density and air resistance.[1141] The efficiencies are not high enough; with removal rates of 50–95 percent (depending on the life stage of the weevils), the insect populations can rebound in continued storage to original levels in two generations unless the grain is cooled.

Farmers and researchers have tried some "natural" solutions; leaves of various aromatic herbs, plant extracts including the use of neem extracts[128,134,316,806,882,1040] to some minor avail, but without data on effects of their residues on humans. The concept of "its natural, therefore it must be safe" seems pervasive among the proponents. Basically, they are adding back many of the undesirable flavors and odors that kept insects at bay in crop progenitors, which were selected away or bred out.

Fumigants such as banned methyl-bromide had been used, as have some insecticides. The latter are not of consequence for seed stored for planting, but their use and their residues for grain to be consumed by humans is regulated. The one safe "natural" insecticide pyrethrum can be used, but it is expensive, even when formulated with synergists to prolong activity.[122] In this respect, the insect growth regulator insecticides seem to be more promising[740] as they typically have no effects on mammals, other than on their wallets; they are not inexpensive. Bruchid weevils infesting legumes can be held in check by application of insecticides to the pods, preventing weevil establishment prior to harvest.

The evolution of insecticide resistance is a well-known pest management problem common to many species of storage insects and to a wide range of insecticides.[198] Resistance stems from careless use, which is common among resource-poor farmers. As a consequence, the only available methods of control of grain weevils involve the use of ineffective and expensive chemicals to which the target species have evolved high resistance.

Cultural control strategies have been employed to control grain storage pests. Grain moisture content considerably affects pest status. This is not a factor that can be cost-effectively manipulated by artificial drying to achieve sufficient control of insect pests in most humid tropical situations, especially with farmer-stored grain for household use. Insect development and growth rates are dramatically enhanced by warm tropical temperatures and the pests seem to develop much faster under storage conditions. The only applicable conventional concept to control insect infestation depends on sealed (hermetic) storage to reduce oxygen availability and infestation.[518]

6.2.2. Breeding for Resistance to Grain Pests

Genetic variability related to weevil damage in maize has been known for more than three decades.[1150] Research groups continue to report on progress reducing weevil damage in maize[271, 272, 374, 390] and wheat.[614] Some green peri-carp properties were positively correlated with a modicum of resistance to weevils,[374] as are diferulic acid phenolics (some of which are probably not de-sired by consumers), grain hardness, and hydroxproline-rich glycoproteins. This indicates that the genetics of resistance should be multifactorial, as has been borne out.[272] Similarly, resistance in wheat is quantitative, and there is combining ability.[614] Still, in decades of effort no varieties have been released from a breeding program devoted to resistance to weevils.

Conversely, no genetic diversity was available in peas (*Pisum sativum*) for resistance to *Bruchus pisorum,* a weevil specific only to peas and the major weevil attacking peas. The needed variability seems to be in the related *P. ful-vum,* which can cross with peas, when *P. fulvum* is used as the pollen par-ent.[466] How hybrids and backcrosses will perform is an open question, as resistance in this case seems to be a matter of preference of pod type for ovipo-sition. When the female does not have a choice, she may well lay eggs on peas, even if they have *P. fulvum* pod types. Interestingly, a single pea gene con-trols resistance to the rice weevil *Sitophilus oryzae,* a grain weevil with a wide host range.[406]

6.3. Potential Biotechnological Solutions and Status

It is fascinating and telling that the biotechnology industry has hardly ad-dressed the issue of storage grain pests. They have dealt so well with root-worms, stem borers, and cotton bollworms and boll weevils. Why have they ignored storage insects? Has the biotechnology industry lost its pioneering spirit? Pioneers see a need and come up with a solution. Industry now sees only existing "markets," not new virgin markets; instead, they covet others' markets and want a share of that existing market. Rootworms, stem borers, bollworms and boll weevils represented large markets for insecticides in the developed world and the biotech industry could assess it, and covet it they did. One may add in defense of the biotech industry, they clearly replaced organophosphate and other not too healthy insecticides with safer ones made *in planta* to control these pests.[918] But why not realize that grain storage is a potential market? There are two reasons. Firstly, the biotech industry is slow to realize that the world's poor, in the parts of the world where storage pest problems are the worst, are a huge market. Other industries have realized this and are profiting, but not agricultural biotechnology.[413] Secondly, no one had been able to make safe and efficacious insecticides for grain storage pests, so there was not another's market to be jealous of.

In biotechnology for storage pests, most of the publications delineating potential solutions have come from the public sector; universities, public research institutions, and government laboratories. They are doing the pioneering work, but they do not have the economic incentives to get the products to market nor the financial means to deal with the regulatory hurdles.

Not all genes that control storage pests would be useful, because some have contraindicated effects on mammals. A biotechnology company was making avidin in transgenic maize, a protein normally found in low quantities in chicken eggs and isolated for use in diagnostic kits.[510] They found that the avidin-containing maize seeds withstood attack by several storage pests. Avidin binds the vitamin biotin, so avidin would pose a risk of avitaminosis to livestock consuming large quantities of avidin-containing grain. It would also pose a risk to humans eating large amounts of the whole meal, as avidin is not denatured by heat (cooking). The transgenic maize was never destined for anything other than use in pharmaceutical kits and was produced in isolation. The avidin was produced in the embryo (germ), which is removed before making corn meal. This production is often cited as an example of commercializing dangerous transgenic crops, which is clearly demagoguery.

A setback recently occurred in the use of transgenes to control pea weevils. The alpha-amylase inhibitor-1 gene product that is expressed without issue in cowpeas, for reasons unknown, is somehow posttranslationally modified in pea to a form that is immunologically inflammatory and thus may be allergenic.[867] This was caught way before a commercial product was released.

6.3.1. The Use of Bt Genes

As with stem borers in maize, genetic engineering of insecticidal proteins into the crops of interest could effectively control grain weevils. Unlike the case of stem borers, suitable Bt Cry proteins have not yet been identified for grain weevils. Cry3 proteins are known to have coleopteran activity, as are the binary proteins in classes Cry34 and Cry35.[317,954] In addition, some Vip proteins have broad coleopteran activity.[193,954] Lepidoteran pests such as the grain moth are largely controlled by low levels of *Cry1Ab* expression in grain.[964] Thus, screening of several proteins from *Bacillus thuringiensis* for weevil activity would be a logical first step. Unfortunately, when the owners of the five largest Bt libraries, containing thousands of different natural and man-made Bt genes were asked whether they would offer their libraries to public institutions for screening against storage pests, they all refused.[431] If a suitable Bt protein could be identified, then currently used expression systems that have already been shown to produce levels of expression in the grain capable of protecting the grain against storage pests could be used. As with stem borers, the use of highly specific, plant-expressed insecticidal proteins would remove the human health risks posed by fumigants and insecticides, or by the insect detritus common in the flour and food.

Genes encoding toxic proteins are available that would control both the grain weevils and the moths. These genes come from projects where hundreds of proteins were fractionated from venomous scorpions and spiders.[661] Many of these proteins have no measurable mammalian toxicity when tested at orders of magnitude higher concentrations than are toxic to insects. The genes encoding many of them have been isolated. In the present regulatory climate where the illogic of "perceived public opinion" prevails, it is unlikely that one would even dream of making and animal testing a cereal or pulse crop with such genes. Still, it should be possible to use such genes in an industrial crop such as cotton, freeing Bt genes for wider use against grain pests.

6.3.2. Using Plants Own Tricks—Disrupt Insect Metabolism

More than a decade ago a pioneer thinker/researcher, Don Boulter, proposed using plants own enzymes to transgenically ward off grain pests.[136] The natural response of plants to grain weevils is either to poison them by synthesizing phytoalexins[234] or lectins, to form galls around them to limit their growth,[960] or to prevent them from digesting the insides of the seed. The prevention of digestion is attained by amylase inhibitors and/or protease inhibitors (many of which are also lectins), which prevent degradation of seed starch[34,261,274,358,360,436,719,754,875,1132] and protein reserves[288,295,359,462] by the insects (Table 9). Still, as peas are attacked by weevils, can pea proteases be of much value, or have the pea weevils evolved resistance to them? Although pea proteases may not have value against pea weevils, members of the three known classes of pea protease genes[288] might have value in other crops against other insects that have not come into contact and evolved resistance to these specific proteases. They may also be used for gene shuffling (section 3.5) to generate new protease-inhibiting peptides. Still, there are cross purposes here. Amylase and protease inhibitors are not a problem for humans when the grain

Table 9. Exogenous Genes or Products for Suppressing Storage Pests

Type of gene and source	Target crop	Target insect	Level tested	Reference
α-Amylase inhibitors				
Tepary bean	Azuki bean	None tested	Functional	1166
Rye	*E. coli*	*Acanthoscelides, Zabrotes, Anthonomus*	On larvae, NME[a]	274
Bean	Chickpea	*Callosobruchus* spp.	Transgenic, MI[a]	940
Bean αA1	Tobacco and peas	*Zabrotes* and *Bruchus pisorum*	Transgenic, MI	436, 595, 754, 875
Bean αA2			Transgenic, NME	957
Wheat	Bean	*Acanthoscelides obtectus*	In vitro	360
Protease inhibitors				
Black-eyed pea and chickpea	Cotton	*Anthonomus grandis* (boll weevil)	In vitro	359, 397
Prosopis juliflora	—	*Tribolium* and *Callosobruchus* weevil	In vitro	991
Hyptus suaveolins	—	*Prostephanus* weevil	In vitro	11
Barley	Rice	*Sitophilus*	Transgenic seeds against larvae	28
Barley	Wheat	*Sitotroga* grain moth	Transgenic seeds	32
Maize	Rice	*Sitophilus*	Transgenic seeds against protease	526

[a]NME, no mammalian effect when tested on mammalian enzyme(s); MI, inhibits a mammalian enzyme.

products are cooked, but protease and amylase inhibitors have been bred out of grains for animal feeding, as they prevent digestion of these key components in animal feed. Perhaps gene shuffling can give rise to amylase and protease inhibitors specific to insects (as well as to fungal pathogens; see Chapter 7) that do not affect mammals? Indeed, there is evidence for cross-species specificities. A more intellectual approach has been to compare the crystal structure of wheat amylase inhibitors that are specific to insects with those specific to human alpha-amylases.[358] Whether the differences found can lead to site-directed improvement of activity and target specificity, or whether random shuffling can do it better and quicker are open questions. Researchers hopefully will take both directions, in friendly competition, to see which works best.

The bean amylase inhibitor αAI 1 when transformed into peas provided excellent protection against the pea-specific weevil under field conditions,[754] but inhibited mammalian amylases. A second inhibitor αAI 2 has no mammalian activity but is not as active against insects. These transgenic plants did not produce as much inhibitor when the pods matured at high temperatures (32°C),[261] again suggesting that a better inhibitor is necessary. An amylase inhibitor from rye seemed active against a variety of insects, where it did not affect the mammalian enzyme tested.[274] Nature typically uses combinations, and such combinations might be valuable in the future. The wheat α-amylase inhibitor is not as active as expected against cowpea weevils because the weevils have the ability to degrade them with weevil proteases, probably as an evolutionary response. Black-eyed pea (*Vigna unguiculata*) protease inhibitor was quite active in inhibiting the cotton boll weevil, but no information has yet been published on its activity on grain pests. Many of these solutions are only effective on first instar insect larvae; later instars become resistant because of the production of other proteases in the larvae that degrade the transgenic amylase and protease inhibitors.[1202] This too can be overcome biotechnologically. The soybean soya cystatin N, a serine protease inhibitor, synergized the activity of wheat α-amylase inhibitor against the cowpea weevil in feeding trials.[34] It is hoped that the same synergy will extend to transgenic plants, without ill effects in animal-feeding trials. Some plants contain bifunctional enzymes having both α-amylase and subtilisin (a protease) inhibition activities.[788] Such bifunctional activities could be an excellent target for gene shuffling.

Many of the legume responses to grain weevils are turned on by brucins, a group of chemicals that emanate from the weevils themselves.[960] One of the genes turned on by brucins is a cytochrome P_{450} *CYP93C18*, which is thought to be an isoflavone synthase gene, especially as its up-regulation coincides with an increase in pisatin, an isoflavone phytoalexin.[234] Not only is this gene important, but there is great promise in isolating the regulatory elements that recognize brucins and control up-regulation. The next generation of transgenic peas should have the amylase and protease genes under control of such an element. This would greatly decrease the amount of amylases and proteases in stored grain, because their synthesis would be limited to only infested pods and maturing legume seeds. Such a concept would not work as well against insects attacking dry grain.

For reasons not clear, the approach of using amylase/protease inhibitors has been applied mainly to legumes[859] but not to cereals, even though heterologous genes/enzymes have been employed. This is surprising, as natural tolerance to *Sitophilus zeamais* in maize has been correlated with the level of amylase inhibitors.[693] Similarly, a protease inhibitor from *Hyptis suaveolens* inhibits a protease from the larger grain borer in vitro.[11] Two cases are reported where protease inhibitor genes were transformed into rice: cystatin, which inhibited proteases in vitro,[526] and the CME trypsin inhibitor of barley, which reduced weevil survival rate.[28] No reports were found on the use of amylase inhibitors in other cereal grains, or if they actually worked in the field. One hopes that both would be tried more widely with cereal grains, especially when broad insect spectrum, low mammalian effect bifunctional genes are synthesized.

The same approach is being tested to protect stored root crops such as sweet potatoes,[891] so far with exogenous amylase and protease inhibitors, where some had a protective effect, and their genes could be used to generate transgenic sweet potatoes.

It would be nice to report that large-scale trials in farmers' home granaries as well as in commercial silos validated some of the preceding approaches. Public research proves concepts; someone must deal with the regulatory hurdles and bankroll the field trials. Then, there is the disconnect between the plant protection community and the biotechnology community. This was brought home at a Pan African meeting convened by the United Nations Industrial Development Organization (UNIDO) a few years ago. The entomologists in Africa score the needs for dealing with grain pests near or above dealing with stem borers as the main problems in the continent. There are chemicals that can cost-effectively deal with stem borers, but there are no solutions for grain pests. The biotechnologists at the same meeting did not list grain pests as a priority in any but a few national programs. The existence of the problem of grain pests and its magnitude were a revelation to most of the biotechnologists in attendance.[431]

Kwashiorkor, Diseases, and Cancer

Needed: Food without Mycotoxins

The average life expectancy in the developed world is about seventy-five years but in most of the developing world it is closer to forty-five. What are the reasons for the difference? Factoring these to a specific number of years, each for malaria, dysentery, HIV(AIDS), measles, starvation, and so on, would allow setting health care research priorities, but only if the numbers were known. The World Health Organization has no such data available to factor the discrepancies of life expectancy, while such data are available for cigarettes, obesity, and so on, in the developed world. The media do not help; Ebola virus, avian flu virus, and severe acute respiratory syndrome (SARS) virus, which probably knock just a few seconds off of the average life expectancy, make front-page headlines. Why is this medical question in a book about continued crop domestication?

One of the factors contributing to or exacerbating many diseases is the lack of food, or access to bad food. Current statistics claim that hundreds of millions lack food, with most of them living in rural environments that are supposed to grow the food. Most people think of malnourishment as just the lack of food (calories) or the lack of critical components in food (specific amino acids or vitamins), which is not untrue, but is far from the whole picture. Still fewer think about the derivation of the word malnourishment: mal = bad nourishment, that is, a definition that tells us that there are components in the food chain that are bad for those who eat it. In other sections we discuss various toxic compounds (Chapters 16 and 19), allergens (Chapter 12), and goiterogenic compounds (Chapter 18) that need to be removed from crops. Our ancestors selected out many such components, but only those *mal*nourishing compounds that tasted bad or caused an acute reaction. Many tasteless compounds are left and perhaps none are as insidious as the mycotoxins produced by fungi that contaminate foodstuffs.[108]

We usually hear about them only when they are in ultra high doses, wiping out French villages when all the villagers ate baguettes from the same bakery, or of outbreaks in Africa or in Asia where many people suddenly die. It is the low levels that are often chronically in foods, and these low and medium levels might put mycotoxins on a list of factors limiting life expectancy, probably knocking off quite a few years from lives in the developing world,[687,688,854,1080] but real statistics are not available. Today this can only be estimated through rough statistics of chronic toxicoses: liver ailments, specific types of cancer, lack of food adsorption, weakened immune systems, and so on. Perhaps, if we had a medical estimate of years lost to mycotoxins there would be more than the small, ultra-dedicated band of scientists dealing with them.

7.1. Incidence and Effects of Mycotoxins

Mycotoxin constraints in staple foodstuffs in the tropics are caused mainly by two carcinogenic mycotoxins: aflatoxin and fumonisin. Aflatoxin was first identified in peanut meal contaminated by *Aspergillus flavus*[631] and was subsequently shown to cause outbreaks of acute hepatitis in animals and humans, to cause liver cancer in animals, and to be associated with liver (hepatocellular) cancer in humans,[435] in particular, in sub-Saharan Africa and southeast Asia.[1123] More than 200,000 deaths are caused annually by this cancer in China,[1123] probably because of the high incidence of hepatitis B, which is also correlated with high aflatoxin in the diet. Lower doses of aflatoxin in food are more insidious, because this mycotoxin impairs liver function, and the efficiency of food conversion is lowered. Weight gain per unit calorie ingested decreases in farm animals and in people fed contaminated food. Impaired childhood growth,[399] and the seasonal appearance of Kwashiorkor syndrome in children has a strong association with the seasonal presence of aflatoxins in the diet.[493] A child can be eating 2,000 calories a day, but if the food contains aflatoxins, the child develops as if he/she had received 500 calories. Susceptibility to disease in children is also enhanced by aflatoxins because of a suppression of immune function.[1079] Thousands of publications on aflatoxin have appeared during the past four decades.

This chapter will deal with these two most severe mycotoxins in developing countries. Tricothecenes, HC-toxin, ochratoxins, and others are more common in temperate climates, or are less important and are not discussed in this chapter. Further information on biotechnological approaches of dealing with other toxins is discussed in an exhaustive review.[565]

Fumonisins were first isolated in South Africa in 1988 from cultures of *Fusarium verticillioides* (= *F. moniliforme*)[381] and their structures were elucidated.[119] The symptoms of acute poisoning vary by animal: there were widespread outbreaks of leukoencephalomalacia in horses[571] and pulmonary edema syndrome in pigs[476] fed corn screenings containing fumonisin B_1.[856] Fumonisins in maize have been correlated with esophageal cancer in humans in South Africa.[1036] Fumonisins cause liver cancer in rats.[379] Fumonisin B_1 inhibits folic acid transport and the deficiency causes neural tube defects and other birth defects in humans.[1015]

Stem borers directly act as the vectors, or cause the lesions through which the fungi enter the crop plant or seed. The *Fusarium* species that produce the mycotoxins infect the crops systemically, spreading throughout the plant, including into the developing seed. Fumonisins are produced in the developing seed while it is still on the plants, and the amount at harvest is the amount later, as fumonisin production ceases upon harvest.[298] The *A. flavus* producing aflatoxins continues to develop and produce toxins during damp storage. Indeed the grain can be infected during attack by grain weevils, a double whammy for the subsistence farmer with poor storage facilities. Indeed the term "storage facilities" often means some sacks under a dirt floor or in the rafters under a leaky roof of a flimsy dwelling. Not only have the insects eaten the farmers' grain, infested grain is likely to be infected with fungi and contaminated with aflatoxins.

A maximum tolerable daily intake of 2 μg of fumonisins per kg body weight was provisionally set for fumonisins.[1149] When such regulations are enforced, they only protect urban populations, and not the 80 percent of the population who are subsistence farmers storing their own grain for household usage. Grain stored by the farmer invariably contains much higher mycotoxin levels than material sold commercially and is not included in legislation regulating maximal tolerable levels of fumonisin B_1 and aflatoxin in commercial grain or processed food and drink in the market place. For example, in one study a third of rural samples tested positive for fumonisins, whereas only 6 percent of urban market samples had them, and the rural samples had a seven times higher level of these toxins. As expected, fumonisins were also found in feces of a third of the rural population and in only 6 percent of the urban population.[206] To make things worse, a synergistic interaction occurs between fumonisin and aflatoxin in eliciting cancer.[380] Both the total intake and the level of contamination determine the risk from mycotoxins in grain, and it is the farm families with the least diverse diets that also consume the most highly contaminated home-grown grain.[382]

Fumonisins and aflatoxins occur in maize and other grains worldwide, including throughout Africa (Table 10). They occur in processed products, from those purchased by normal households but also in commodities such as heroin, as the mycotoxins are not completely degraded by heat. These toxins even make their way through the food chain into beer[961] and aflatoxins (but not fumonisins) carry over into milk,[372] including the milk of nursing mothers.[1189] Thus the preparation of beer or the feeding of contaminated grain to farm animals can be counterproductive, but farmers often separate grain before food preparation, giving the contaminated or weevil-infested grains to their animals, or use them to prepare home brew.

The World Bank has estimated an annual loss of two thirds of a billion dollars in exports from Africa to Europe due to aflatoxins because of the newly reduced European Union (EU) aflatoxin thresholds.[830] This would reduce health risks by approximately 1.4 European deaths per billion population per year, in contrast to international standards. These data have been extrapolated globally.[1157] Aflatoxin-contaminated foodstuffs rejected by Europe will be consumed in Africa, which is more serious than the monetary loss. These products would be a small fraction of the total food intake in Europe, diluting the consumed levels to well below thresholds. They would represent a larger part of the food intake in Africa and in the developing world where the populations are at much higher risk for primary liver cancer because of nutritional deficiencies and hepatitis B virus infections.

Table 10. Postharvest Levels of Mycotoxins in Grains

Locale	Commercial/ farmer storage	Main findings[a]	Reference
Argentine	Farmer	Fumonisins 0.5–12 μg/g across field locations	260
Benin		Parasitic insects on grain weevils reduce infestation by 10%	966
Benin	Farmer	Heat/humidity/insects/correlated with aflatoxin.	488, 489
Botswana		Aflatoxins in 40% sorghum and peanut samples, fumonisins	980
Brazil	Commercial	Fumonisins in all samples, 0.7–23 μg/g (mean 10) and remained constant 2 months	822
Burundi		High levels of fumonisins in maize and sorghum	765
Egypt		Fuminosin producing *Fusarium* spp. in maize ears	31
Kenya	Farmer	Fumonisin levels above 1 μg/g in 5% of samples	570
Nigeria	Markets	High aflatoxin in 40% cassava samples	519
Philippines	Farmer	Fumonisins 0.3–1.8 μg/g across locations	260
South Africa (and USA)	Supermarkets	Fumonisins low in commercial maize	951, 1035
Zambia	Farmer	1.8 μg/g aflatoxins in maize, 14.8 μg/g in peanuts	795
Worldwide		High incidence of fumonisins in maize	1035
Africa (and Europe)		High fumonisin contamination Italy, Portugal, Zambia, Benin maize samples	286

[a] In many cases the sample sizes are small. Mycotoxin levels worldwide have been reviewed by Shephard et al.[973] and by Placinta et al.[855]

The social costs of the impact of aflatoxins in maize and peanuts in Indonesia, Philippines, and Thailand have been estimated by using economic models that calculate the cost of disability due to aflatoxin-related primary liver cancer in humans and suppression of growth of poultry and pigs.[662] The total annual social cost in these three countries due to aflatoxin in maize was Australian $319 million (Table 11). The estimate for aflatoxin in peanuts was Australian $158 million. Unfortunately, no such estimates have been attempted in other underdeveloped tropical regions.

Table 11. Social Costs of Aflatoxins in Maize in Indonesia, Philippines, and Thailand (1991)—The Kinds of Data Needed Elsewhere

Sector	Impact factor	Social parameter estimated	Cost (M Aus$)/yr[a]
Commercial grain	Spoilage effects	Wastage/postharvest costs	71
Household	Human health	Disability from liver cancer	64
	Human health	Premature death from liver cancer	113
Poultry, meat	Reduced feed efficiency/ increased mortality	Increased cost vs. aflatoxin-free feed	29
Poultry, eggs	Reduced feed efficiency/ increased mortality	Increased cost vs. aflatoxin-free feed	7
Pig meat	Reduced feed efficiency/ increased mortality	Increased cost vs. aflatoxin-free feed	36
Total			319

Source: Condensed from Lubulwa and Davis.[662]
[a]Australian dollars

7.2. Conventional Technologies Have Failed Outside the Developed World

Fungicides have been widely used in the developed world to control the toxigenic fungal infections of grains, but are only affordable for export crops in the developing world. Thus, the local population does not benefit from their use. The effectiveness of fungicide usage in many cases is limited[242] because of a lack of consistency as well as the evolution of fungicide-resistant pathogens. Fungicides are not used in organic agriculture, and it is very hard to obtain organic peanut butter that is below the FDA threshold for aflatoxin. Steeping and washing peanuts and maize can reduce aflatoxin levels sixfold.[795] This is insufficient with heavily contaminated material and probably impractical in many places where water is limited or contaminated with other agents. Conventional peanuts are grown with a strict fungicide regime to prevent infection. Organic corn (maize) meal imported into Britain in 2003 was removed from the shelves when it was found that ten brands had a average of seven times the EU threshold for fumonisins, yet all corn meal from conventionally cultivated maize was below the threshold, probably due to insecticide use to control the stem borers that act as vectors. Various studies (discussed below) show that transgenic Bt maize on average has less fumonisin than conventional maize.

Mycotoxins disappear during ensilage of contaminated material, suggesting that they are degraded by microbes. Various pure cultures of microorganisms that degrade mycotoxins have been isolated, and it has been proposed that these organisms or enzymes from them be used to detoxify mycotoxins in feedstuffs and food.[565] Positive results with one commercial preparation, Mycofix Plus, could not be replicated in another laboratory[565]. There are reports though of cell-free preparations detoxifying aflatoxins, but fumonisins could only be de-esterifed in vitro.[298] The remaining backbone retained some toxicity. Many mycotoxin-detoxifying reactions require cofactors or energy (ATP or NAD(P)H) and cannot be performed outside organisms.[300] This precludes using the biotechnology of immobilized enzymes to clean up liquids (milk, soft drink bases, malt) or adding enzymes to feedstuffs.

An antischizistosomal drug, oltipraz, accelerates aflatoxin detoxification in humans. It was tested in China where it was efficacious,[1123] but about half of the smokers who took it had severe gastrointestinal problems.[572] This alone precludes the development of such a drug as long as there are still smokers. Clearly it would be better to prevent aflatoxin from entering the food chain, and it would probably be cheaper than supplying a drug to such a large population, although the pharmacological approach continues to be studied.

Folic acid reduces the risks associated with fumonisins, and fortification of all grain products with 1.4 pg/kg folic acid is now compulsory in the United States, and soon will be in South Africa. It is not certain that folic acid fortification will reduce the risk to all population groups.[1022] The incidence of mycotoxicoses in South Africa is three- to sixfold higher in rural than in urban blacks, but urban blacks have been reported, paradoxically, to have lower plasma folate concentrations than rural women.[1082] Household-stored grain will not be fortified with folate, further exacerbating the problem to subsistence farmers. The incidence of neural tube defects and folate status in rural and urban Africans needs to be determined before the introduction of folic acid fortification requirements to fully determine efficacy in counteracting fumonisins, although there may be other valid reasons to add this vitamin. Clearly it is more important to prevent mycotoxins from forming in grain than counteracting them in the food chain, as precluding them will also benefit subsistence farmers and their livestock.

7.3 Potential Biotechnological Solutions and Status

The solution to the problems caused by mycotoxins in staple foodstuffs is not governmental regulation, but reduction of fungal infection and mycotoxin levels.[687,688] The most promising approach is to provide innovative solutions through biotechnology. Various biotechnological approaches are available to deal with mycotoxin contamination of grain, and as none will be sufficiently effective, it may be best to use a combination of measures. Both the conventional and novel approaches to mycotoxin management have been recently reviewed[298,766] and an earlier review[121] exclusively deals with molecular approaches.

The first line of defense is the control of the vectors carrying the fungi that produce mycotoxins, both the stem borers that cause a large proportion of the systemic infections by endophytic *Fusarium* spp., and the grain weevils, and especially the lepidopteran earborers that carry the *Aspergillus* spp.[184,966] The second line of defense is to suppress fungal attack, either by biocontrol or by engineering resistance to the fungi that produce the mycotoxins. This is needed to deal with infections not vectored by insects. The third line of defense against mycotoxins is to prevent their biosynthesis, and a fourth line is to degrade them in the grain before they enter the food chain. These lines of defense are described below.

7.3.1. Excluding the Vectors

Genetic engineering of insecticidal proteins accomplishes this in maize to some extent vis-à-vis fumonisins, but not to aflatoxins. Transgenic maize expressing the Cry1Ab protein strongly reduces the mycotoxin levels in conventional maize where insect pest pressure is significant.[766,767] The effects have been variable in different locations. Mycotoxin levels are also highly dependent on environmental factors such as humidity, and weather may be more important than insect damage in some circumstances.[260] Studies in Canada and northern Europe did not show great reductions in mycotoxin levels,[766] but again the basal levels were also low in these northern climes. However, a large difference, indicative of the tropical situations comes from experiments where Bt and non-Bt maize were artificially inoculated with corn borer. The mycotoxin levels in the non-Bt maize were exceedingly high, and in the Bt maize tenfold lower. Strong reductions in fumonisins were naturally found in Bt maize in the warmer climes of Spain and France, where the Bt maize always had less than 0.5 μg/g and the normal maize had levels as high as 10 μg/g. The results with Bt maize and aflatoxin were not as consistently good, even in warmer climates.[766] This is probably due to the continued growth and toxin production by the *Aspergillus* in warm humid storage, as well as the ability of grain storage pests to transmit the fungus, whereas the fumonisin producing *Fusarium* spp. cease toxin production at harvest.

A large study was performed in Argentina and the Philippines to try to differentiate between the factors causing the variability of fumonisin levels in maize. About half of the variability of total fumonisins among maize grain samples was explained by location or weather, followed by insect damage severity in mature ears (17%), and which hybrid was cultivated (14%) and with the use of Bt hybrids (11%).[260] Still, across all locations in Argentina, Bt-containing maize hybrids reduced fumonisin concentrations from 6.3 to 2.5 μg/g in the year 2000 and from 3.1 to 0.6 μg/g in 2001. In the Philippines, where the basal mycotoxin levels were lower, there was only a significant reduction at one of two sites in one of two years.[260] The data crunching may not give a full picture in blaming the presence on location and weather. It seems from all studies that where there were ultra-high levels of fumonisins in conventional maize, they were greatly reduced in Bt-containing isolines.

The use of biotechnology to control these insect vectors carrying mycotoxin-producing fungi is described below, but a 90 percent reduction of fumonisins from a high level may not be sufficient where maize is a large part of the diet. Thus, the engineering of Cry proteins into maize at present represents only a partial solution to the mycotoxin problem, and it will remain so until it is possible to use Bt or some other solution to also control grain storage pests that transmit aflatoxin-producing fungi. Plants will have to be engineered with stacked Bt genes, with separate specificities to coleopteran and lepidopteran vectors of pathogen-producing mycotoxins, which are active in the grain as well as stems. Other approaches must be simultaneously employed if mycotoxin levels are to be significantly reduced in all locations, as not only insects transmit these fungi. Maize is not the only tropical crop with mycotoxin problems. These issues have not been, but should be addressed in other tropical grains.

7.3.2. Controlling Mycotoxin-Producing Fungi

Biotechnology has been used to control fungi. Two forms of biocontrol have been tested in general. The first has been to find and then enhance the activity of mycopathogenic fungi that will compete or kill the mycotoxin-producing fungi. The many approaches for doing this are described in a recent book.[1117] Although this approach has been used to control various fungi, it is typically used to control them before they enter the plant, either in the soil or on the surface of plants. Such biocontrol agents have been engineered to degrade their host before entry. Some of the organisms are in the same genus, e.g., *Fusarium* species that control other *Fusarium* species. In the past, researchers have had some success using organisms related to the fumonisin and aflatoxin toxigenic fungi to compete with them, though not to control them.[266] The nonproducing strains did provide a modicum of fumonisin or aflatoxin reduction, but these are still organisms living at the expense of the crop plants, reducing yield.

The advantage of biocontrol agents that control other microorganisms is that they are nonspecific as to crops, and they can typically be used on a large number of crop species. The approaches of directly engineering crops are variety specific, although it is possible that a construct used to engineer one crop variety or species can be used successfully with most others. Transformation is still not trivial, and many transformants must be made to ultimately choose one to be released.

7.3.3. Controlling Toxigenic Fungal Growth in Plants

Plants have been engineered with a large coterie of antifungal agents to prevent fungal growth including phytoalexins (e.g., stilbene) and enzymes (e.g., chitinases and glucanases). Few reports have been found where this strategy was used to prevent attack by mycotoxigenic fungi. An amylase inhibitor from a legume inhibits fungal growth and aflatoxin production,[328] and one from maize works against *Fusarium*.[345] An antifungal trypsin inhibitor protein was isolated from maize kernels and is associated with a modicum of tolerance to *A. flavus* and, to a lesser extent, to *Fusarium verticillioides*.[213] Thus, some gene "candidates" for engineering suppression of mycotoxigenic pathogens are available. They could be used as such, or "enhanced" by selection or shuffling (section 3.5) of their genes for greater activity.

7.3.4. Preventing Mycotoxin Synthesis and Degrading Mycotoxins in Plants

The approach of preventing mycotoxin synthesis has been discussed for nearly a decade. This could be done by preventing the plant from producing (the elusive) signals that initiate toxin production by the fungi, but there are no reports that it has been done. Microarray profiling might assist in finding the genes in the plant that trigger this response in the fungi.

A strategy further along has been to interfere with the metabolic pathways of mycotoxin production. The pathways have been elucidated and the genes are clustered on the fungal genomes.[1142] The genes for aflatoxin biosynthesis have also been cloned.[1185] Microarrays were used to compare a fumonisin minus mutant of *F. verticillioides* with the wild type to elucidate the candidate genes in fumonisin production.[852] A zinc finger regulatory gene controlling fumonisin production has already been discovered.[352] The information from these studies has already been used to generate a reporter gene to screen plant lines for resistance[160] and for qualitatively screening grain and food samples for the presence of mycotoxin-generating fungi.[1172] How precisely this information will be used to generate resistant plants is an open question. One could envisage (science fiction?) cloning an antisense/RNA interference (RNAi) construct encoding one or more of the mycotoxin biosynthesis genes into a phage (virus) that parasitizes the mycotoxin-producing fungus that will live systemically in the crop. Another approach would be to have plants synthesize small RNAi constructs. Such constructs go from cell to cell, and perhaps will penetrate the fungi. Such an RNAi approach has been used to reduce nematode attack, with the transgenic soybean RNAi entering and suppressing the nematode.[515]

The degradation of mycotoxins seems to be the route being investigated by several private enterprises, based on the number of recent patents issued. An esterase that degrades fumonisins has been isolated and cloned from a black yeast[298] and other microorganisms.[298,299] Similarly, an amino-oxidase has also been isolated and cloned from black yeast,[125] and its activity enhanced by mutagenesis[301] and gene shuffling.[203,1194] The patents claim that transgenic plants degrading fumonisins "can be made," and not that they have been made, nor is it fully clear that they have actually cloned or synthesized all the genes that are exemplified in the patents.

Two human genes that degrade aflatoxin were discovered epidemiologically, and then were isolated and cloned. The genetic susceptibility to cancer from aflatoxin was traced to a lack of either a glutathione transferase or an epoxide hydrolase that degrade aflatoxin due to a mutation in the genes encoding either of them.[707] This led another group to find a human aflatoxin aldehyde reductase gene.[88] In a patent application they propose its use only for gene therapy in humans who are supersensitive to aflatoxins, and not for degrading aflatoxin at or near the source. Engineering such genes into plants could answer the basic question of whether the aflatoxins are necessary for the virulence of the *Aspergillus* vector that produces them. On the practical side it could be used to rid grain of the aflatoxins, and depending on the answer to the basic question, might possibly reduce the damage caused by the mold.

Another patent proposes to use gene shuffling to enhance the mycotoxin-degrading activity of several mycotoxins, including aflatoxins and fumonisins.[1023] Thus, there seem to be a number of patented paths to mycotoxin degradation, but hard data are sparse to suggest which path may actually work.

As mycotoxins are a major constraint to human health in the tropics, more effort should be made in further developing these and other lines of protection. This would effectively increase grain yields in tropical developing countries, both by decreasing yield loss to the fungi and by increasing the efficiency of caloric utilization by humans and livestock, besides precluding disease. It would also lengthen life expectancy. Industry has avoided the tropics as a market and should reevaluate its stance, partly due to climate change with its concomitant increase in tropical areas and markets.

CHAPTER EIGHT

Emergency Engineering of Standing Forage Crops to Contain Pandemics— Transient Redomestication

Potential flu and other viral or bacterial pandemics are a source of both media and actual scientific concern. This is independent of whether this is a natural pandemic such as severe acute respiratory syndrome (SARS) or avian flu or an artificial epidemic due to a bioterrorist attack. The expensive response to most such possible pandemic spreads is to develop and stockpile huge amounts of drugs to combat the disease. Besides the direct expense, there is the added expense of work days lost while people recuperate on the drugs, or if the pandemic does not materialize, of destroying the drugs. A more logical approach is to develop vaccines to immunize animal vectors so that the disease does not get to humans and that livestock is not lost. Despite the hype, immunizing humans far from the initial source of the disease need not be top priority, as developing a vaccine considered safe for humans takes longer than one for livestock. Countries at the source of the disease are expected to immunize unaffected animals with relatively expensive vaccines, slaughter livestock and flocks, and decimate the wild life and domestic vectors of a disease to prevent its spread,[1137] to prevent the development of a global pandemics.[173] This has a huge cost to local and world economies, food security, and the environment. With avian flu, for example, most Asian and African countries do not have the resources to immunize their flocks (if vaccines were available), nor the will to slaughter (yet) unaffected birds. The utility is also questionable, considering the reservoirs of migratory birds already affected.

8.1. Present Solutions—Avian Flu as an Example

Most of the effort to deal with avian flu is in stockpiling antiviral cox-1 neuroaminidase inhibitor antiviral medicines in the developed world in waiting for the global pandemic to reach their shores from the developing countries of southeast Asia.[5] Developed countries stockpile vaccines in anticipation of pandemics, mainly for the human population,[333] because of the expense.

It should be possible to inexpensively immunize both the wild and the domestic livestock and populations of vector mammals and birds, for example, the chickens and wild and domestic ducks and geese and other birds along with pigs affected by avian flu. The standard vaccines take an inordinate amount of time to produce, at a great expense, in a large amount of tissue (usually eggs) to grow the live attenuated viruses. The cost of producing such vaccines renders them mainly appropriate for human use. Even if they could be used to immunize livestock, the wild vector populations cannot be easily trapped and immunized, and will spread the pandemic even if domestic flocks are immunized. An inexpensive oral vaccine is needed that can be quickly produced and administered to both livestock and to the wild vectors. This could conceivably come from vaccines being developed based on antigens, but produced in standing fodder species instead of by expensive systems.

8.2. Solutions by Transient Further Domestication

It is suggested here that the most rapid and efficient method to disseminate an oral vaccine is through forage crops. Once such an antigenic protein is developed, its gene can be cloned into a disarmed plant-pathogenic virus and the virus can then be used to infect standing fodder crops. Such plant-pathogenic viruses have been used to provide single-generation transformations of both grain[218] and broad leaf[975] crops as described in section 3.8. Transgenic plants are an ideal means to produce oral vaccines, as the rigid walls of the plant cells protect antigenic proteins from the acidic environment of the stomach, enabling intact antigen to reach the gut-associated lymphoid tissue.[935] People and mice have been thus immunized against rabies with an oral vaccine composed of raw spinach leaves, after the spinach had been transformed with a chimeric gene encoding 22 amino acids from a rabies glycoprotein along with 14 amino acids from a rabies nucleoprotein, all fused with alfalfa mosaic virus coat protein[1187]. Other antiviral and antitumor proteins have similarly been cloned into plants using viral vectors (e.g., references 58, 848, and 858).

8.2.1. Problems with Field-Scale Inoculation

A major problem to be solved is how to inoculate large areas with such viruses. Traditionally, inoculation of viruses has been done by rubbing the leaves with carborundum and the virus or its nucleic acid. It has been proposed to inoculate young plants in greenhouses with a rotary roller with abrasive and virus,[942] or by abrading the leaves with a brush.[943] A tractor -mounted rig system has been described using a liquid virus mixture sprayed at high pressure with an abrasive. Six people could treat a hectare in a hundred minutes with more than 95 percent of the plants infected.[858] Similarly, a multibarrel plant inoculation gun was developed that gas impels fast-discharges of viral solutions, for greenhouse use.[207] A group that considered the viral replicon inoculation to be too inefficient put the replicon in *Agrobacterium,* which they then vacuum infiltrated into leaves of potted plants.[692] None of these systems are appropriate for rapid, wide-scale field implementation, as they cannot easily be scaled up to inexpensive technologies.

8.2.2. Inoculation by Sandblasting

Inoculation could possibly be made more efficient on a field scale using commercially available industrial sandblasting equipment mounted on tractors in a similar manner as standard spraying equipment, which should sufficiently abrade leaf cuticles to allow viral penetration. It is highly likely that this will work, as it has long been known that windblown sand can transmit virus particles, causing disease in plants.[981] Applying the virus dry on particles has many advantages, especially the lower volumes to be transported to the field and the higher equipment speeds. A broad-spectrum virus, such as wheat streak mosaic virus[218] for grain fodder crops, would allow wheat, barley, oats, maize, and pasture grasses to be inoculated. Cowpea mosaic or alfalfa mosaic viruses[181] would infect clovers, alfalfa, berseem, and other forage legumes. The inoculation of standing fodder/ forage crops precludes having to plant seed and waiting for the crop to grow, reducing the lag time between seeing the need for such vaccines and being able to harvest oral vaccinating crops.

Pesticides must be spread evenly over treated fields by careful, uniform spraying. Crop dusting plant viruses with encoded vaccines with sandblasting equipment needs only to establish a few infection sites per plant, as such virus infections typically spread systemically throughout a plant. As animals browse many plants, or eat chopped, mixed fodder, a proportion of misses during field inoculation would be inconsequential if the average titer is sufficient. Thus, equipment could be operated at much faster speeds than with typical spray equipment. The systemic viruses would be spread throughout the plant tissues within a few weeks, producing the antigens. The wild avian and mammalian disease vectors all too often avail themselves of farmers' fields, and would be immunized, as would grazing livestock. Chickens, geese, and ducks fed chopped fodder would also be immunized along with penned livestock. This technology would not immunize the feral cats and dogs that are now a new worry, because they become violently ill from avian flu after eating infected dead birds. Some fear that they may become the bridge to human infection.[617] Still, if domestic and wild fowl are immunized to avian flu, there will be fewer infected birds for the cats to eat.

The first areas to be crop dusted by sandblasting should be around known areas of disease infestation, and the crop dusting could then be continued along the known migrational paths of mammalian and avian vectors. The fodder crops in the treated areas could be used to extract the viral nucleic acids needed to infect further areas. Because plants could produce the vast majority of the vaccine, such a process requires much less industrial production of the viral inoculum than would the conventional production of oral vaccines. Immunization in a given area would not have to be complete. Modeling suggests that disease epidemics can be contained by immunizing only about 80 percent of the target species.[264, 768] Fewer herds or flocks would have to be slaughtered if such technologies are readied.

Disarmed vector viruses that are not seed borne would provide single-generation vaccine production, that is, they would not persist in annual forage crops and would not likely persist for too long in perennial crops. Previous evidence[218, 975] suggests that the yield loss due to such viral infection can be minimal, although the cucumber virus expressing an active epitope against hepatitis C virus did cause its host plants to bleach somewhat.[780] Thus, if there is a yield drag, the loss would be compensated by not having to slaughter livestock, and the yield drag would only occur in the season the technology is used. Should the disease pathogen mutate, and the vaccine become less effective, forage species can be infected with new virus-bearing genes encoding better vaccines. Thus, it is envisaged that such or similar technologies can provide the maximum coverage against disease pandemics with minimal economic impact.

Although little chance exists that this technology can be ready for the present strains of avian flu for both technical and regulatory reasons, it would seem logical to develop this sort of rapid response for future outbreaks of such diseases. The technology is also obviously highly amenable to rapid response to bioterrorist attacks that are expected to use livestock disease pathogens.

CHAPTER NINE

Meat and Fuel from Straw

Demographers have revised scenarios of population growth down from a predicted thirteen billion mouths to about eight billion,[665] predominantly middle-class people, most of whom will desire a diet containing animal products. The same amount of grain is needed to feed these eight billion people and their livestock as for thirteen billion impoverished people who eat only grain. With almost all arable land under cultivation, and more land leaving production than entering, where will the food needed to feed an expanding humanity come from? To make matters harder to contemplate, this huge mass of middle-class people will not only wish to eat meat, they will want to go places, putting far more pressure than at present on fossil fuels, further raising prices to far higher levels. The present technologies to replace a small part of the fossil fuels with bioethanol and biodiesel from feed grains has quickly absorbed the recent oversupply of grain, with dire consequences to the developing world (section 2.11). The present yet-crude technologies for biofuel production are economically viable, according to economists, when crude oil is sold at fifty dollars a barrel in the United States, where fuels are sold at half the price that the rest of the world pays (stimulating profligacy). The fuel prices will go up as grain supplies decrease, and biofuel costs will go down as the technologies become more sophisticated, both in production efficiency and economy of scale when cloned production plants will make biofuel factories much cheaper. Additionally, changed automotive engine design to use 95 percent (undried) ethanol in blends with gasoline will reduce the break-even point.

How much can the world expand feed-grain production to satisfy the needs for both meat and fuel? Despite the economics of biofuel from grains, the grains cannot be available to replace more than a minuscule part of the petroleum used, but the grain available for food and feed will be limited, at a price that reflects petroleum prices. Thus, alternative feedstocks are needed for fuel and meat production, without expanding the land under the plow.

9.1. Straws as a Substrate for Producing Meat and Fuels

About half of the above-ground biomass of grain crops is wasted: the straw that bore the grain. Most of the nearly two billion tons of cereal straw produced annually in the world (Table 12) has a negative economic value. (The thick straw of maize and sorghum is often termed "stover," but it will be referred to as straw). In years past, much of the cereal straw had been burned after harvest to kill pathogens, and since banning, fungicide use has increased. Plowed-under straw temporarily binds mineral nutrients while being degraded by soil microorganisms, often requiring additional fertilizer in the following crop, with negative economic and environmental consequences. Small amounts of straw are fed to ruminants as roughage or as an extender to animal feeds, but very little caloric value is derived from it. This lack of feed value results from an export of the sugars and amino acids digested from polysaccharides and proteins in the leaves and stems to the developing seed during grain filling. By senescence, most of what remains in straw is polymeric

Table 12. *Grain Production: Near the Amount of Wasted Straw*

	Wheat	Rice	Maize	Sorghum	Millet	Total
			(Million metric tons)			
Sub-Sahara	3	11	27	20	14	**74**
Africa	17	17	43	21	14	**112**
World	568	579	602	55	26	**1,830**

Source: FAO statistics for 2004.[329]

hemicelluloses (mainly xylans) and cellulose, but their biodegradation by carbohydrases, whether by ruminant bacterial cellulases, or by those in commercial bioreactors for ethanol production, is heavily prevented by a smaller component of lignin. Very small amounts of lignin intercalate into and around the cellulose and prevent biodegradation due to a steric hindrance to the cellulolytic enzymes.[430] Thus, companies developing bioethanol from straw achieve only a 20 percent efficiency of conversion (250 liters of ethanol, equaling 200 kg of ethanol per ton of straw),[523] despite all but a few percent of straw being organic carbon compounds that theoretically can be metabolized.

9.1.1. Breeding for Better Digestibility by Ruminant or Industrial Cellulases

Breeders have endeavored to breed higher straw digestibility within the limited variability of the genomes of the various crops. Brown midrib mutations in maize and sorghum (Table 13) that have a lower lignin content and much higher digestibility have been isolated. They have been used to breed forage (silage) maize and sorghum, invariably with somewhat lower yields, which can be economically compensated for by the greater digestibility.[94,454,629,733] The brown midribs are due to mutations at various loci in lignin biosynthesis, which lead to slightly less lignin and a modified lignin

Table 13. *Biochemical Specificities of the Brown Midrib Mutants of Maize in an Isogenic Line*

Mutant	Fiber in straw[a]	Lignin in fiber[b] percentage (% of cv)	Digestibility[c]	Deficiency[d]
cv[e]	56.2 (100)	16.4 (100)	15.9 (100)	None
bm1	54.9 (98)	14.6 (89)	19.2 (121)	Cinnamyl alcohol dehydrogenase (CAD)
bm2	51.3 (91)	13.6 (83)	20.2 (127)	No information
bm3	52.2 (93)	13.0 (79)	23.1 (145)	5-Hydroxyconiferaldehyde O-methyltransferase (CaldOMT)
bm4	53.4 (97)	14.3 (87)	22.8 (143)	No information

Source: Collated from data in Barriere et al.[94] and references cited therein.

[a] Measured as neutral detergent fiber.

[b] Percent of Klason lignin in neutral detergent fiber.

[c] In vitro neutral detergent fiber digestibility.

[d] Mutated enzyme where known.

[e] Cultivar F92, into which all brown midrib mutations were backcrossed, until near isogenicity.

subunit composition. These modifications lead to the signature brown midribs that characterize such mutations and allow for easy mutation screening. There may well be mutations in lignin composition/quantity that do not have this signature, but they would be too hard to discern as there would be no phenotype to be detected. In looking through the results there are always unclear correlations with yield and other parameters. For example, in a multiple-variety sorghum field test, the brown midrib variety had a high frequency of lodging, but so did one of the conventional varieties,[733] so it is incorrect to say that the brown midrib per se increased lodging. Whether it is the total lack of a critical enzyme in the brown midrib varieties (instead of a genetically engineered modulated level), a yield reduction that will always be present, or a linkage disequilibrium, is yet unclear.

9.1.2. Should We Switch to Switchgrass?

There is considerable discussion (including by politicians on the highest possible podiums) of cultivating crops such as switchgrass (*Panicum virgatum*) for production of biofuels or for burning in power plants. Switchgrass has the same problems as straw; lignin limits digestibility. The cost of producing a ton of switchgrass is infinitely more than producing straw, as straw is a by-product with negative value, while inputs must be invested to cultivate, fertilize, and harvest switchgrass. Thus, the cost of a ton of baled hay or stover will always be considerably less than a ton of baled switchgrass. The lignin content of mature switchgrass is close to 17 percent, compared with about 12 to 14 percent in wheat straw and maize stover. Without modification of lignin, switchgrass will be far more appropriate for burning than for bioethanol, unless the lignin content is modified, as far less cellulose will be available. The simultaneous saccharification and fermentation of switchgrass to fuel ethanol was less efficient than maize cobs and stover, wheat straw, and even wood residues.[1164] This necessitates pretreatments with either hot dilute acid[222] or ammonia explosion.[29] The latter technique raised the ethanol yield two and a half times to 20 percent,[29] the same as with wheat straw.[523]

Those touting switchgrass as a new form of snake oil make it sound simple; a perennial grass that is drought tolerant and requires few inputs, even on the poorest of lands, overlooking laws, including laws of conservation of matter, and those of basic plant physiology, as well as economics. Just sow and reap forever is proclaimed. They forget that even the poorest of lands may well be supporting wildlife or have a finite value, as the farmers' time and energy do. As this book is about overcoming limitations, we will leave the positive aspects of switchgrass to others, and highlight its problems, some of which can be overcome transgenically (section 9.5). The first problem is that stand establishment is not easy, and frequencies of 25 percent establishment are not uncommon.[952] In such situations it is unwise to harvest for the first few years. After establishment, yields can vary more than fourfold from less than two to more than nine tons per hectare (Fig. 19A). It may surprise some, but 90 percent of this yield variability could be explained by the amount of rainfall.[637] Switchgrass is drought tolerant, but that does not mean that it will yield heavily without water (Fig. 19A). Switchgrass actually consumes more water than traditional crops under all climatic conditions (but it is also better at preventing erosion than many other crops except winter wheat).[159] Companies will not like to construct huge bioreactors for processing switchgrass unless there is a guaranteed steady supply of feedstock. Will they construct facilities where the rainfall is as variable as that described in Fig. 19A? Because of transportation costs, bioreactor facilities will have to be constructed near the source. Is this a repeat of the years when investors were suckered into planting jojoba in the desert, where it also lived, but was not productive without water?

Switchgrass was highly responsive to nitrogen fertilizer in a multiyear, multisite study (Fig. 19B). Yields increased linearly to more than twenty tons per

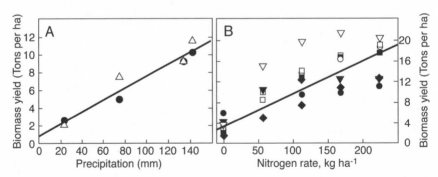

Figure 19. Switchgrass does not defy the laws of conservation of matter; it has high yields only in response to water (A) and nitrogen fertilizer (B). (A) Yields at two locations (different symbols) in response to April-May precipitation in South Dakota are shown (replotted from data in Lee et al.[637]). (B) Yields at one location in Texas in response to fertilization during different years (different symbols) (replotted from data in Muir et al.[761]).

hectare at rates from a quarter ton per hectare. Alas, at high nitrogen rates, lodging becomes a problem with this two-meter-tall species.[761] It can join the crowd of other grass species with that problem during domestication.

Switchgrass is envisaged for use in direct production of energy and/or burnable gasses as well as for bioethanol production from digestible materials in the cut material. The energy would be produced by cofiring with coal, and the gas would be produced by pyrolysis. In both cases considerable ash is produced, as with all plant material, but part of the particulate material is potentially dangerous silicon particles.[126,1060] Healthy plants of other species in general can have as low as 2 percent ash, but switchgrass has almost 5 percent ash,[126] and in some soils up to 10 percent.[542] This ash contains more than 60 percent silica. Fifty percent more of this potentially dangerous compound (when in the form of microscopic particles) is emitted on burning than by coal.[126] As silicon is not an essential element for plant growth (although small amounts might be of some value), there is reason to consider lowering its presence. When switchgrass was mixed with coal, the fine-particle concentrations were much higher than with dedicated coal combustion.[126] These fine particles are a concern because they are not captured by electrostatic precipitators or other devices used to lower particulate emissions. The microparticles of switchgrass ash should be captured, as besides the silica they contain large amounts of phosphorus and potassium salts, which should be recycled back to the field by farmers, not by wind. This is far more important than with coal, where most of the particulates are aluminates of no agricultural value.

Breeding switchgrass for anything but dominant traits is not easy, as it is an autotetraploid, with a high degree of preferential pairing.[735] Still, there has been success in varietal selection from among the natural variability of this U.S. native species.[710] Any transgenic improvements of switchgrass will have to consider the necessity of curtailing gene flow to remnant wild populations and of interbreeding related species.[1112]

9.2 Possible Biotechnological Solutions

The basic solution to the environmental problem of straw waste and the immediate potential utility of using straw to produce fuel, and the future needs for grain to feed people instead of automobiles, is to transgenically modify straw so that it can be utilized more efficiently in biofuel production or fed to ruminants.[93,423,453] Ruminant animals can digest cellulose and hemi-celluloses, unlike monogastric animals. Plant material containing more cellulose or less lignin, or with modified lignin composition, is more digestible by the carbohydrases of ruminant animals or in fuel bioreactors. For each percent less lignin, two to four times more cellulose is available to bioreactor carbohydrases or to ruminants. Wheat and rice have very little genetic variability in straw composition, so it is doubted that classical breeding can provide a solution, especially in hexaploid wheat where recessive mutations are hard to find.

Chemical and physical treatments to enhance digestibility (e.g., reference 1101) have not been cost-effective. Increasing cellulose-digestible material by 20 percent would upgrade the immediate economy of fuel production and the future nutritional value of straw to that of hay, and meat could be produced using straw as a major carbon source, considerably decreasing the amounts of feed grains that must be used to feed cattle. This can be achieved transgenically by creating transformants with increased cellulose having a more open (biodegradable) structure using the *CBD* gene to up-regulate cellulose synthesis, and separately or together using RNA interference (RNAi) techniques to modulate the lignin content.

Rice straw and switchgrass will require more modifications for commercial bioreactor use or for ruminant feed than those discussed below as the general case, because they also typically contain silicon inclusions, which can render the straw unpalatable and possibly undesirable. They also produce a potentially dangerous ash on burning. Biotechnological solutions to this silicon problem are discussed in section 12.4.

Considerable efforts to decrease chemical wastes during paper pulping by transgenically reducing the lignin content or composition of trees by affecting the genes controlling biosynthesis of lignin monomers have led to a beginning of understanding of lignin biosynthesis and its relationship to cellulose availability.[749] Yet many basic compositional differences exist between tree and other dicot lignins and those of grasses, and it is thus not clear how much one can extrapolate from dicots to grasses. Genetic modification/reduction of lignin by classical breeding using the brown midrib types enhanced digestibility of some forage crops. Similar mutations that would allow breeding decreased lignin have not been identified in small grains such as rice, wheat, and barley, probably because the genes for lignin biosynthesis are in multigene families in grains, which are not amenable to single mutations. Most sources of variability would probably be quantitative, where more than one isozyme may have to be suppressed, requiring extensive breeding to modify lignin without modifying other grain quality characters.

9.2.1. Lodging and Lignin

Partial (but not major) reduction or change in straw lignin composition should leave dwarf and semidwarf wheat and rice with sufficient strength to resist lodging. It is a general misconception, without proof, that lignin is singularly responsible for the structural stability that prevents lodging, the propensity to keel over in windstorms, which can be a major impediment to mechanical harvesting. A comparison of lignin content in straw and the susceptibility to lodging showed no significant correlation.[1073] Indeed, it is not easy to understand the chemical and structural factors involved in lodging, but attempts are being made, both genetically and physically.

Lodging is typically precipitated by wind combined with rain or irrigation and may result from buckling or partial breaking of the lower stem, or from the roots twisting out of the soil.[1014] The latter should be more correctly termed "dislodging." The driving force of both is the wind drag exerted on the grain head. In practice, wheat varieties with short stiff stems are often more resistant to stem buckling, whereas wheat varieties with compliant stems may be less susceptible to root lodging in wet or water-saturated soils. Breeders and agronomists as well as genetic engineers must balance the competing constraints imposed by stem rigidity and flexibility to select the best varieties for wet versus dry environments to deal with stem buckling versus anchorage. Most discussions of lodging do not differentiate between stem-buckling and anchorage-susceptible varieties despite there being two very different physical reasons for each one.

Whereas dwarf and semidwarf varieties have less stem to lodge, they have larger grain heads, acting as bigger sails in the wind. As drag is presumably a function of sail size, some if not much of the drag should be on the awns that spread out from the heads, unless they act as some sort of damper due to their spiky structure. No comparisons of lodging resistance have been published using awnless versus awned isogenic lines of wheat or barley. The question of awns has also not been addressed by the physicists or agronomists dealing with lodging. The turbulent wind flow at the surface of a field is dominated by intermittent horizontal eddies, and the top of each stalk is subjected to successive impulses from varying directions. The physics of these wave-like motions, and their possible relationship to lodging, are not fully understood. A group of physicists (including one from an aviation company) and biologists followed the effects of artificial wind gusts on wheat plants by video photography.[330] Unfortunately they tested only two unrelated varieties, one lodging and one nonlodging. They developed a highly esoteric nonlinear model that mimicked the results using four parameters (possibly oversimply interpreted herein) as two different directions of stiffness, linear viscous drag and stem torque.[330] They do not state whether this relates to stem breakage or root twisting lodging. In a later, somewhat inconclusive paper, they try to model whether there can be an ideal wheat stalk that will be resistant to both stem buckling and root dislodging.[331] Basically, the authors claim that because of the competing requirements of the two types of lodging, the optimal wheat stem geometry has "a nonlinear dependence on the intensity of gravity and the frequency spectra of wind."[331] The results and conclusions would clearly be more important if more varieties had been tested. They pointed out (but did not measure) that the collisions between stems buffeted by wind could have a domino effect in the field, while they measured only single plants.[330]

Another group built a portable wind tunnel, which they took to the field.[1014] They used only one variety of wheat, but measured lodging under a variety of soils and soil wetness conditions. They found that stem lodging occurs instantaneously, but root lodging takes a few minutes at a lodging wind velocity, suggesting that there may be an element of plant and/or soil fatigue occurring during failure.[1014] If only their wind tunnel had been used on different varieties, with different lignin contents and compositions, with and without awns, we might better know the directions to follow to genetically engineer a more lodging-resistant cereal.

Another cause potentiating lodging that was not discussed by the physicists is stem borers. Stem-borer damage renders the stems far more susceptible to injury by wind because of the structural damage, just as termites weaken a house. There have been claims that low/modified lignin maize varieties are more susceptible to lodging. Is that because the plants are inherently less strong, or because the more digestible plants are more appreciated by stem borers? An analysis of quantitative trait loci (QTLs) for lignification shows that many colocalize with those of resistance to corn borers[885]; that is, borers are not stupid; indigestible cell walls are less tasteful to the corn borers. Rootworm damage is also highly correlated with susceptibility to root

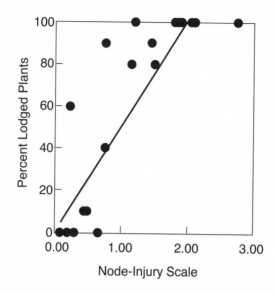

Figure 20. Rootworm injury correlates with lodging under high environmental stress. Relationship between percent lodged plant and root injury evaluated on the node-injury scale under a high level of environmental stress levels. The correlations between lodging and injury were significant but less so at low ($r^2 = 0.28$) and medium ($r^2 = 0.46$) levels of environmental stress). *Source:* Redrawn and modified from Olefson et al.[816]

lodging, especially under environmental stress (Fig. 20), but this was not correlated with lignin content or composition. The stem borer and rootworm should not be ignored when modifying lignin or the highly digestible crop could remain a laboratory curiosity.

9.3. Transgenically Enhancing Straw Digestibility

A solution to increasing digestibility without affecting important varietal traits is to transform elite material to have modified lignin and cellulose contents. Partial silencing of the phenylpropanoid pathway enzymes leading to lignin, encoded by whole gene families (Table 14) can be achieved by antisensing or other RNAi strategies using small interfering RNAs (siRNAs) that conform to consensus sequences of the gene family. Most of these genes have already been partially silenced in dicots[55] changing monolignol levels,

Table 14. *Enzymes of the Phenylpropanoid Pathway in Cereals are Encoded*
by Small Gene Families—How to Suppress?

Rice gene[a]	Type	No. copies identified	Sequence identity (%)			
			Barley	Wheat	Maize	Dicot[b]
PAL (AK067801.1)	FL-cDNA	At least 5[c]	86	85	86	<76
C4H (AK104994.1)	FL-cDNA	At least 2	89	89	87	<80
C3H (AK099695.1)	FL-cDNA	At least 2	ni	89	79	<80
4CL (AK105636.1)	FL-cDNA	At least 3	83	ni	76	<80
CCoAOMT (AK065744)	FL-cDNA	At least 2	ni	93	90	<82
F5H (AK067847)	FL-cDNA	At least 2	ni	84	ni	LS
COMT (AK061859.1)	FL-cDNA	>1	71	86	87	LS
CCR (AK105802)	cDNA	At least 3	88	85	90	<75
CAD[d] (AK 104078)	FL-cDNA	At least 12	ni	ni	83	<71

Source: Updated from Gressel and Zilberstein.[423]

Note: Gene copy and nucleotide homology estimations are according to the rice and other cereal genome data currently available on the NCBI and Gramene websites. FL, full length; ni, not identified; LS, low score.

[a] Rice gene reference for homology comparisons.

[b] Highest identity with a dicot ortholog.

[c] Number of copies as of December 2005.

[d] According to Tobias and Chow.[1065]

increasing cellulose levels and digestibility.[93,423] Decreasing transcript levels of gene families may suffice, but inhibiting more than one gene type may be necessary because of biochemical compensation by parallel pathways producing monolignols (Fig. 21). A decrease in function of a single gene provided sufficient down-regulation and modification of lignin structure and enhanced the digestibility in maize (cf. reference 454), sorghum and pearl millet,[629] poplar[1076] and pine,[674] but little evidence has been published that affecting more genes each to a lesser extent can increase digestibility with fewer side effects.

Partially suppressing shoot lignification by ectopic antisense (RNAi) based on the desired phenotype is unlikely to affect mechanical strength. The compressed internodes of semidwarf and dwarf wheat and rice should maintain structural integrity with somewhat less lignin. It is unlikely that selected modulation of lignification would affect defense mechanisms deriving from phenylpropanoid intermediates, as the gene encoding (at least one) isozyme involved in defense lignification had a quite different sequence from the isozyme for xylem lignification.[511] Still, the task will not be easy as it is still unclear which lignin modifications / reductions will do so without affecting yield.[840] The typical gene jockey who performs the transformation, regeneration, and greenhouse analyses might think that a good product has been achieved because no growth or yield differences were observed with the small number of regenerated plants. The statistics to ascertain whether there has been a yield reduction require multiple field testings, over a few growing seasons, at many locations. It is unlikely that farmers will cultivate a grain crop with even a five percent average yield reduction to gain a more digestible straw. This is why the brown midrib maize and sorghum mutations are presently used only to develop varieties to be used specifically for forage or silage and are not found in varieties to be used for grain production. Thus, it will be essential to incorporate many lignin-reducing/modifying gene constructs with different tissue-specific and expression levels promoted into numerous genetic backgrounds and have agronomists and biofuel and animal nutrition specialists evaluate the lines.[840]

A courageous attempt has been made with maize to define the ideal ideotype with optimal stover digestibility, which should be of use to both the breeder and the genetic engineer, as well as to facilitate their interactions.[715]

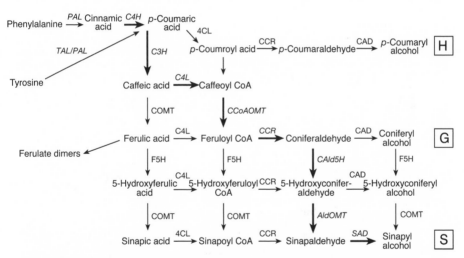

Figure 21. Convoluted pathways of lignin biosynthesis (modified from references 55, 647). The typical monocot-specific features are noted in gray and the pathway suggested to dominate in angiosperm trees is emphasized by the thick arrows. Abbreviated enzyme names are indicated above the corresponding catalytic steps. Details of the cereal pathways await clarification. *Source:* Gressel and Zilberstein,[423] copyright 2003, by permission of Elsevier.

Twenty-two inbred lines of maize were sectioned and stained for different properties. The microscopic results were correlated with lignin and p-hydroxycinnamate contents as well as cell wall degradability. Various combinations of the results could then genetically define 90 percent of the results. The ideal ideotype contained less lignin with a higher ratio of syringyl to guaicyl subunits, which were preferentially located in the cortex and not the pith tissues of the maize stems.[715] They modeled in vitro digestibility based on the two histological and two biochemical variables and were able to obtain a highly significant regression correlation with observed digestibility of thirteen inbreds. But on what interests us the most; what will happen with mutants and transformants, the model is inadequate. Brown midrib *bm3* mutant maize was actually far more digestible than the model predicted.[715] Their results must yet be compared in the type of wind tunnel described above to add a correlation with lodging. The results should also be econometrically correlated with added/reduced yield and the added value of the digestible stover. Despite there being so much more wheat and rice straw than maize stover, few publications deal with modified lignin in these small grain crops. If this is because maize is sold as hybrids, with economic advantages to seed companies, then it is time for the public sector to increase its involvement with these more important, nonhybrid crops.

Another interesting challenge has been raised to the molecular biologists from a study of twelve lodging-resistant and -susceptible varieties of wheat. In the analysis of all their data, it appears that a higher fiber (including lignin) content in the second and third internodes correlates with resistance to lodging (with a correlation coefficient of ca. 0.6).[1075] Would it be possible and invaluable to increase the fiber and lignin contents of these two internodes and lower it in all other ones using tissue-specific promoters? Having high lignin content in the lower internodes and less elsewhere may not affect biofuel/ruminant digestibility. Soil scientists are adamant that straw for such purposes be cut high, leaving 20–30% of the straw in the field so as not to overly affect soil organic matter and tilth. This would leave residues with the most lignin, the best for generating humic materials.

In generating more digestible cereals, one must not forget the stem borers and possibly rootworms, which prefer the digestible lines.[885] One must consider that Bt or some other transgene that will control stem borers and rootworms must be cotransformed with all genes of choice, or that genes of choice be transformed into lines that already contain the Bt (or other insecticidal) gene(s).

9.3.1. Genes Affecting Lignin Composition in Dicots and Their Cereal Orthologs

Lignin is a matrix of copolymerized hydroxyphenylpropanoids (monolignols) covalently linked via a variety of bonds. The monomeric composition and the types of bonding vary among species, with a high content of *p*-coumarate and ferulates covalently linked to monolignols in grasses. Genes encoding the shikimate and phenylpropanoid pathways are activated during cell wall lignification, along with up-regulation of transcription factors[133,567] having an unclear role in cereals. Ferulate and diferulates cross-link cell wall components to xylan α-L-arabinosyl side chains during grass lignification, leading to further copolymerized lignin complexes.[402]

Gradual elucidation of lignin monomer biosynthesis pathways and related genes (Fig. 21) resulted from efforts to decrease lignin content in trees and model dicots and from genomic projects. A high level of sequence homology with more than 70 percent identity exists among gene families encoding enzymes of the phenylpropanoid pathway in dicots and cereals (Table 14). Cinnamate, the first phenylpropanoid pathway product, is a precursor of lignin, but also of pigments, phytoalexins, and flavonoids, suggesting that its production must not be modified, and only later enzymes should be modulated.

The brown midrib maize mutants have recently been characterized biochemically in isogenic backgrounds, which had not previously been done (Table 13). There is up to 45 percent greater digestibility of the fiber with *bm*3, which is now thought to be mutated in 5-hydroxyconiferaldehyde *O*-methyltransferase (CaldOMT).[94] Yield or lodging information was not presented, but if acceptable, this is a good target for modulation in cereals. It is hoped that mutant *bm*4, which also has a very high digestibility, will also be characterized soon.

9.3.2. Engineering Decreased/Modified Lignin

As discussed above, down-regulating expression of genes encoding various enzymes of the lignin biosynthetic pathways (Fig. 21) decreases or modifies lignin, with preliminary evidence for enhanced digestibility without overly modifying structure.[629] Antisense repression of 4-CL in aspen lignin biosynthesis was compensated by increased cellulose.[512] Antisensing with dicot COMT cDNAs caused decreased tobacco COMT activity and altered lignin composition.[302,787] Transgenic plants with dual antisensing of COMT and CCoAOMT had less lignin content, but each alone was ineffectual.[1196]

No field studies have yet been reported with transgenic annual plants (let alone dwarf grains) with decreased/modified lignin, where lodging was compared with the original variety. No lodging problems have been reported with decreased lignin transgenic poplars, except for the dislodging of trees by ecoterrorists in a European field trial. Still, in some cases where lignin was heavily modified, the changes could be ultrastructurally visualized,[926] but there is no need to engineer such vast changes, minor modulation or down-regulation may be sufficient to economically enhance digestibility.

A considerable amount of possibly useful information has been obtained from model plants such as *Arabidopsis*. Still, *Arabidopsis* is a dicot, and may have it wrong for cereals, which have totally different morphologies and needs. Tiny *Arabidopsis*, growing among wind-breaking trees, near the forest floor, is unlikely to have had a need to evolve resistance to the type of wind turbulences hitting a wheat field in the plains. The rice genome sequence, additional maize and wheat expressed sequence tag (EST) sequences, and genomic data have allowed identification of cereal orthologs of genes encoding lignin precursors, including sinapyl alcohol dehydrogenase (SAD),[647] which is highly homologous to cinnamyl-alcohol dehydrogenase (CAD) genes (Table 14). This, and the ease of transformation of wheat, maize, and rice precludes the continued reliance on models such as *Arabidopsis*, when the real targets can be used. This genomic information from the target species paves the way for extending dicot modifications to cereals, as a first step.

9.4 Modulating the Quantity and Structure of Cellulose

Increasing the amount of cellulose (especially at the expense of lignin), or modifying its structure such that more is available to cellulases could also increase the feed and bioethanol value of straw. This has been done transgenically with two genes: *CEL1* and *CBD*. *CEL1* is an endo-1,4-glucanase gene from *Arabidopsis thaliana*. Transgenic poplar plants overexpressing *CEL1* were taller, had larger leaves, increased stem diameter, wood volume index, and dry weight, and a higher percentage of cellulose and hemicellulose than control plants.[968,969] Transgenic poplar overexpressing the poplar endo-1,4β-glucanase gene also produced more cellulose.[808] *CBD* encodes the cellulose binding domain of cellulase on cellulose. Transgenic plants of tobacco *Arabidopsis*, potato, and poplar were generated that overexpressed CBD.[970] Expression of the CBD protein in the cell wall of these species resulted in an increase in growth rate and enhanced biomass accumulation.[970] Two-year-old field-grown poplar had a threefold increased volume index over the non-transgenic controls, as well as an improvement in fiber quality.[642] There are no extant published reports of modulating these genes in grains.

9.5. Special Biotechnological Possibilities for Switchgrass

Switchgrass and the other grass species envisaged for biofuel production have problems that might be addressed by transgenic approaches in addition to the ones above. The problem of too much lignin and the problem of too much lodging at high nitrogen fertilization rates might be partially solved in a single step by transgenic dwarfing. Dwarfing has always worked in the past to prevent lodging, but why should this decrease the lignin content? With dwarfing, much more of the biomass will be in leaves, and leaves of switchgrass have been shown to contain a lower proportion of lignin than stems.[555] Another approach that is not part of the typical straw approaches is to delay flowering, as has been done in a forage grass (*Festuca*) to provide more available feed to ruminants by having it grow longer before reaching sexual senescence. Indeed, switchgrass has less lignin during the vegetative phase,[555] but also less cellulose, as it has starch and protein, both of which can be utilized by the yeast making bioethanol. It would be less advantageous if the switchgrass were to be pyrolyzed or burned. Flowering was delayed in *Festuca* by transforming the plants with a flowering suppressor gene from *Lolium*.[544.] These approaches might be quicker to lower the lignin content, than the direct antilignin approaches described above, and of course successes from both approaches could be stacked.

Decreasing the silicon content in the manner outlined in section 12.4 would substantially decrease the problems from the most dangerous of the microscopic particulates released in burning for energy production or pyrolysis for biogas production,[126] the particulates most likely to cause silicosis.

Switchgrass has been stably biolistically transformed.[895] More recently the same group using *Agrobacterium* with the *bar* gene along with acetosyringone to overcome the inherent problem of infecting monocots with this bacterium that normally infect only dicots, used the herbicide glufosinate as the selectable marker.[1004] The *Agrobacterium* transformation was most efficient using highly proliferative somatic embryos giving rise to many transformed plantlets each.

The problem of gene flow from switchgrass to related *Panicum* species[1112] could be obviated by many of the mitigation strategies outlined in Chapter 4, but the delayed flowering discussed above for keeping lignin lower could overcome it as a mitigator, vastly decreasing the possibilities of overlapping flowering times and conferring unfitness on any hybrids that do form. It will be harder to produce switchgrass seed, but as this is a perennial crop once established, the added price of the seed should easily be offset by the quality of the product.

9.6 Lignin and Cellulose Modification/Reduction May Not Be Enough—Integrating Approaches

Reducing or modifying the lignin content of straw would render far more carbohydrate available to ruminant animals, but this would still not be hay containing 10 to 20 percent protein, and animals must be fed proteins, or must they? Only some of the protein fed to ruminants is directly digested to amino acids. Much more is recycled through rumen bacteria, which eventually die, releasing amino acids to the animal. Bacteria need not be fed proteins to make amino acids; the bacteria can utilize and reduce inorganic nitrogen sources. When Europe was severed from feed grain imports during World War II, beef cattle were fed ammonified waste paper (cellulose).[1008] Ammonifying straw also separates some lignin from cellulose, rendering more cellulose digestible, even more so if done with heating. Urea, which releases ammonia under heat or water, has the same effect. Thus, there can be multiple effects of injecting plastic-wrapped bales of lignin-reduced/modified straw with liquid ammonia or urea. If the wrapped bales are left in the sun, solar heating assists in delignification. Small-scale subsistence farmers can use solar heaters manufactured from old barrels, aluminum foil, black paint, and polyethylene sheeting to convert chopped straw and a handful of urea fertilizer into daily fodder.[729]

If straw is to be heat pasteurized and a nitrogen source added, another biotechnology can add to the nutritional quality of the product; short-term fermentation with preferentially ligninolytic microorganisms such as *Aspergillus japonicus*. Unlike ligninolytic Basidiomycetes, *A. japonicus* degrades straw lignin in the presence of nitrogen sources.[730] Ruminants can then utilize the lignin biotransformed into fungal biomass, instead of excreting the lignin in the manure. In the distant future, a second fermentation with a cellulolytic fungus could then convert the cellulose to utilizable biomass, good enough for supplementary feeding to monogastric animals such as pigs and poultry, and if with mushroom flavor, to humans.

9.7 Potential Impact

A mixture of these transgenic with other technologies could yield nearly two billion tons of inexpensive, high-quality hay or silage from rice, wheat, maize, and sorghum straws around the world, replacing about half a billion tons of feed grain. Switchgrass will mainly be useful if it is not grown on land taken from cereal production. Treated straws from lignocellulose-modifying technologies should somewhat replace the use of concentrated feeds in mega-feedlot cattle production where high feed value grain is brought to the cattle. Instead it should support a more diffuse, farm-based feeding operation, with the farmer receiving considerable value added from the straw, having to purchase less grain for feeding. The cost of the ammonia/urea for upgrading the straw is mainly offset by using the resulting animal manure as a slow-release fertilizer that is superior to urea or ammonia, which readily leach and run off the soil. Whereas large feedlot operations cause considerable water pollution, the on-the-farm use, with the returning of the wastes to the fields or paddies, will vastly reduce such problems. As less straw is incorporated into the soil, fewer of the mineral-binding problems occur during the initial microbial degradation of fast-degrading components, which requires additional fertilization in spring. As plant-disease-carrying straw is removed from the field, there could be less need for fungicide application the following season. The soil-improving humic compounds are not lost by harvesting the straw (as they were when straw was burned), because they are returned to soil in the manure.

There had been a reluctance to deal with lignin in straws because of the present slight oversupply of grain that has now disappeared as more grain goes to biofuels. Policymakers did not consider future needs, nor the time needed to develop the biotechnologies. If the lignin reduction technologies had been developed, straw could now be used as a better, efficient feedstock for ethanol production, far better than with the present technologies that utilize only a small proportion of the cellulose to produce ethanol.[523,931] It will be necessary to allay public fears relating to scientific/technical issues, although it is likely that by the time the technologies are ready, transgenics will probably no longer be a public issue, as generating the ideal straw will take time.

Very poorly digestible straw with its very low nutritional value potentially can be economically converted into hay-quality material having both highly enhanced caloric value as well as being a nitrogen source using biotechnologies described above, especially when used together with physical and chemical treatments. If successful, the Americas and Europe could produce another 200 million cattle a year, 35 percent more than at present; Asia, 250 million cattle, 50 percent more than at present; Africa, 170 million more goats per year (or half a billion more goats per year if grain yields were increased to match world averages), 80 percent more to triple the present number; and Australia, 30 million more sheep, 25 percent more than at present. Genetic engineers should turn straw into milk and meat, or into biofuels instead of feeding grain to ethanol factories while straw rots in the field. More of the digestible straw can be fed to (castrated male) steers than milk cows, as steers do not need to be grown intensively with as much concentrated feed as dairy cows. Grass-fed ruminants produce a leaner meat, containing less fat and less cholesterol, than animals fed on maize and soybeans. This should have a considerable positive impact on consumer health.

The redundancy of the monocot phenylpropanoid pathway in homologous small gene families with sequence homology to dicots (Table 14) indicates that RNAi/antisensing is the most feasible strategy for down-regulating expression and changing lignin/cellulose ratio. Moreover certain highly homologous short consensus sequences may serve as RNAi machinery initiators for down-regulation of similar gene families in both monocots and dicots.

The Kyoto carbon dioxide emission accords have many countries interested in obtaining carbon credits for not releasing or for delaying the emission of carbon dioxide into the environment. All manipulations of straw, natural or artificial, bring about return of the CO_2 fixed by photosynthesis to the environment. The key questions are which processes delay CO_2 return the longest, keeping the carbon fixed and not released as CO_2, and which replace fossil fuels most efficiently. The fastest return of CO_2 is burning straw after harvest, followed by burning as a biofuel, followed by the presently mandated soil incorporation and microbial degradation. Still, because biofuels will replace fossil fuels, full credit will be deserved for using a resource that in any case would soon be released as CO_2 due to rot. Using transgenic straw as feed for ruminants keeps carbon fixed the longest among the potential major uses, as part goes to animal products, part is stored for long periods, and the un-degraded lignocellulose is returned to the soil where it slowly degrades as hu-mic material. The only commercial use that might keep the carbon fixed longer is the use of straw in construction materials (a minor market nearly saturated at present) and possibly paper manufacture, which most experts feel is not feasible because of the varying fiber length. The paper industry of ne-cessity produces other pollutants.

It is conceivable that when the technologies for using straw as a higher-grade feed or as a renewable fuel are in widespread use, it will be possible to receive Kyoto carbon credits for the differential period that the carbon is not released as carbon dioxide, between this process and those processes presently in use. The best solutions to the worlds' most abundant agricultural waste is to recycle it through energy or through ruminants, producing more food. These technologies have elements that render them ideal to demonstrate why/where genetic engineering can benefit farmers, the environment, and consuming humanity as a whole. It will require more than a decade to iso-late the genes, transform the plants, analyze each series of transformants, and fine tune the levels of expression such that sturdy, high-yielding cereals will result, with more digestible cellulose. It will take years more to either cross and backcross the genes into more varieties of the crop, or to transform each variety. The subsidiary technologies of processing the straw to fuel or feed-ing the straw will also have to be developed. If the research would commence now, with luck it would be ready soon after there are perennial food short-ages due to overpopulation. Petroleum-based fuels are already in deficit when compared with the rate of new discoveries. Straw utilization will be environ-mentally beneficial, compared with present uses, but it will not provide fuel or feed value equal to grain. The best-quality hay or silage, a viable target, has only a quarter the feed potential of grain, but a 25 percent increase in agri-cultural efficiency is also exceedingly valuable, especially when the environ-mental worth is added to the equation. Agricultural productivity in much of the developing world is less than half of the world averages. This 25 percent advantage will remain as productivity increases.

The final products will not be simple to obtain, they will not result from engineering a single gene with a nonspecific promoter. They will surely con-tain a large number of transgenic modulations of cereal genes, with tissue-specific promoters, as well as the addition of genes, based on the needs documented above.[55,647]

CHAPTER TEN

Papaya

Saved by Transgenics

Dire predictions have forecast that major crops such as bananas (and their starchy sisters, the plantains) could go extinct because of pandemics of diseases such as sigatoka.[857] This has not yet occurred, but papayas have been on the way to local extinction, especially in Hawaii where 80 percent of the orchards had been decimated by Papaya Ringspot virus, and it was clear that the rest would follow. The disease is rampant elsewhere, but not quite to the same extent, yet growers worldwide are worried. If endogenous resistance genes exist, they have not been found. Even if they did exist in some rare germplasm, the need to backcross such gene(s) into commercial varieties would be daunting.

10.1. Biotechnological Further Domestication

Papayas have early on become an easy species to transform and regenerate.[175,350,678,1171] Standard virus resistance techniques of engineering coat protein genes into papaya were used to confer resistance,[343] and the Hawaiian industry has been rejuvenated; other countries are following suit.[98,341,655] The ringspot virus is so devastating that transgenically resistant trees yield more than fifty times more fruit in the field than their untransformed cohorts.[98]

Papaya ringspot virus varies considerably around the world, being very similar to a potyvirus attacking cucurbits. Another sequenced potyvirus, the leaf distortion mosaic virus, also has a similar distribution between cucurbits and papayas.[686] Indeed, some virologists believe that papaya ringspot virus is a series of concurrent mutations from the cucurbit virus in various locales around the world, with the papaya pathogen also attacking cucurbits. Thus, the transgenic Hawaiian papaya ringspot-resistant varieties are not necessarily resistant to other ringspot accessions and vice versa,[72,97,492,534,1051] requiring that papayas must be transformed locally, with resistance to local viral strains, unless a consensus sequence can be found that will confer universal resistance. Attempts have been made to do so, with a modicum of success.[1052] With viruses, a 5 percent genomic variability is common within the same species. There is a 2 percent genomic difference between humans and other primates, yet they are classified as different species. Quite a bit of esoteric discussion is going on among virologists about the cause of the lack of cross resistance among transgenics using different coat protein sequences to confer resistance.[1074] Consumer groups intent on detecting transgenic papayas have had to develop multiplex systems, reflecting this genomic variability.[1122] Perhaps using the replicase gene[208] instead of coat protein genes may confer a broader resistance. Surely the use of stacked coat protein/replicase/other genes will confer a more robust resistance that is less likely to break down due to evolution of the virus.

Ringspot disease is not the only problem of papaya. The solution of this problem then shows that the crop is susceptible to other pathogens, including *Phytophthora palmivora*. Plants often use disease-induced phytoalexins to naturally ward off pathogens. Most of these complex chemicals, which vary from species to species, are synthesized in a series of steps from primary metabolites, which would render the problem of genetic engineering to be a tough problem. Getting a series of genes into the plant and having them be correctly expressed is still a daunting task. One phytoalexin is an exception to this generalization, stilbene (a.k.a. resveratrol) is synthesized in a single step from the common metabolite naringenin (one step from the amino acid phenylalanine), by stilbene synthase. When the *VST* gene encoding stilbene synthase was transformed into papaya, a modicum of resistance to *P. palmivora* was achieved.[1200]

10.2. Lessons to Be Learned from Papayas

Many valuable lessons can be learned from this short case history.

1. The number of person-years required to solve the problem was really minimal, demonstrating that transgenic engineering is not just for the high-volume bulk crops, but for "minor" crops as well. (Considering the place of papayas in the microcosm of Hawaiian agroeconomics, papaya is not really a minor crop.)

2. The regulatory costs for obtaining approval to field test, grow, and then market were nearly two orders of magnitude less than the figures typically cited by the biotechnology industry for such a registration at that time. Such a relationship between relative costs may remain although total costs are higher. This suggests that growers' groups can deal with the regulatory approval at a much lower cost than the commercial agricultural-biotech industry, either because industry has too high an overhead, industry inflates their costs, and/or growers groups find a more responsive and less bureaucratically restrictive regulatory environment. Still, the growers groups in other countries have not been able to get their transgenic-virus-resistant papayas through the regulatory regimes, for whatever reasons, and the growers are losing their trees.

3. "One size does not fit all." One transgenic line does not give universal resistance to all strains of the same virus disease. It will thus be necessary to transform papayas with different coat protein constructs in various places. The disease virus may mutate to resistance and it may also be necessary to retransform resistance to newer strains. "Event-based" regulatory regimes (section 4.1) will clearly hamper these processes.

4. Unlike the debacle of Flavr-Savr tomatoes (Chapter 21), where a good gene was put in a tasteless background by a company without good, on-the-ground agricultural knowledge, the viral resistance was put into excellent papaya varietal material, which was the main market determinant for its success.

5. The rhetoric of consumer marketing boards has been that, while it may be acceptable to have genetically engineered crops fed to animals, consumers are not ready to eat transgenic food. This rhetoric proved to

be clearly incorrect, because the consumer acceptance has been excellent. Such self-proclaimed experts on consumer habits have not learned their lesson. They had previously claimed that consumers would not drink milk from cows injected with transgenic bovine somatatropin to increase milk yield, increase feed efficiency, increase protein content, and reduce butterfat. Consumers have shown no reluctance to eat the tasty transgenic papayas from Hawaii, where available, through consumer choice, even though they contain an "alien gene" encoding a "virus coat protein." European importers of gourmet foods were recently accused of importing transgenic Hawaiian papayas, in contravention of European regulations.[170] It is most interesting that importers and their gourmet customers were willing to take these risks. It is common in Europe that the major retail food chains declare in advertising and banners to consumers that they will not market transgenic products while saying, without irony, that they support "consumer choice."

10.3. The Future of Transgenic Papayas

The consumer acceptance of transgenic papayas, which flew in the face of conventional wisdom, will clearly stimulate groups to further improve the quality of this tropical fruit. It is an ideal for transgenic intervention because of the short (often less than a year) juvenile period, and then nearly a decade of bearing fruit. Its color and aroma could be enhanced, and adverse flavors such as those coming from benzyl isothiocyanates can be removed from already superior varieties, without the necessity of crossing and multiple backcrossing. A massive cataloging of papaya fruit-ripening genes has begun by isolating expressed sequence tags (ESTs) from ripening fruit and comparing their sequences with those in gene banks.[267] Little more information is needed to start using these ESTs to modulate pathways and further modify this fruit.

CHAPTER ELEVEN

Palm-Olive Oils

Healthier Palm Oil

The various tropical oil palms (mainly *Elaeis guineensis*) are the highest yielding crop species in terms of yields of oil per hectare. Their plantings have been increasing over the past decades, and production has zoomed (Fig. 22), especially as oil yields can be ten times those of soybeans. The profits are not ten times higher because the cultivation, picking, and processing costs are less for soybeans, and the soybean meal left after oil extraction is an important feed material for livestock and poultry, so that palm oil is only a third cheaper than soybean oil. Palm oil is the lowest-price edible oil on the market, and it is often hydrogenated to make margarine and frying fats and is heavily used for frying by fast-food outlets. Demand for palm oil is bound to decrease further as a food oil. This is due to both the high production and the consumer realization that the quality of the oil is contraindicated on a health basis, despite many counterclaims by proponents. The oil is naturally highly saturated,[306] to the point that it is considered to be more cholesterogenic than animal fats; it is claimed that coffee whitener from palm oil is worse for the human system than whole dairy cream, and margarine from hydrogenated palm oil is worse than butter. Some claims against palm oil may be overstated

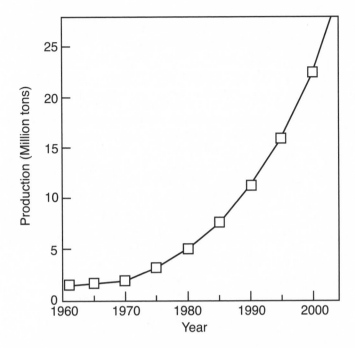

Figure 22. World palm oil production (data from FAOStat).[329] Who will use all this cholesterogenic oil in the future, unless the composition is changed?

in this era where to pump one product, it is deemed necessary to trash the competition, instead of standing on ones' own virtues. The situation is exacerbated by those who cite literature inaccurately and/or selectively to strengthen their points against palm oil, e.g. reference 157, which is unnecessary, as the oil is clearly nutritionally the worst major plant oil for human health.

Nutritionally, saturated palmitic acid (16:0) from palm oil may actually be a reasonable alternative to trans hydrogenated fatty acids from partially hydrogenated soybean oil in margarine, if the aim is to avoid trans fatty acids.[763] The health situation is such that the Danish government has banned oils and fats containing more than 2 percent industrially produced trans fatty acids,[1012] and New York City has recently instated a similar ban. A palm oil–based margarine is less favorable, however, than one based on a more unsaturated vegetable oil,[763] especially some of the transgenic ones that are not hydrogenated,

A Reminder about Lipid Biochemistry and Notation

For example, what does 18:2 $\Delta^{9cis,11trans}$ mean?
The first number refers to the number carbon atoms in the acyl fatty acid chain, the second number (after the colon) refers to the number of double bonds. This can be followed by a Δ and a superscript denoting the position and configuration of the double bonds. Various elongases add 2 carbons at a time to 4-carbon malonyl-CoA, to get the 18-carbon chain, and a Δ^9 desaturase made the double bond in cis configuration after carbon 9. The chain can be omega (ω, but sometimes written n), 3, or 6 denoting an unsaturated double bond between carbons 2 and 3, or 5 and 6 of the fatty acid chain.
Lipid chemists love abbreviations; most are eschewed here but some are used:
PUFA and MUFA = poly- and mono-unsaturated fatty acids, with FA = fatty acid (obviously) and
VLCFA = very-long-chain fatty acid, all of which are made on:
ACP = acyl carrier protein (upon which the 2 carbon pieces get added stepwise) by:
KAS = β-ketoacyl synthases until the chains are long enough and become condensed, forming a:
TAG = triacylglycerol = fat or oil, on the:
ER = endoplasmic reticulum by a:
GPAT = a membrane-bound glycerol-3 phosphate acyltransferase,
but we won't go into abbreviations LPAAT, DAGAT, TE, PDAT, DAG, KCS, FAR, CPT, PC, ALNA, EFA, EPA, ARA, DHA, DPA, KA, ad nauseam.

but why eat margarine? Trans fatty acids are produced during the transformation of palm oil to cooking fats by heating the liquid oil in the presence of hydrogen and a catalyst. They are worse for cholesterol levels than saturated fats because they not only raise low-density lipoprotein (LDL) (bad) cholesterol, but also lower high-density lipoprotein (HDL) (good) cholesterol.[1030] A somewhat better view of palm oil can be found in the review of Edem[306] who states:

> Although palm oil-based diets induce a higher blood cholesterol level than do corn, soybean, safflower seed, and sunflower oils, the consumption of palm oil causes the endogenous cholesterol level to drop. This phenomenon seems to arise from the presence of the tocotrienols and the peculiar isomeric position of its fatty acids. The benefits of palm oil to health include reduction in risk of arterial thrombosis and atherosclerosis, inhibition of endogenous cholesterol biosynthesis, platelet aggregation, and reduction in blood pressure. Palm oil has been used in the fresh state and/or at various levels of oxidation. Oxidation is a result of processing the oil for various culinary purposes, and a considerable amount of the commonly used palm oil is in the oxidized state, which poses

potential dangers to the biochemical and physiological functions of the body. Unlike fresh palm oil, oxidized palm oil induces an adverse lipid profile, reproductive toxicity and toxicity of the kidney, lung, liver, and heart.

Edem[306] is the only reviewer differentiating between the types of palm oil marketed, and proposes no solution for preventing its oxidation, or for labeling the oil "fresh" or "oxidized." The bottom line is that most consumers are getting an oil that is not the most healthy, and its transgenic transformation to "tropical olive oil" is called for.

A comparison of the effects of various dietary fats on the ratio of total cholesterol over HDL (good) cholesterol in human serum is shown in Fig. 23, based on a meta-analysis of sixty studies. A positive bar indicates that the relative amount of HDL (good) cholesterol has decreased. The predicted changes are calculated based on 10 percent of the energy in the "average" U.S. diet

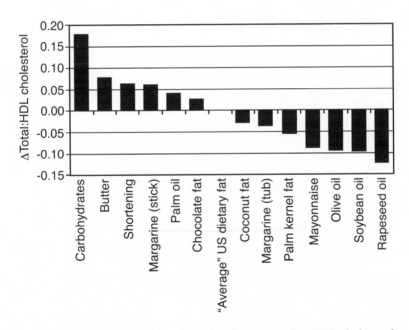

Figure 23. Predicted changes (Δ) in the ratio of serum total to HDL cholesterol when mixed fat constituting 10% of energy in the "average" U.S. diet is replaced isoenergetically with a particular fat or with carbohydrate. *Source:* Redrawn from Mensink et al.[722]

being replaced by one particular fat source or by carbohydrates.[722] For animal fats, they adjusted for the slight effects of dietary cholesterol on this ratio. The epidemiological studies suggest that a change of one unit in total:HDL cholesterol is associated with a 53 percent change in the risk of myocardial infarction, and palm oil has near the risk of butter, but surprisingly, so would replacement with carbohydrates (Fig. 23). The largest reduction of risk to heart health is seen with unhydrogenated oils, such as rapeseed, soybean, and olive oils,[722] especially those predominantly containing mono-unsaturated fatty acids,[709] so these oils would be the role model for modifying palm oil. The relative saturation levels of oils are shown in Fig. 24 and Table 15.

Epidemiological studies associated higher intakes of saturated fatty acids with a higher incidence of colorectal cancer and breast cancer,[709] but these are based on limited studies. Still, palm oil approaches animal fat in just about all compositional categories (except taste) (Table 16, Fig. 24), and high intakes of animal fats are eschewed by most medical authorities. The small amount (10 percent) of omega-6 18:2 linoleic acid is lower than that in many other oils, but it is associated with a greater risk of cardiovascular disease, whereas a higher level of omega-3 α-linolenic is considered preventative. The omega-6 fatty acids

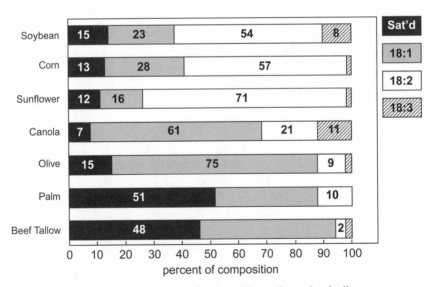

Figure 24. The fatty acid saturation of palm oil is similar to beef tallow, not commercial vegetable oils. *Source:* Redrawn and modified from Ohlrogge et al.,[807] with permission of the authors who are copyright holders.

Table 15. *Dangerous Saturation of Vegetable Oils and Animal Fats*

Oils and fats	Fatty acids		Total unsaturated fatty acids	Saturated fatty acids
	Polyunsaturated	Monounsaturated		
Vegetable oils and shortening				
Safflower oil	75	12	87	9
Sunflower oil	66	20	86	10
Corn	59	24	83	13
Soybean oil	58	23	81	14
Cottonseed oil	52	18	70	26
Canola oil	33	55	88	7
Olive oil	8	74	82	13
Peanut oil	32	46	78	17
Margarine, soft tub[a]	31	47	78	18
Margarine, stick[a]	18	59	77	19
Vegetable shortening[a]	14	51	65	31
Palm oil	9	37	46	49
Coconut oil	2	6	8	86
Palm kernel oil	2	11	13	81
Animal fats				
Tuna fat	37	26	63	27
Chicken fat	21	45	66	30
Hog fat (lard)	11	45	56	30
Mutton fat	8	41	49	47
Beef fat	4	42	46	50
Butter fat	4	29	33	62

[a]Made by hydrogenating soybean plus cottonseed oil

are precursors of arachidonic acid, which in turn is a precursor of necessary prostaglandin hormones and throboxanes, which both promote blood clotting,[807] for better (clotting after a cut) or worse (thromboses).

Thus, perhaps despite the increasing production of palm oil, the medical/dietary community is pressuring that palm oil be abandoned by the discerning, overweight consumers in the developed world. This oil does not present as much a problem to those still on (perforce) low-calorie, high-fiber diets in the developing world, but obesity is becoming a problem in large parts of Asia.

Table 16. Fatty Acid Composition of Vegetable Oils Compared with Butter

Fatty acid	Cotton seed	Linseed/flaxseed	Peanut	Rapeseed	Safflower	Sesame	Soya	Sunflower	Olive	Palm	Butter[a]
					g/100 g oil						
Saturated											
8:0	0	0	0	0	0	0	0	0	0	0	1
10:0	0	0	0	0	0	0	0	0	0	0	2
12:0 lauric	Tr[b]	0	Tr	0	0	0	0	0	0.1	0.1	3
14:0 myristic	1	0	Tr	Tr	0.1	Tr	0.1	0.1	0.1	1	8
16:0 palmitic	22	2	11	4	7	9	11	6	10	42	22
18:0 stearic	2	1	3	2	2	5	4	4	3	5	9
20:0	0.3	N	1	0.6	0.3	0.3	0.4	0.3	0.4	0.3	0.1
22:0	0.1	N	3	0.3	0.3	0.3	0.5	1	0.1	0	<0.1
24:0	0	N	1	Tr	0.1	0.1	0.1	0.3	0.4	0	<0.1
Monounsaturated											
16:1 palmitoleic	1	0	Tr	0.2	0.1	0.1	0.1	0.1	1	Tr	1
18:1 oleic	17	7	43	58	11	37	21	20	72	37	30[c]
Polyunsaturated											
n-6 18:2 linoleic	50	22	31	20	74	43	52	63	8	10	1
18:3 γ-linolenic	0		0	0	0	0	0	0	0	0	<0.1
1-3 18:3 α-linolenic	0.1	>50	0	10	0.1	0.3	7	0.1	0.7	0.3	0.5

Sources: various

[a] Butter also contains 3% 4:0, 2% 6:0, 1% 8:0, and 2% 10:0.

[b] Tr. trace

[c] Includes all forms of 18:1; cis, cis/trans, n = 7, and n = 9.

The producers of palm oil are expected to have a temporary market respite due to increasing fuel prices; much palm oil is being diverted to produce biodiesel, and many new biodiesel oil refineries are being constructed. The predictions that petroleum will never again be sold for less than fifty dollars per barrel due to rising demand and less production coupled with political instability render palm oil very profitable for such conversions. This must be causing some second thoughts for those who championed the use of re-newable resources such as palm oil. The higher prices paid for palm oil will put more land into production, causing destruction of rain forest and wildlife[157] as more plantations are carved out to meet increasing demand in southeast Asia. The same would refer to the increased production of soy-beans in Brazil for producing biofuels, also profitable, but not as much so as palm oil.

New technologies will eventually fuel vehicles and a healthier food oil will then be sought. Additionally, a healthier oil should demand a higher price on the market. Furthermore, palm oil producers should worry that regulations promulgated by health authorities will ban import of unhealthy palm oil, a prerogative health authorities have under international trade agreements. Obesity is increasing in the developing world and obesity is exacerbated by cholesterol and cholesterogenic oils. Thus, in its present form, palm oil will be relegated to the lower price end of food oils, never commanding a price that it could attain if the quality was better.

From all the considerations the most desirable palm oil would be one coming from another tree, and the other tree is not coconut, which possi-bly has a worse oil than palm oil. The best all-around oil for salad and fry-ing is from olives. It is mainly mono-unsaturated, which seems to be the best of both worlds. Poly-unsaturated oils oxidize quicker during deep frying than the mono-unsaturated acids and go rancid with bad flavor and odor.

11.1. Genetic Solutions

It has been possible to modify oil content of oilseed crops by breeding, where the variability exists. Linseed contains over half 18:3 α-linolenic acid, which gives it drying qualities that make it good for paint. By good selective breeding it was possible to select for the two mutants in linseed, which ren-dered the oil palatable, and was named "Linola," high in 18:2 γ-linolenic acid.[405] Similarly, oilseed rape had a very high concentration of erucic acid (22:1), a long-chain fatty acid considered good for industrial uses, but poisonous to humans. This and a glucosinolate antifeedant have been bred out to give the modern-day Canola oilseed rape. These properties both come from loss-of-function mutations; specific desaturase and elongase functions were lost. Oil palm needs to gain at least two functions; an elongation and a desaturation. It seems unlikely that it would be that easy to attain such gains-of-function steps by breeding, although it is possible for a species to lose biochemical functions by breeding. One cannot gain functions that do not exist within the genetic makeup of a species.

Despite most oil palm trees in southeast Asia having been derived from four trees introduced, probably from the same parent, an intensive breeding effort has been coming from this narrow base because of the heterogeneity of those trees.[999] The breeding effort has been directed to oil yield and bacterial and fungal disease resistances, and not to oil composition.[999] No published information could be found about the genetic variability inherent in oil composition in this species. The variability is probably not much, as other palm species also have highly saturated oils.

11.2. Biotechnological Solutions

Oil palms should clearly be redomesticated to produce a more healthy oil. Their genomes probably possess the genes to do so, as the desired fatty acids are present at low levels. Still, breeding a tree crop such as oil palm is almost impossible when compared with the ease of breeding in linseed or oilseed rape. It would require mutations in regulatory genes to increase the needed elongase and desaturase activities. It is even easier to breed dicot trees, where once a genotype is found, it can be grafted on orchards of mature trees. Still the needed genes are known from work with other oil crops, overcoming one hurdle. They come from a wide variety of sources and they have been used in other species (Table 17). This demonstrates the value of knowing comparative biochemistry at the outset to ascertain where to look, together with genomics to perform the final search. Before setting out to engineer oil palm, it is wise to see what has been done in other species, what the sources of the genes were, and how they were used (Table 17), as described in the next section and reviewed at much greater length in several recent publications.[303,709,807,989,1111]

Table 17. *Transgenes Presently Being Used to Modify Oil Crops*

⇑(elevated) ⇓(suppressed) gene[a]	Source of gene[b]	Effect	Reference
Edible oils			
⇑ acetyl-CoA carboxylase	p, *Arabidopsis thaliana*	5-fold increase in oil in potato tubers	591
⇑lauroyl-acyl carrier protein thioesterase	p, California bay tree	Increased lauric (12:0) acid in oilseed rape	305
⇑acyl-CoA:diacylglycerol acyltransferase (AtGAT)	p, *A. thaliana*	Increased triacylglycerol content	138
⇑site-directed mutagenized Garm FatA1,acyl-acyl carrier protein (ACP) thioesterase	p, *Garcinia mangostana* (mangosteen)	Rape plants accumulate 55–68% more stearic acid than plants expressing the wild-type enzyme	325
⇓ Stearoyl-acyl carrier protein (stearoyl-ACP) desaturase	p, *Brassica rapa* Oilseed rape	Up to 40% increase in stearic acid levels in rape seeds	597
⇓ ghSAD-1-stearoyl-acyl-carrier Δ9-desaturase	p, cotton	Increased cotton seed stearic acid to up to 40%	658
⇓ ghFAD2-1-oleoyl-phosphatidylcholine ω6-desaturase	p, cotton	Increased cotton seed oleic acid up to 70%	658
⇑Cpa2 Δ12-epoxygenase	p, *Crepis palaestina*	Suppresses endogenous Δ12-desaturase, increase in oleic acid in *Arabidopsis* seed	988
⇓ FAD 2 desaturase	p, soybean	Suppresses oleic acid desaturation in soy, giving high oleic oil	587
Specialty oils and waxes			
⇑Ch FatB2 acyl-ACP thioesterase	p, *Cuphea hookeriana*	Up to 75% 8:0 caprylate and 10:0 caprate in canola seeds	265
⇑lauroyl-acyl carrier protein thioesterase (MCTE)	p, California bay	Express high levels of lauroyl-CoA oxidase activity but not palmitoyl-CoA oxidase activity in rape	305
⇑ fatty acyl hydroxylase	p, castor bean	Accumulation of ricinoleic, lesquerolic, and densipolic acids	155
⇑MomoFadX oleic acid desaturase	p, *Momordica charantia*	>15% α-eleostearic acid in soy	176
⇑ImpFadX oleic acid desaturase	p, *Impatiens balsamina*,	>15% α-parinaric acid in soy	176
⇑Pt∆6 desaturase + ⇑PSE1∆6 elongase + ⇑Pt∆5 desaturase	a, *Phaeodactylum tricornutum* m, *Physcomitrella patens* a, *Phaeodactylum tricornutum*	Decrease in 18:2; increase in ∆6 C18 and C20 PUFAs	4

[a]⇑Expressed positively; ⇓ expression suppressed by using antisense or RNAi.
[b]a, alga; m, moss; p, plant.

211

11.2.1. Engineering Oil Composition in Plants

Oilseed rape has already been engineered to produce virtually any oil composition that could be of value (Table 17), from the industrial equivalent of sperm oil, to higher-quality frying oil (to fry at higher temperatures without oxidation), to fatter (more solid) types for margarine manufacture, to mono-unsaturated oils with a composition similar to olive oils.[156,540,1054] Similarly, soybeans have also been reengineered to provide an oil with a high omega-3 fatty acid content to provide this key component, which can replace fish oils in human diets.[1087] The engineering effort used two approaches: antisense or RNA interference (RNAi) to close pathways that lead to the wrong products, and new genes from distant sources (so as not to get cosuppression with native underexpressed genes) with strong promoters to push metabolites in the desired direction (Table 17). The genomics researchers dealing in this area annotate more than six hundred *Arabidopsis* sequences as possibly encoding genes for more than one hundred twenty reactions in fatty acid biosyntheses, but 80 percent of these putative annotations have yet to be experimentally verified.[807]

Engineering Edible Oils

Many technical successes have been achieved in engineering oil seeds to modified fat content. Some of the successes are summarized in Table 17. Soybean is a good example as it naturally produces an oil rich in polyunsaturated fatty acids; about 50 percent linoleic acid (18:2) and 10 percent linolenic acid (18:3) (Table 16). In particular, 18:3 makes the oil unstable and easily oxidized, and as noted above, the oil develops objectionable flavors and odors when heated. Thus unprocessed soybean oil is unsuitable for many applications, and therefore, for many edible uses it is chemically hydrogenated, giving rise to trans-fatty acids.[807] The biosynthesis of polyunsaturated fatty acids naturally begins with the enzymatic conversion of oleic acid (18:1) to linoleic acid (18:2). The gene (FAD2) encoding this enzyme was isolated from *Arabidopsis* by screening mutants generated by T-DNA insertions.[814] Shortly afterward, molecular biologists succeeded in isolating and suppressing the expression of the gene in soybean.[587] This strategy led to a major decrease of the 18:1 to 18:2 conversion step, and almost completely eliminated polyunsaturated fatty acids in the soybean oil. The new transgenic soybean oil has 85 percent oleic acid, one of the highest oleic acid contents found in nature,

more than olive oil and other high-oleic oils, which are considered to provide health benefits, compared with other plant and animal oils. An unantic-
ipated benefit of the oleic increase was that the saturated fatty acid content
of the oil fell from approximately 15 percent to less than 8 percent.[587] With
the resources of a major corporation, genetic engineers only needed less than
five years from gene isolation to a field-tested transgenic soybean crop ready
for commercialization of a new product.[587]

The primary approach for enriching plant oils in stearic acid has been to
enhance metabolic flow from palmitic to stearic acid in developing seeds.
There are several enzyme targets that had been manipulated to achieve this
goal (Table 17) including: suppression of FatB thioesterase activity (preventing cleavage of 16:0-acyl carrier protein (ACP); increasing KASII elongase activity (enhancing conversion of 16:0-ACP to 18:0-ACP); increasing FatA
thioesterase activity (promoting cleavage of 18:0 stearic acid from ACP and
thereby preventing desaturation by stearoyl-ACP desaturase); and/or suppressing stearoyl-ACP desaturase activity (blocking conversion of 18:0-ACP
to 18:1-ACP).[303] Although a slight increase in stearic acid content was obtained by overexpression of KASII in transgenic plants,[587] a much larger increase in stearic acid was achieved by increasing FatA thioesterase activity.
FatA activity was increased in canola by transgenic expression of the FatA
thioesterase gene from mangosteen, a tropical fruit that accumulates high
amounts (56%) of stearic acid. The 9*cis*,12*cis*-stearic acid content in the seed
oil of the resulting transgenic oilseed rape plants increased from 2 to 22 percent.[480] The stearoyl-ACP level was later increased thirteenfold by DNA site-
directed mutagenesis, resulting in transgenic plants that accumulated up to
70 percent more stearate than plants expressing the wild-type enzyme.[325] A
yet higher stearic acid content was obtained by also down-regulating stearoyl-
ACP desaturase activity. The highest levels of stearic acid were eventually
achieved by antisense suppression of stearoyl-ACPase activity, which elevated
stearic acid from 2 to 40 percent in oilseed rape and cotton.[597,658]

High-stearic and high-oleic cotton seed oil genotypes were generated using such technologies (Table 17). By intercrossing them it was possible to simultaneously down-regulate both ghSAD-1 and ghFAD2-1 to the same degree
as observed in the individually silenced parental lines. The silencing of
ghSAD-1 and/or ghFAD2-1 to various degrees allows the development of cottonseed oils having novel combinations of palmitic, stearic, oleic, and linoleic
acid contents that can be used in margarines and deep frying without

hydrogenation and also potentially in high-value confectionery applications without the need for hydrogenation to stabilize the oil.[658]

There has even been an effort to engineer oilseeds to resemble palm oil for the specialty markets of shortening for baking, and for soaps. This makes little economic sense as palm oil is significantly less expensive, but the research is scientifically interesting. Palmitic acid content was increased in soybean oil from 10 to 50 percent by suppressing KASII activity and increasing FatB thioesterase activity.[588]

Alterations of seed oil fatty acid composition have compromised other agronomic traits. *Arabidopsis* seeds containing high oleic acid did not develop properly at low temperatures, resulting in lower total seed oil content and lower germination than wild-type seeds.[732] Cotton and oilseed rape engineered for high stearic acid content in seed oil exhibited poor germination and reduced survival of seedlings,[597,658] so the level of manipulation that can be performed may well be limited by environment.

Engineering Specialty Oils

The greatest interest at present is not in the mundane changes in edible oil composition of standard crops, but in the synthesis of specialty oils and waxes that are expensive but do not have large markets. Palm is probably not a target of such fascinating modifications, so our discussion will be brief, but some of these syntheses are summarized in Table 17. Various polyunsaturated long-chained fatty acids (PUFAs) are very important for human nutrition for the synthesis of various hormones and regulatory compounds. Humans do not perform the elongation and desaturation reactions when there is some fat in their diet. Neither do fish, but they are the major source of such compounds in our diet. Fish bioaccumulate these fatty acids from their diet of marine algae. Some hard-to-cultivate plants also accumulate these fatty acids, but the best sources are all algae or mosses, neither of which we normally eat. The genes from one moss and one alga were found to be optimal in producing these very-long-chain fatty acids in seeds of tobacco and linseed.[4] The authors enjoy pointing out that producing such very-long-chain PUFAs transgenically in plants can only be good for the rapidly depleting wild fish stocks,[287] although they should realize that getting one's PUFAs is not the main reason, or even a reason, that most people consume fish, and many consumers are happy eating less expensive penned, cultivated fish.

11.2.2. What Genes to Choose for Oil Palm?

Palm oil is meant for bulk commodity purposes; it will not be easy to precisely engineer and tweak the composition. The vegetative propagation will take more time than for an annual-seed producer that can be induced to produce a few crops a year. Thus, it is probably not best to focus on the boutique-market designer oils such as sperm oil equivalent and fish oil. Boutique markets are more amenable to much more rapid turnaround time, seed-propagated annual crops, and not vegetatively propagated perennials. Clearly there is a need for a general-purpose healthy household edible oil that has a composition approaching that of olive and sunflower oils and can also be used for frying. Unfortunately for the obesity of the world, a major part of the oil goes for industrial-scale deep frying, where the longest possible continuous use of the oil at the highest possible temperature without oxidation is desired. That does not mean that one cannot or should not fry with liquid cooking oil. The Italian and Spanish cuisines, which do not lead to obesity, use mono-unsaturated olive oil for quick frying, even if it is not McDonalds' industrial-strength saturated and artery clogging.

A perusal of the types of genes that have been used to modify oils (Table 17), and the compositions of the various oils (Table 16), allows developing a scheme of what might be useful for palm oil (Fig. 25). The general concept should be to drastically lower the 42 percent of saturated palmitic acid (16:0) by pushing it through the system to mainly become monounsaturated oleic acid (18:1), while preventing it from becoming further saturated to linoleic acid (18:2). This might require most if not all of the following genes: a down-regulation of the acyl-carrier protein that stops elongating at 16:0, replacing

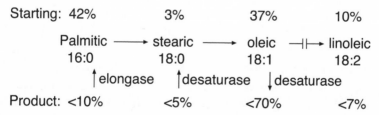

Figure 25. Engineering needs of palm oil. ↑ The genes/enzymes needing up-regulation; ↓ The genes/enzymes needing suppression.

it with an up-regulated, site-directed mutationally enhanced mangosteen FatA1 acyl-protein thioesterase that drops its load as stearic acid (18:0). This must be augmented with a desaturase to keep the stearic acid low and move the fatty acids to oleic acid (18:1). A gene encoding a very active, highly promoted stearoyl acyl carrier protein desaturase is called for. A few possibilities are already available to prevent further desaturation to linoleic acid: cosuppression the native gene by overexpressing a gene such as the Cpal2Δ12 epoxygenase, by RNAi of palm FAD2 desaturase.

11.2.3. Shortening the Chains to Prevent Global Warming

While ruminant animals are more feed efficient than monogastrics, they have the annoying habit of releasing large amounts of methane into the atmosphere while chewing their cuds. Methane contributes more to global warming (per mole) than carbon dioxide, and the more than eighty million tons of methane released annually by livestock has a greater effect on the environment than the carbon dioxide emitted by all vehicles worldwide.[1010] The metabolism of good food in the rumen by methanogenic organisms can cause a 2 to 15 percent loss of gross energy intake,[672] which is quite a loss to the farmer. It was astonishing that a 3 to 5 percent addition of fatty acids or esterified oils containing 12:0 (lauric) and 14:0 (myristic) saturated fatty acids to the diet could halve methanogenesis and increase general feed efficiency, especially with the low fiber content diets of the types fed to dairy cattle.[672] The lauric and myristic acids directly inhibit the methanogenic protozoans in the rumen. Thus, there is a good chance that there will be a new market for such medium-chain fatty acids, which could be obtained in palm by preventing the elongase reaction leading to 16:0 stearic acid, and palm oil would be as good a source as any for this. This has already been accomplished in oilseed rape by increasing the expression of lauroyl-acyl carrier protein thioesterase from the California bay tree.[305]

11.2.4. How to Transform?

The beautiful job of modifying oilseed rape to obtain whatever oil composition was desired was a first. Until then, genetic engineers resembled primitive tribes that could count: one, two, . . . ? many? The concept of a multigene transformation was an anathema, a far-off dream. Suddenly, here were oilseed rape plants containing five or more transgenes. Much of this was not performed in a single sophisticated step, as it superficially seemed. Much was sequentially performed, often with the help of the breeder. The genes were separately transformed into different plants, the plants crossed, and F_2 families selected that were homozygous (nonsegregating) for the complex of traits. Such an approach is appropriate for crops that can have a few generations per year, but not for a tree species such as oil palm. Here it will be necessary to cotransform all the genes that are expected to be needed, and then to obtain a large number of transformants. These will have to be probed to ascertain which have the whole gene complex introgressed into the genome, with part of each transformed clone regenerated, and part continually propagated. As soon as nuts are available, the oils will have to be analyzed and upper and lower limits set for each fatty acid, such that many different clones within the limits can be regenerated in quantity, while not limiting numbers to too few clones, to establish a modicum of diversity. If all clones are within the broader limits, the blended oil from the resultant mixed plantations will also be within the newly desired narrower limits, to be sold at a premium, as tropical "palm-olive" oil.

11.2.5. Transformation and Regeneration

Palm is a species that has and can be biolistically engineered[836] with targets of obtaining high-oleate and high-stearate oils, and the production of industrial feedstocks for production of such as biodegradable plastics. Because of the long life cycle of the palm it is envisaged that commercial planting of transgenic palms will not be widespread before the year 2020. This should be at about just about the time when the bottom drops out of the biodiesel market as new technologies for transportation come on line. Without a transgenic, new high-quality edible oil, the bottom could drop out of palm oil.

11.3. Biosafety Considerations

Oil palm is a species introduced to Southeast Asia, where most are grown, and no wild species are known to cross with oil palm in these areas. Should there be a cross with a tree and the nuts germinate and grow to a tree (highly unlikely in a plantation), then the oil can only be healthier. The same would go for pollen from a transgenic tree pollinating another tree, the endosperm would be expected to have the dominant traits conferred by the transgene, bettering the oil. Indeed, one way to improve the oil of old orchards would be to artificially pollinate them every season with transgenic pollen. It is assumed that many of the nonscientific regulations will be off the books by the time such a technology is closer to coming on line. Thus, the hindrance of event-based regulation, that presently would force all growers to a single clone, will hopefully no longer be in effect. Still, it will be necessary to perform food safety analyses to ascertain that no known novel products are made. As this has not been a problem with the other oil crops using the same genes, it is unlikely that there should be such problems. If the engineering is successful, the product should be far healthier than previous palm oil, providing an inexpensive, healthy oil to the Asians who are presently the major market for the present unhealthy, untransformed palm oil.

CHAPTER TWELVE

Rice

A Major Crop Undergoing Continual
Transgenic Further Domestication

The reader might consider it most peculiar that rice, the largest volume crop for human consumption (more wheat is produced, but less wheat than rice is consumed by humans) is considered as a case for needing further domestication. Yet there are two issues with rice where further domestication is called for, based on human evolution with rice. First, the green revolution, by dwarfing rice, vastly increased its yield because grain is now produced at the expense of straw. This trait was crossed into a limited number of varieties, greatly reducing the biodiversity of rice produced, and desirable landraces and varieties are being abandoned, lessening biodiversity of this crop (section 12.2). Transgenic technologies can preserve the genetic diversity of the landraces.

Second, major changes have taken place in the technology of rice cultivation. World industrialization has led to workers abandoning agriculture, the poorest paid employment on earth, yet among the most back breaking. Less labor is available for transplanting rice, the time-honored method of controlling the major weeds of rice. Economic globalization might well force the abandonment of the expensive and subsidized machine transplanting common in Japan and Korea. Inexpensive direct seeding replaced transplanting and has brought a series of new problems to rice, requiring further modification of the species. These include a necessity for herbicide resistance while preventing gene flow to feral forms, as discussed below (section 12.3). Rice has disease, insect, and mycotoxin problems that are not tractable to breeding and are discussed generically with other crops (Chapters 5 and 6). Its straw is wasted, and some possible solutions amenable to rice are described in Chapter 7, but rice has special needs (described in section 12.4). Despite rice being a major crop, it has a multitude of varieties and landraces. Useful genes should be engineered into rice (section 12.2), and new transformation technologies are discussed (section 12.5) that will facilitate this. Rice has been the leading crop in plant genomic research, and the first crop to have its genome completely sequenced, aiding the discovery of genes needed in rice as well as in other grain crops.

12.1. Varietalization of Rice

Unlike other major crops that are considered bulk commodities, without varietal separation, rice is valued by the consumer who recognizes types and varieties by name. Even wheat is not separated by variety, just by type. Varietal name recognition is considered a key reason for slow acceptance of new rice varieties,[772] especially in epicurean countries such as Japan. Genetic engineering of single traits such as dwarfing, or herbicide resistance, or disease resistance does not substantially change a variety (except in the particular engineered trait), which is a distinct advantage, so transgenics should not be a problem in impeding varietal name recognition. Transgenes can ease the addition of needed traits to specific varieties without laborious backcrossing. Thus, it makes sense to engineer each variety with the transgenes of choice, but this will necessitate not adopting an "event" based regulation system, so prevalent at present, as discussed in section 4.1.

There are many political and socioeconomic reasons why transgenic rice varieties are late in coming. The major ones stated—varietal nature of rice production (no single variety covers large areas) and the fear of insufficient profit to the private sector (e.g., from selling only farmer-saved seed)—have delayed development except in China, where transgenic rice is being accepted, and Iran, where the transgenic rice came from government research. The acceptance in China was initially demonstrated by farmers, who illegally cultivated "experimental" Bt varieties. The situation is now being legitimized. It is because of the public need of national food security that the public sector has become heavily involved in rice transformation in China and India. The only problem is that the decision makers in the public sector are not always cognizant of farmers' needs, as is evidenced by the paucity of research and development of herbicide-resistant varieties, the major need emanating from the continuing switch to direct seeding.

12.2. Easily Domesticating the Multitudes of Rice Landraces

A method of preserving landraces based on genomics but not transgenics has been instituted in Yunnan province of China. It was designed to deal with two issues: the propensity of the tall varieties to lodge and the typically high susceptibility of the local landraces to rice blast disease. In dealing with rice blast disease, the scientists followed the reasoning and models of Wolfe[1161] who dealt with disease prevention in barley. The concept was to take strains with different susceptibilities and tolerances to different pathogen strains and to mix them, making it much harder for the pathogen to spread among compatible strains. In the case of rice landraces, they are interplanted by transplantation in fields of hybrid, dwarf, blast-tolerant rice. The landraces are chosen to have a more than 70 percent genetic dissimilarity by amplified fragment length polymorphism (AFLP) and by having resistance to different rice blast strains.[1199] Because the landraces are tall, they can be harvested separately from the hybrids. Their high value justifies the considerable added cultivation costs. Clearly this system has had some success, considering the adoption rate in the region it was pioneered (Fig. 26). It also demonstrates the value of academic/agronomic interchange. Still, this is a stopgap, and it is valuable mainly for high commercial value landraces, not for the varieties that provide most of the rice eaten.

The breeding of green revolution dwarf rice, with its high harvest index (the ratio of grain to straw) and resistance to lodging, even with high fertilizer rates, tripled rice yield in India and China, countries that had been on the brink of starvation and possibly nuclear warfare. The varieties developed were those with the highest yield potential, and the crossing and backcrossing with the initial dwarf mutants meant losing important traits relating to quality and desirability. Traits such as the fragrance and flavor of a basmati rice variety are polygenically controlled, and thus the necessity to cross the recessive genetic dwarfism was too daunting to local breeders, for fear of los-

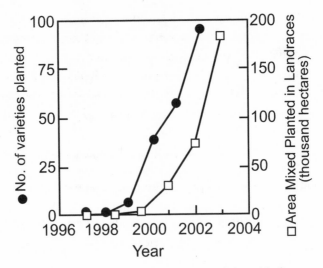

Figure 26. Adoption of mixed planting of landraces of rice with disease-resistant rice in one region of China to obtain more landrace production. *Source:* Plotted from data in Zhu et al.[1199]

ing these quality traits. The same is true for any tall, low-yielding landrace with desirable characters. If one wants an otherwise identical high-yielding dwarf, it is an arduous task by classical breeding.

The gene constructs are readily available that provide the same effect as genetic dwarfing, yet as a dominant transgene, and without bringing in alien genes during the initial breeding. The concept of engineering dwarfism into a large number of landraces was unthinkable a decade ago. Now that the constructs are available, and many a lab technician can engineer genes into rice, it should be possible to increase the number of dwarfed landraces with improved yield. The dwarfing genes could be stacked in tandem with genes for disease, insect, and/or herbicide resistances, which would add value. Additionally, the dwarfing would not just lead to a higher yield, it would provide mitigation for gene flow, as discussed in later sections.

Some varieties of rice are easier to transform than others (section 12.5), but it is worth enhancing the biodiversity of this crop by transforming the landraces. The economics are there, for example, an increasing middle class in India is willing to pay much more for a basmati rice. This tendency is continuing wherever a middle class is developing.

The biolistic techniques currently used for rice allow mixing the DNA of various gene constructs and performing cotransformation. The vast majority of transformants will bear all the traits in the DNA cocktail,[599] inherited as a block, as the traits seem to integrate together at the same locus. Thus, a favorite landrace can be transformed with herbicide resistance[194] as both a selectable marker and a useful trait, a dwarfing gene as a needed gene and as a mitigating gene,[459,811,1083] disease resistance genes,[351,548,776,794,947] insect resistance genes,[96,149,548] and so on, in the cocktail, along with other mitigating genes (section 4.2.3) to deal with gene flow. While dwarfing itself is an excellent mitigating gene to preclude establishment of transgenes in feral rice or other varieties,[19–22] a gene that should be able to prevent shattering has also been described[644,645] and could be used in tandem, both for its positive effect on many landraces and for mitigation. Instead of less than half the yield of a good hybrid, the newly transformed landrace will have all the properties of its parent, but at a much higher yield, and can be direct-seeded or transplanted alone, without the added cost and bother of mixed planting and separate harvesting. This is the kind of active landrace preservation that is compatible with the farmers' desires and market forces. This active landrace preservation also assures continual positive "Red Queen" evolution of the landrace, the type of evolution that has occurred for millennia with landraces and contributed to their continued survival.

12.3. Direct Seeding Necessitates Further Domestication— Better Not Red or Dead

According to the Food and Agriculture Organization of the United Nations (FAO), rice is clearly a crop with problems.

1. Rice has geographic problems; areas are being taken out of production faster than are being brought into cultivation due to urban encroachment, competing crops, soil degradation and salinization, and lack of water; *transgenics are needed to increase yield on the remaining land and for land reclamation.*

2. Rice culture has demographic problems because of the continuously increasing average age of rice farmers. Their children, and often their husbands, flock to slightly higher paying, less arduous employment.

The remaining older farmers and the women are tired and more likely to take on direct seeding. Unless agronomic strategies change, paddies will become weedier, and rice yields will drop; *herbicide-resistant transgenics are needed to replace labor.* Groups such as the International Rice Research Organization (IRRI) are promoting direct seeding without forethought to the consequences of having feral rice.[551]

3. Rice culture has agronomic problems that relate to the demographic problems. Direct seeding has alleviated the labor problems of transplanting while exacerbating the weed problems; *transgenics are needed to overcome the weed problems.*

When surveyed, farmers claim that the major constraint to rice production is weeds[499]; insects and pathogens together account for only 12 percent of losses. Much of the loss due to insects was later attributed to uncontrolled weeds that act as secondary hosts for the insects.[499] The weed issues do not get the attention due them and are rarely addressed by the international rice community, whose members are typically trained to deal with every aspect of rice culture other than weeds.

The shift to dwarf green revolution varieties greatly enhanced yield, but it was counter to millennia of continued selection for taller varieties that would outcompete weeds. The weeds followed suit and evolved taller biotypes matching the selection by the farmers. The cultivation of dwarf varieties necessitated a reliance on herbicides to control the weeds, and herbicides were extremely successful and economic for the farmer.

Herbicide use in rice inexpensively facilitated the newly necessary weed control in low-stature, high-harvest index "green revolution" rice, allowing fewer farmers to harvest far more rice. Herbicides also facilitated the ecological and evolutionary changes that are just beginning to appear; these tell us that the chemical answers that allowed cultivation of direct-seeded short varieties are not forever. An ecological vacuum will not remain, and weeds continue evolving just as they have before, but now they have evolved resistance to the herbicides that could selectively control them in rice. Thus, major weeds such as various *Echinochloa* spp. have evolved resistance to all rice herbicides, but not everywhere to all herbicides. The fact that somewhere there is resistance to each herbicide suggests the inevitability of the phenomenon wherever the same herbicides are used.

There is one grass weed that is naturally resistant to all selective herbicides that can be used in standing rice, and has been held in check by a mixture of techniques: the use of certified seed, transplanting, and water level management. That weed is the feral (weedy or red) rice strains that are mainly the same genus and species as rice (*Oryza sativa*) or its immediate progenitors (the perennial *O. rufipogon* or the annual *O. nivara*, which are botanically all the same), or crosses with very closely related species.

There can be no return to the labor-intensive, herbicide-free, back-breaking, transplanted, and midseason hand weeded rice. Japan and South Korea use machine transplanting and then cocktails containing a slew of herbicides to effect complete chemical control of weeds, much more than anywhere else. Other vast rice-growing areas exist where just one or two herbicides are often used: a grass killer and sometimes a broad leaf killer. The nature and economics of farming in this vast middle realm necessitates the use of cheap generic herbicides; 2,4-dichorophenoxyacetic acid (2,4-D) for broad leaf weed and either propanil, pendimethalin, butachlor, or thiobencarb for grass weed control. The latter four compounds are not rotated; each is typically the sole herbicide used in a given region. This use of single compounds has led to resistance problems.[412] Broadleaf weeds have not escaped the adequate control with 2,4-D, except in isolated instances.[419]

Major Weeds in Rice

There are three weed groups that somewhat arbitrarily fit the designation as major weeds, that is, are globally distributed, pernicious, hard-to-control weeds that have become acute problems because of recently instituted cultural practices:

1. *Echinochloa* spp. are always problem weeds, but are now evolving resistance to the rice herbicides used for their control;

2. The sedges (*Cyperus* and other) that were never well controlled by any herbicide chemistry and the areas infested are expanding;

3. The feral (red, weedy *Oryza* spp.) rice types that were never selectively controlled in rice by herbicides.[87,797,798] Their control is especially amenable to biotech solutions that confer selectivity between domestic and feral rice. The same solutions are frightening and futile if the crop transgenes for herbicide resistance (or other traits) introgress into

the feral rices enhancing their competitive ability and obliterating the possibility of selective control.

The *Echinochloa* spp. are major weeds wherever rice is grown.[509] Their distribution is truly global from temperate to tropical areas in a wide variety of crops. The *Echinochloa* spp. were major targets for graminicide development, and *Echinochloa* spp. were excellently controlled for a very long duration. The excellent control was the key to resistance problems. If there is excellent kill through all weed germination flushes, the only survivors are individuals that are resistant to the herbicide doses used, that is, the selection pressure for evolution has been intensive. Propanil provided excellent grass weed control throughout the Americas and in Europe until resistance evolved.[410,419, 1089,1091] Propanil resistance in *Echinochloa* was not at the photosystem II target site in the chloroplasts. Two *Echinochloa* spp. evolved elevated levels of the same acylamidase enzyme system that rice uses to metabolize the herbicide propanil to nonphytotoxic compounds.[186,636]

Large-scale butachlor/thiobencarb resistances evolved in southern China,[363,514] and interchemical group metabolic cross resistances evolved in two *Echinochloa* spp. to a variety of herbicides in California.[348,349] The number of sites and the area infested with acetolactate synthase (ALS) inhibitor-resistant weeds of rice have steadily increased. More than half of the twenty-four reported herbicide-resistant biotypes in rice are resistant to ALS inhibitors.[483]

The Sedges

The newer direct-seeding cropping systems for rice favor sedges. The excellent control of grass weeds left an ecological vacuum, which nature abhors. The sedges *Alisma plantago-aquatica* in Italy and Portugal, *Cyperus difformis* and *Sagittaria montevidensis* in Australia and the United States, *Scirpus mucronatus* in Italy and the United States, and *Lindernia* spp. in Asia are of particular concern. The lack of good alternatives for control of some of these species in rice heightens the concern of growers. The only good chemical way to kill sedges is with systemic herbicides that will penetrate to the storage organs of these pests, preventing their regrowth. Genes are available to confer resistance in rice to a few systemically translocated herbicides that kill sedges.

The Feral/Weedy Rices

Feral, conspecific (same species) weedy/red rice and other *Oryza* spp. have also filled a vacuum and are much harder to deal with in direct-seeded rice where the cultivated rice does not have the head start due to transplanting from nurseries.[87,509,797,798] Their genetic, morphological, and phenological similarities to domestic rice kept them as minor camp followers, until cultivated rice was dwarfed to increase the harvest index. The taller feral rice strains now have a competitive advantage over domestic rice, but transplanting left them behind. Two feral rice plants per meter square can give a measurable yield loss[624] and the numbers that can be found will reduce rice yields by as much as 85 percent.[315,323 347,625]

In addition to the United States, Spain, Central America, and Italy, where only direct-seeded rice is grown, feral rice is now becoming an acute problem in Malaysia, Thailand, Vietnam, and the Philippines where direct seeding is becoming popular because of increasing labor costs. It had not been a major problem in South Korea and Japan where transplanting rice is mechanized and rice production is heavily subsidized. South Korea is now having feral rice problems because their direct-seeded rice areas are increasing. There is a move toward direct-seeded rice in China; the area increased nearly sixfold from 1995 to 1999 to nearly a quarter million hectares in Zhejiang province alone. The trend is expected to continue, and feral rice species could become a problem as they have everywhere else where direct seeding is practiced.[1125] Feral-rice may be less of a problem in China as the farmers continue to increase the use of hybrids, and feral rice is carefully kept from hybrid seed production areas, so that there is little contamination of seed.

Some of the conspecific feral rice biotypes that have appeared can be considered as progenitors to, or as recently evolved feral forms of, domestic rice (Fig. 27). Other *Oryza* spp. have weedy characters, as well as wild species that are not competitive at present in agroecosystems.[1102] The various feral rice strains shatter most of their seeds before cultivated rice is harvested, so the farmer loses rice yield while filling the soil seedbank with weed seeds.[87] Enough weed seed is left in the harvested crop to further sow farmers' fields with this problem if contaminated saved seed is used. This weedy rice seed mimics rice seed; and it is nigh impossible to mechanically separate it from

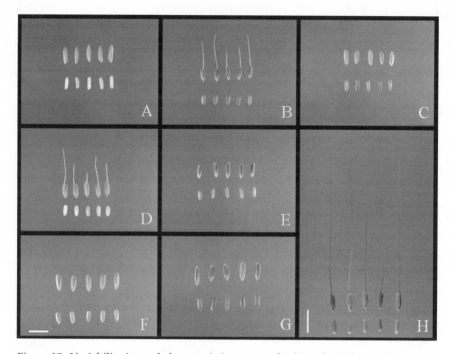

Figure 27. Variability in seed characteristics among feral/weedy rice types found at a single rice field in Vietnam (Long Thanh village, Giong Rieng district, Kien Giang province at 09°51.64′ N and 105°13.74′ E). (A) Cultivated local commercial variety; (B) through (G), weedy rice accessions: (B) gold-hull, short-awned, red-pericarp weedy rice; (C) straw-hull, awnless, red-pericarp weedy rice; (D) straw-hull, awned weedy rice with white pericarp; (E) gold-hull, awnless with brownish pericarp; (F) shattering "varietal" type with white pericarp; (G) gold-hull, awnless type with long, red-pericarp kernels—grew in association with the local commercial variety; (H) wild perennial rice, *Oryza rufipogon*. H was collected at a nearby field. Bar = 10 mm (magnification slightly different for H, which has its own bar). For more information on the problem see Valverde.[1090] *Source:* Photo and legend courtesy of Dr. Bernal Valverde.

rice seed. Recently Vietnam lost a third of its annual export sales of high value aromatic rice because of contamination with feral rice seed.

Another aspect of feral rice can be prolonged dormancy, germinating over several years, resulting in a prolonged problem. Without these two qualities, shattering and dormancy, the feral rices would almost be domestic rice. The feral rice strains have become greater problems since farmers became more reliant on chemical means to control other weeds in rice. It should be no surprise to the geneticist or biochemist that the domestic and feral rices are nat-

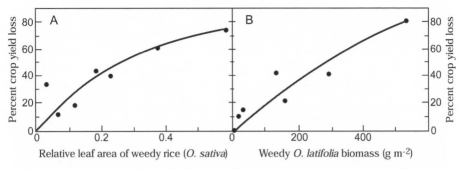

Figure 28. Damage to rice yield from small amounts of competition by weedy rice: (A) by conspecific feral forms and (B) by the related wild species *Oryza latifolia*. *Source:* Replotted from data of Valverde and Madsen cited in Valverde.[1090]

urally resistant to the same selective herbicides,[798] being the same species. Herbicides that controlled other weeds allow feral rice to thrive and fill an empty ecological niche. The feral rice strains are exceptionally competitive with rice (Fig. 28), especially as they mutate to taller forms than the cultivated dwarf rice varieties.

Presently, the best farmers can do to deal with feral rice is to delay seeding their rice crop until after the feral rice has germinated, and then control it with nonresidual graminicides to which they are still susceptible. The return to transplanted rice is generally not in the cards as the "fe"manual (predominantly female manual) labor for planting is missing. Delayed planting shortens the season, reducing the yield. Rotating rice with other crops and instituting strong control measures to reduce the seedbank of feral rice also can be effective where there is no permanent paddy.[66,356] Italian farmers have had to resort to ancient herbicides such as dalapon to control feral rice before delayed planting of rice (A. Ferrero, pers. commun.).

The easiest way to obtain selectivity among closely related species such as rice and feral rice strains is to engineer herbicide resistance into the crop, as has been done with glufosinate resistance.[802, 803]

12.3.1. Molecular Differentiation of Rice Strains: Historical Evidence for Gene Flow

Molecular techniques for distinguishing rice genotypes have been widely published, but they are predominantly used for breeding of cultivated types. Only recently have there been a few studies relating to the feral rices, mainly those thought to be *O. sativa,* but there have been some surprises (Table 18).[1103]

Based on randomly amplified polymorphic DNA (RAPD) analyses as well as morphometric and isozyme analyses of numerous weedy rice accessions, Suh et al.[1024] present a most comprehensive story. They concluded that one group of feral rices in Korea probably originated from gene flow between japonica and indica, another between wild and cultivated indica, one group comprised old rice varieties gone feral, and one group arose because of gene flow between wild and cultivated japonica types. Thus, gene flow has been rampant in the eight millennia of rice cultivation, and one can expect it to remain so. Gene flow was related to proximity; in Korea the short-grain weedy rices were related only to the short-grain japonica types. Longer grain weedy types were similar to longer grain indica types grown in the south.[1024] Similar results were obtained in China using simple sequence repeat markers.[182] Many of the wild relatives of rice are somewhat weedy.[1102] It was a surprise that some feral rices in the United States, far from the center of rice origin, were related to perennial *O. rufipogon* and the annual *O. nivara* progenitors of rice that are conspecific with it.[1104] How they arrived in the United States is open to question. These weedy accessions were found near breeding stations where, in the past, breeders had used such strains to increase diversity

Table 18. Molecular Genetic Characterizations of Feral Rice—Selected Examples

Molecular method	Locale	Relatedness	Reference
Allozymes and morphometry	Asia	Indica-like mimics, indica-like self-propagating, and japonica-like self-propagating types	656
RFLP	Korea	Both indica- and japonica-like forms found	216
RAPD	Asia	Forms of indica and japonica and intermediates	1024
SSLP	USA	Mainly indica but japonica-like and *Oryza nivara* and *Oryza rufipogon* found	1104
SSR	USA	Distinguish domestic from groups of red rice, and hybrids of domestic and red rice	377
SSR/AFLP	France	Map weediness genes related to *O. rufipogon*	150

Abbreviations: RFLP, restriction fragment length polymorphism; RAPD, randomly amplified polymorphic DNA; SSLP, simple sequence length polymorphism; SSR, simple sequence repeat; AFLP, amplified fragment length polymorphism.

for disease resistance in rice. Further analysis will be needed to ascertain whether the wild rice species have introgressed genes from cultivated rice, information that is needed as part of risk analysis.

Most importantly, by using controlled crosses between cultivated rice and feral rice it was possible to distinguish hybrids by using molecular techniques.[376,640,1103] This is of utmost importance as it allows both the establishment of a baseline frequency of hybridization before transgenic rice is cultivated and before a strong selection pressure exists, which could favor hybrids carrying the novel transgenic traits of rice. Of course, multitudes of laboratory studies with transgenics have reinvented the wheel; they show that hybrids can occur, but they do not allow accurate prediction of the frequency at which they will occur in the field. Most modeling studies have predicted "slowly," and regulators have permitted commercial release. This does not match practice, as will be discussed in the section on introgression of herbicide resistance genes. Herbicide-resistant rice can be of great benefit, but only if it is used with care to prevent or mitigate gene transfer to widely distributed related feral and wild rices. More sophistication will be needed than is presently being used with transgenic rice to delay or mitigate such introgression, which can provide fitness advantages to hybrids and their progeny.

In the United States, where transgenic herbicide-resistant rice varieties were released (but not commercialized), gene flow issues were only dealt with on paper. Statements were required by U.S. regulatory authorities about the possibility of introgression during the registration process. The authorities did not deny registration based on the possibility of introgression, nor do they demand that monitoring for such introgression be initiated, nor do they require that fail-safe mechanisms to prevent or delay introgression be tested or instituted. Presumably the regulating authorities will do so after the first introgression of transgenic herbicide resistance into weedy rice is widespread and they are forced to remember that their responsibility is to farmers and the common weal, not to the marketers of herbicide-resistant rice, or to the producers of herbicides. Resistance may appear in feral rice on a large scale in the United States due to introgression, well before such rice is released elsewhere. The United States provides a large-scale testing laboratory to determine the rate of introgression, so that authorities and farmers elsewhere can learn.

12.3.2. Implications and Risk Analysis

It has already been shown that transgenic glufosinate resistance can be transferred genetically from rice to conspecific feral rice[802,938,939]; how easily this will occur in the field is unclear, as rice is predominately self-pollinated, before flower opening.[71,148,566,633,681,690] Rice is a member of a genus where there is far too little field information on natural introgression with other weedy and wild *Oryza* spp., despite considerable information in the breeder' and cytogeneticists' laboratories from trying to bring wild traits into cultivated rice. Cultivated rice *O. sativa* has an AA genome as do the many feral forms of *O. sativa*.[10,582,772] Genes readily move between the cultivated and feral forms. The perennial rice progenitor *O. rufipogon* has an AA genome and is considered a weed of rice.[508] Another weedy rice *O. officinalis* has a CC genome. The ease of homoeologous gene transfer from the AA to genome CC is unknown. Many other diploid and tetraploid wild (but not weedy) rice species bear genomes through the alphabet from AA to HHJJ, but of them only *O. latifolia* (CCDD) is an important weed in areas of Central America. Breeders have transferred new traits from many of these wild and weedy species to rice, despite the chromosomal incompatibilities.[148,582] This often requires embryo rescue and/or intermediate crosses through bridge species. The significance of this to field problems is unclear. The taxonomic differentiation among these species is also rather unclear. In many places where weedy rices are problems, it is not known whether the weeds are the conspecific feral forms or the other species.[87,230] Indeed, recent molecular studies have shown that the material in genetic resource depositories is often misclassified.

12.3.3. Assaying Introgression in the Field

Few contemporary studies dare to comparatively estimate how long it will take to have resistance introgress and predominate in field weed populations versus how long it would take resistance to evolve by natural selection versus the expected commercial lifetime of the herbicide. This can be done with marker genes that allow for quick plant-by-plant analysis for its presence. The marker that allows large-scale field epidemiology is a gene conferring resistance to a rarely used herbicide.[422] This strategy proved unnecessary because of the release of a nontransgenic mutant herbicide-resistant rice. Field data with the ALS level mutant (nontransgenic) rice dispelled the myth that the breeding systems of rice would delay gene transfer. In the first year after the release of this "Clearfield^{TM}" rice for commercial use, the researchers gathered seed from many plants that withstood the herbicide that year. The ALS target site is highly mutable (typically one in a million) and some of the feral rice plants were ALS resistant without any cultivated rice morphological traits or molecular markers.[376] Ten times more resistant types had morphological and molecular signs of being hybrids, despite an outcrossing rate of only one in ten thousand.[377,884] The hybrids were probably the result of crossing the resistant trait into feral rice, as feral rice shatters, leaving behind seed. Previously it was only possible to morphologically detect hybrids in a sea of nonhybrids. The new ability to control most nonhybrids with herbicide made detection simpler because of the strong selectable marker for discerning introgression. The hybrids may be unfit in the Haldanian sense (when the selector is not used), and had the imidazolinone herbicide not been used in the second year, the resistant hybrid might have succumbed due to competition from their susceptible siblings. Their Darwinian fitness (resistance to the herbicide selector) predominated because of the continued selection by repeated use of herbicides. This is no longer an experimental system. The imidazolinone-resistant rice has been commercialized in the United States and throughout Central America, and the gene has widely introgressed into feral forms of rice and is an increasing problem.[376,1090] That has not stopped the manufacturer from releasing the same material in Thailand. If the imidazolinone resistance had been transgenic instead of mutant, the transgene could have been coupled with mitigating genes, precluding these introgression problems. Instead, the material was marketed accentuating that it is nontransgenic. This is a classic case of doing a disservice to the long-term interests of agriculture, as well as the long-term interests of the company stockholders.

There will be far greater possibilities of transgene introgression into feral forms of rice where hybrid rice is grown. Hybrid rice cultivation is on the rise in China, and its use will surely spread because of the much higher yields, even when accounting for inputs. The male lines used to obtain the hybrids were bred to overcome the cleistogamy (self-pollination in the flower before it opens) typical of rice. Instead, their anthers protrude, shedding vast amounts of pollen, increasing the likelihood of pollinating feral rice. Thus, introgressional risks will be greatly elevated where hybrid rice is cultivated and feral rices are present.

There has been a single attempt to model what might happen by computer simulation.[676] The simulations suggest that hybrids between rice and weedy rice would be a major problem after two to four years of herbicide use if 5 percent outcrossing occurred, resulting in hybrid formation (the variation depends on the rate of predators devouring seed). At 1 percent outcrossing, the problems from hybrids would only be acute after four to seven years.[676] Alas, the actual field data, with a far lower rate of outcrossing (one in ten thousand), have demonstrated that hybrids are an acute problem after two years.[376,377,1090] No attempt was made to ascertain the effect of rotating two different herbicide-resistant cultivated rice varieties, where the rotated herbicide would kill hybrids, resistant individuals, and volunteer rice. Crop rotation previously has been found to be a better tool for delaying resistance than even the most optimistic models had predicted.[832] The effectiveness of rotations would be predicated on domestic × feral hybrids not having weedlike secondary dormancy, that is, having a seedbank that will germinate uniformly. This is unlikely, as secondary dormancy is typically dominant.

Thus, the present transgenic, herbicide-resistant rice varieties may be a very temporary answer to the feral rice problems, but cannot be envisaged as sustainable single-gene products unless containment and/or mitigation technologies are instituted. Herbicide-resistant rice can be of great benefit if used with care; farmers must delay or counter the further inevitable evolution of resistant weeds. They must also prevent the introgression of genes into conspecific feral rice. More sophistication will be needed than proposed by manufacturers for the glufosinate-, imidazolinone-, and glyphosate-resistant rices that have been extensively field tested for commercialization, because of the Darwinian fitness advantage of hybrids over their susceptible cohorts. Transgenic insect and disease resistance can also potentially provide a fitness advantage to feral rices.

More herbicide resistances are needed. One can even choose bacterial genes for inexpensive old herbicides such as dalapon, modify them for plant codon usage, and transform them into grains.[427] Even if the dalapon dehalogenase does not confer total resistance, it can lessen the delay in the planting time discussed above with nonresistant rice. Such genes have the advantage that it is harder (but not impossible) for the weeds to mimic bacteria than plants in evolving resistance.

Conversely, perhaps herbicide resistance is not needed in rice. In a long review on the needs for transgenic rice varieties from the International Rice Research Institute (IRRI) there is no mention of weeds being a constraint to rice production.[253] The only (passing) mention of herbicides is their use as selectable markers. Or perhaps IRRI is missing out on dealing with the major needs of the aging rice farmer, abrogating dealing with them to industry. Industry does not seem to believe that there is a market for, and has not commercialized transgenic herbicide-resistant rice with resistance to inexpensive generic herbicides, with fail-safe mechanisms to prevent introgression.

12.3.4. Biosafety Measures to Prevent Gene Flow

It is best to assume that if there has been a proven field movement or unassisted laboratory movement of any genes (transgenes or others) between rice and related weeds in the past, then it will also occur at some time in the future. Thus, if herbicide resistance in the weed will be a problem, then it is obligatory to consider ways to delay the transfer from crop to related species and to mitigate the effects of transfer. In some cases it is clear that introgression will happen, sooner probably than later; that is, transgene movement from rice to feral rice. In these cases, the consequences are great where the weeds exist.

Of all the systems for minimizing gene flow described in section 4.8, the following seem most appropriate for rice:

1. *Single generation transformation by seed companies.* The herbicide resistance is transformed into certified or hybrid rice seed using vacuum/pressure infiltration by the seed company of a transformed, appropriate disarmed virus. This is untested so far, and the infiltration technologies must be developed. If it works, and it is demonstrated that viruses really do not move through pollen/seed, as tested/monitored during large-scale release, it could be allowed. The present viruses available cannot carry large exogenous genetic loads, that is, usually one small gene. This could change as it would fit well with hybrid seed technology, giving a higher value to the hybrid seed.

2. *Seed-inducible systems.* An elite variety, possibly even a tall basmati or excellent landrace, is transformed with a tandem construct (or cotransformed in a manner that inserts in a single site), the following genes:

   a. *herbicide resistance* under an inducible promoter that is turned on by the seed company prior to sale to the farmer. The duration that the gene remains induced should be limited to expression up to a certain growth stage (e.g., four to five leaves), limiting the window where the herbicide can be used to early season for early grass weed control, but preventing multiple applications during the season. Such an inducer could be:

i. *chemical* (Syngenta has a patent for glyphosate resistance under a chemically inducible promoter.)[546]

ii. *stress/chemical* and also conferring stress resistance, for example, the inducer is a very low level of paraquat[1177] or salt.

iii. *physical,* for example, heat shock or cold shock.

Induction guarantees control by the seed company. If the genes move to weedy rice, or to wild rice, they would be inactive, because they are not turned on. The induced expression of the gene is to be used as the selectable marker during transformation/selection of transformants.

b. *other valuable genes,* for example, for pest or a biotic resistances, each also under an inducible promoter to preclude a selective advantage should the construct introgress into feral/weedy rice. If these are coupled with herbicide resistance, and the herbicide is used to control feral rice, genes for pest/disease resistance will not confer any advantage, as it does not help feral rice to be potentially more healthy if it is dead.

c. *transgenic mitigator genes,* which are necessary should the inducible promoter(s) on the primary gene(s) somehow mutate to constitutiveness, and there be a hybrid with the weedy types. These could include one or more of the following: dwarfing, antishattering, antisecondary dormancy, antitillering, and so on. Dwarfing would be useful for a high-value tall rice (e.g., a Basmati) and redundant (neutral) in a dwarf/semidwarf variety. Unlike the semidominant nature of natural dwarfing in a hybrid with weedy rice, the transgenic dwarfing would be dominant, limiting competitive fitness. Antishattering is probably the most important potential mitigator for rice. One major shattering gene *qSH1* has already been cloned from rice.[604] The *qSH1* gene encodes a BEL1-type homeobox gene, and a single nucleotide polymorphism (SNP) in the five prime regulatory region of *qSH1* gene caused loss of seed shattering due to the lack of formation of the abscission layer. The SNP correlated with shattering among *japonica* rice, implying that it was a target of artificial selection during rice domestication.[605] Another shattering gene *sh4* from rice has also been isolated.[644, 645] Both might be useful as mitigator genes in the proper constructs in rice, but their utility in other grass species is in question. No similar genes exist in wheat or maize, and the quite homologous barley gene *JuBel 2* does not map to a region of the barley genome controlling shattering.[650] In the dedomestication leading to ferality, the first trait to appear is a return to shattering, allowing seed to return to the soil seedbank, guaranteeing continuity of the feral line.

These strategies would also preclude a fit establishment of the genetic material in species of wild rice as well as in feral lines. One aspect of both the viral and seed-inducible technologies is that they are especially appealing for areas of intensive cultivation where farmers have not learned the value of certified seed, and continuously sow material that further contaminates their fields with weedy rice and other weeds. The magnitude of the effects of the cavalier disregard of whole regions of farmers for the use of quality seed is well described by Valverde.[1090] For this reason large areas of Central America have feral rice disseminated by heavily contaminated "informal" rice seed. The use of certified or hybrid seed coupled with mitigated transgenic herbicide resistance is the best insurance against feral weed problems.

12.4. Removing Rice Straw-Silicon Inclusions

Rice straw is clearly amenable to the lignin-modifying/reducing tactics discussed in Chapter 9 that would render the straw more digestible to ruminants or generate biofuels. Rice straw has an additional deterrent to ruminant digestibility and to utility as a combustible biofuel, silicon inclusions. The silicon content in plants greatly varies among varieties, ranging from 0.1 to 10 percent in dry weight. The difference in silicon content has been ascribed to the ability of the roots to take up silicon.[1042] Rice accumulates silicon in the stalks to levels of up to 10 percent of shoot dry weight, being the most silicon-effective accumulating crop known. The ascribed benefits of silicon accumulation in shoots include maintaining the leaf blade erect, increasing pest and pathogen resistance, counteracting nutrient imbalances and other claimed beneficial effects,[321,322,667,944] with silicon being termed a "quasi-essential" element.[322] It is taken up by active processes into cereals, with inhibitors of respiration stopping uptake.[881] It is claimed that half the paddy soils in China are deficient in silicon, and hypothesized that the best varieties for such paddies would be those that elsewhere accumulate the most silicon.[243] High silicate in nutrient media also competes and excludes arsenate from rice plants,[441] which if this translates to the paddy, could be important in parts of Southeast Asia where arsenate is an important water contaminant.

Much has been made of studies that show that silicon is effective as a preventative against rice blast disease,[334,458] suggesting that silicon is some sort of signaling molecule, as it became clear that previous hypotheses about its physically preventing fungal penetration seemed to be on shaky ground. A perusal of the data[912] on which the disease resistance hype is based tells a slightly more ambiguous story. Silicon had no effect whatsoever in protecting a blast-resistant variety from the disease on silica-free media, but (only) partially protected a blast-susceptible variety.[912] It is also claimed that silicon facilitates light capture and that it minimizes transpirational water loss.

No quantitative correlations have been found in published journal articles showing how much silicon is actually needed by a rice plant for optimal growth, but by analogy to other species and by analyzing the variance among rice varieties, it is clearly not the 10 percent present in some varieties in some conditions and is probably closer to 2 percent, if naturally or transgenically blast-resistant rice varieties are cultivated. Blast resistance, along with *Xanthomonas* resistance has been achieved transgenically by several groups, with many disparate genes,[228,746,837,945] which probably should be stacked to afford long-term resilience. Would the use of such varieties overcome any need for more than micro-amounts of silicon in rice?

This high percentage of silicon in rice is a major hindrance to ruminant digestibility, and the ash content with a high level of silicon particles during combustion would be an environmental hazard without careful engineering of either the rice or the burners.

Some progress has recently been achieved in elucidating the pathways and molecular mechanisms of silicon absorption and deposition through the use of rice mutants that are deficient in silicon uptake.[669,1045] The ability of rice roots to take up silicon is much higher than that of other graminaceous species; a kinetic study indicated that silicon uptake is mediated by a proteinaceous transporter.[1045] There is no sequence homology between rice genes and the genes of the known silicon transporter from the marine diatom *Cylindrotheca fusiformis*.[500,501] Expression of the diatom gene in a higher plant did not increase the silicon levels,[670] suggesting that it may have relevance only to diatoms. A silicon-binding protein SBP 117 with unknown function has been reported.[974] It has possible homologs in other graminae but not dicots (as determined immunologically). The silicon content of rice is under polygenic control,[243] but a single dominant gene encodes a rice silicon transporter[669] that controls silicon loading into the xylem.[670] The transporter gene was mapped to rice chromosome 2 by comparison with a mutant that did not actively accumulate silicon into the shoots. This transporter gene has just been cloned, and its sequence suggests that it is a member of the aquaporin family of genes encoding water channel proteins.[671] It is constitutively expressed in rice roots near the casparian strips of the endodermis.[671] Up-regulating the gene should provide a benefit to rice in low-silicon-accumulating varieties or in paddies where the soil contains suboptimal levels of silicon.[671] One awaits their results of using the gene in antisense or RNA interference (RNAi) constructs to achieve several otherwise isogenic transformants with different levels of silicon, and testing the transformants in many environments. Transgenic plants with decreased silicon uptake should be compared with the original variety with respect to the digestibility, lodging, arsenate uptake, as well as pest and disease resistance. It will be possible to ascertain the truly needed levels of silicon, and then transgenically reduce the quantity to a bit above those levels needed, so that the straw may be better utilized, without affecting whatever benefits that this element may have for the plant. The optimum levels of expression of the transporter gene may be different in different rice-growing areas, depending on available silicon content of the soil.

12.5. Regenerating Transgenic Rice

Regeneration from cell culture is a critical step in the production of transgenic rice. Regeneration protocols usually used model varieties such as Nipponbare (*japonica*) and Kasalath (*indica*), but leading field varieties such as IR64 in tropical countries and Koshihikari in Japan had low regeneration ability with the protocols used, precluding efficient production of transgenic plants. It was known that regeneration mainly depends on a few key genes,[626] but they had not been identified. A group in Japan has just resolved a major quantitative trait locus (QTL) governing the low regeneration ability of cv. Koshihikari.[793] They identified some QTLs that control the regeneration ability by conventional crosses of low-regeneration rice strain Koshihikari with high-regeneration rice strain Kasalath. They then isolated a main QTL using a map-based cloning strategy. Surprisingly, it encoded ferredoxin-nitrite reductase (NiR), and its low expression correlated with poor regeneration. Clearly nitrate assimilation is an important component, as the wild-type non-regenerable variety secreted about 1.2 μmol nitrite per g of medium, whereas the regenerable variety had more than forty times less, under the limit of detection, suggesting that toxic nitrite was not being detoxified, leading to a lack of regeneration. As proof that enhanced expression of this gene could greatly increase the frequency of regeneration, they used the *NiR* gene under a strong promoter as a sole selectable marker, transforming a gene into rice without typical marker genes.[793] It remains to be demonstrated that nitrate accumulation due to the lack of this gene is the reason for other varieties to be so recalcitrant. If so, there is now a quick fix to a major problem. If not, the group has shown how one might go about remedying the situation with other material.

Many other genes have been transformed into rice, too many to elaborate. A search of the ISI Web of Science database with the key words "rice and transgenic" gives one thousand six hundred hits. Even if only 5 percent are relevant, it is clear that this crop has many other needs for further domestication that are being addressed by biotechnology.

CHAPTER THIRTEEN

Tef

The Crop for Dry Extremes

Tef (*Eragrostis tef*) (sometimes spelled teff) is a traditional crop that grows very well under both drought stress and high water table vertisols in Ethiopia and Eritrea, yet it is little known elsewhere. It is cultivated on a regular basis as part of rotations on about a fourth of the arable land in those two countries, but also as an emergency crop. Farmers always have some seed in storage. When a planted crop such as sorghum fails because of drought, farmers will plant tef and still get some yield 60–90 days later. Its ability to produce a yield of grain, under the driest of agricultural conditions (it does require water; no species can grow without water) suggests that it should be considered further because of the considerable desertification in the world. Tef is the subject of a lengthy monograph,[577] which is the source of much of the background information in this chapter (unless stated otherwise). Another source is a chapter on tef in the book on the *Lost Crops of Africa*,[50] which is well written but less reliable, as it is unreferenced to primary sources.

Its cousin, *Eragrostis curvula* (African lovegrass) is both cultivated as a forage grass and is considered a weed where it has been introduced and has gone feral.

13.1. The Crop

Tef is a C_4 annual cereal with very small seeds, similar to related weedy *Eragrostis* spp. In the early 1990s tef was grown on about 1.3 million hectares, with a yield of 900–1400 kg/ha. Modeling suggests that yields could be tripled with more water.[1182] Comparative experiment station trials have indeed achieved yields of 4.6 t/ha with current varieties and 3.4 t/ha with 35-year-old varieties.[1050] Netting was required to support the plants due to lodging at these high yields, so these yields are more theoretical than practical. FAOStat[329] does not list tef among the crops of the world in their annual statistics, so more up-to-date information is hard to obtain. Tef tolerates anaerobic waterlogged conditions better than other cereals (except rice). Tef can withstand the abuse of these slick-muddy soils when wet, and the cracking clay vertisols that heave, sag and split, and break plant roots when they dry. India, in particular, has vast areas of these "impossible" soils[50] but does not cultivate tef.

The converse property, the ability to grow on very little water is more likely to be a factor in its future distribution compared with its waterlogging tolerance. Most cultivars require at least three good rains during their early growth and a total of 200 to 300 mm of water. Some rapid-maturing cultivars can obtain the 150 mm they need from water retained in soils at the end of the normal growing season.[50] Tef can also be grown in areas too cold for sorghum or maize.

13.1.1. Major Use

Tef is highly prized among Ethiopians (and their diaspora, who are a growing market) and is used to make "injera," a fermented pancake cum pasta cum pita-like, moist, stretchable spongy, bread that is the substrate on which stewlike meals of meat or legumes are served. The dough of ground tef is fermented for about four days with a natural *Lactobacillus* during which phytic acid (a compound that binds phosphate, rendering it unavailable) decreases, inorganic phosphorus, iron, calcium, and zinc increase, considerably increasing its nutritional quality.[1086] The lactic acid bacteria isolated from fermenting tef dough inhibit pathogenic *Salmonella* spp., *Pseudomonas aeruginosa*, *Klebsiella* spp., *Bacillus cereus*, and *Staphylococcus aureus*,[789] rendering the products microbiologically safe with respect to pathogens, when served freshly baked.[790] Sensory panels could not significantly distinguish differences in flavor among freshly prepared injera made from tef, maize, or sorghum.[1191] Still, tef injera keeps its soft, spongy texture for a few days, whereas others dry out. Thus, tef injera is clearly preferred over other injera types,[1191] yet much injera sold is made with mixed grains, because tef is more expensive to produce. Tef grain can also be used to make alcoholic drinks and pastries. A recent survey of farmers showed that they produce white-seeded tef (the whiter, the better price obtained) for the market place, but they prefer and grow brown-seeded varieties for their own home consumption.[106] Why the farmers have different tastes from other consumers was not made clear. Tef does not contain gluten and is thus appropriate for people having the celiac syndrome, who must not have gluten in their diet.[1190]

Unlike other African grains, tef is not attacked by weevils or other storage pests; thus, seed is stored very easily for several years under local storage conditions.

13.1.2. The Species

Tef is a self-pollinated, allotetraploid plant ($2n = 40$) (in *Eragrostis* $x = 10$) with both the stamens and pistils being found in the same floret. The degree of outcrossing in tef is very low, 0.2–1.0%. Most of the Ethiopian farmers use traditional landraces of tef. Early maturing varieties (<85 days) are widely used in areas that have a short growing season due to low moisture, or low temperature, or where double cropping is practiced. In the highly productive, major tef-producing regions, where environmental stress is not severe, local cultivars and modern varieties are cultivated. Very-early-maturing types are ready to harvest in 45–60 days, early types in 60–120 days, and late types in 120–160 days.

Quite a bit of genetic variation occurs among the various accessions of tef available in the germplasm banks (predominantly in Ethiopia) (Table 19). A larger study with a 144 heterogeneous accessions clustered them into eight groups, with significant differences between clusters,[9] providing quite a bit of diversity for breeders. The inflorescences are also quite variable (Fig. 29), with little information on preferred structure to prevent shattering, to preclude bird predation, and so on.

The nutrient composition of tef is similar to that of millet, although, in general, it contains higher amounts of the essential amino acids, including lysine, the most limiting amino acid, more than most cereals and similar to rice and oats (Table 20). A comparison with maize is presented in Fig. 30. Glutelins

Table 19. Variation in Key Tef Characters within Available Germplasm

Character	Minimum	Maximum	Mean	COV[a] (%)
Days to germination	4	12	5	13
Days to heading	26	54	37	10
Days to maturity	62	123	93	8
Days heading to maturity	29	76	56	11
Culm length (cm)	11	82	38	20
Peduncle length (cm)	7	42	19	23
Panicle length (cm)	14	65	41	17
Plant height (cm)	31	155	98	13
Grain yield/panicle (g)	0.3	3	0.9	38
Grain yield/plant (g)	4	22	8	48
Straw yield/plant (g)	20	90	41	39
Total shoot biomass/plant (g)	26	105	49	38
Harvest index (%)	7.0	38.0	17	33
Flag leaf area (cm^2)	2.0	26.0	12	52
Culm of first two internodes (mm)	1.2	5.0	3	37

Source: Condensed from Katema[576] as cited in reference 577
[a] COV, coefficient of variation

Figure 29. Variability in panicle structure of tef. Tall, loose-panicled varieties that are most commonly grown.[50] Many questions about tef remain unanswered: with 0.26 mg seeds at present, what panicle structure will be desired with increased seed size? Which panicle structure has less shattering during mechanical harvest? *Source:* From Ketema,[577] with permission of the publisher.

Table 20. Excellent Amino Acid Composition of Tef Grain Compared with Other Cereals

Amino acid	Tef	Barley	Maize	Oat	Rice	Sorghum	Wheat	Pearl millet	FAO ideal	Hen eggs
Lysine	3.6 *2.3*	3.5	2.7	3.7	3.8	2.0	2.1	2.9	4.2	6.6
Isoleucine	4.0 *3.2*	3.6	3.7	3.8	3.8	3.9	3.7	3.1	4.2	7.5
Leucine	8.5 *6.0*	6.7	12.5	7.3	8.2	13.3	7.0	7.3	4.8	9.4
Valine	5.5 *4.1*	5.0	4.8	5.1	5.5	5.0	4.1	4.5	4.2	7.2
Phenylalanine	5.7 *4.0*	5.1	4.9	5.0	5.1	4.9	4.9	3.5	2.8	5.8
Tyrosine	3.8	3.1	3.8	3.3	3.5	2.7	2.3	1.4	2.8	4.4
Tryptophan	1.3 *1.2*	1.5	0.7	1.3	1.3	1.2	1.1	1.6	1.4	1.4
Threonine	4.3 *2.8*	3.3	3.6	3.3	3.9	3.0	2.7	2.5	2.8	4.2
Histidine	3.2	2.1	2.7	2.1	2.5	2.1	2.1	2.1	—	2.1
Arginine	5.1	4.7	4.2	6.3	8.3	3.1	3.5	3.5	—	3.8
Methionine	4.1 *2.1*	1.7	1.9	1.7	2.3	1.4	1.5	1.3	2.2	3.8
Cystine	2.5 *1.9*							3.2	2.0	2.4

Source: Ketema[577] except that the values for tef that are in italics are from anonymous.[50]

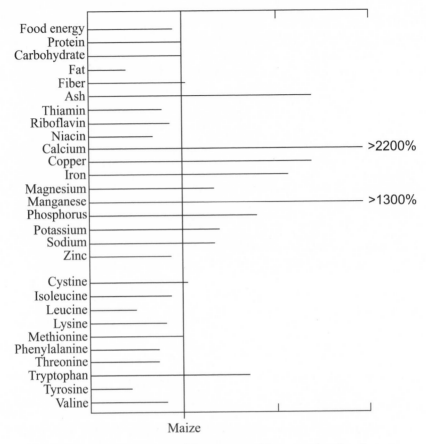

Figure 30. The quality of tef, one of the most nutritious cereal grains. *Source:* Replotted from reference 50, by permission of the National Academy of Sciences.

(44%) and albumins (37%) are the major storage proteins in tef. Tef is lower in total protein content than other wild *Eragrostis* species, although tef grown in the United States is claimed to have about 15 percent protein, not the 10 percent of tef grown in Africa. Tef seeds are tiny and thus have a greater proportion of bran and germ. Tef is almost always produced as a whole-grain flour. Because the germ and bran are consumed along with the endosperm, tef has as much, or even more, food value than the major grains.

Tef straw is used for animal feed, and has a better feeding value than wheat and barley, although the latter usually have higher straw yields. Tef straw is claimed to be 65 percent digestible by ruminant animals,[50] a very high figure, which if correct, would add considerably to its value, as this is twice the digestibility of other straws. If correct, it would also render tef straw far more amenable to bioethanol production than all other plant material tested so far.

13.1.3. Fertilizer Responses

Modern varieties of tef are more responsive to fertilizers than landraces. They have better growth and phosphorus uptake than the corresponding traditional varieties.[685] Nitrogen produces more straw, whereas phosphorus encourages good grain production but also promotes lodging. Potassium fertilization is of minor importance to tef production. Finger millet was more tolerant to salt stress in early growth than tef, which was more tolerant than pearl millet.[569]

13.2. The Domestication of Tef

Tef is assumed to have originated in northeastern Africa, probably in the western area of Ethiopia, where agriculture is precarious and seminomadal. Tef differs from most related species in its complete absence of glands. Progenitors of tef were probably eglandular plants and only eglandular forms of *Eragrostis pilosa*, a wild allotetraploid, are widespread. *E. pilosa* is often considered to be synonymous with *E. tef*, complicating the picture. Phylogenetic analysis of sequence data from the nuclear gene *waxy* and the plastid locus *rps16* strongly supports the widely held hypothesis of a close relationship between tef and *E. pilosa*. *Eragrostis heteromera*, another previously proposed progenitor, is deemed by the waxy data to be a close relative of one of the tef genomes,[522] so tef may well be an allotetraploid with two sets of parents. The level of polymorphism among the wild *Eragrostis* species was extremely high,[77] while low polymorphism was detected among tef accessions using randomly amplified polymorphic DNA (RAPD) markers.[83] This loss of diversity is typical of domestication.

13.3. The Constraints to Wider Cultivation

13.3.1. Photoperiod

Tef is sensitive to day length, requiring approximately twelve hours of daylight. Cultivation was nigh impossible in short-day conditions in British greenhouses. This would prevent the cultivation of tef as a winter crop in dry areas with annual winter rains, such as the arid subtropical Mediterranean and Middle East where cereals are grown in the winter, the wheat-growing areas of India, where irrigation for wheat is becoming a problem, and wheat-growing areas of Australia that often suffer from drought, and where tef could be an appropriate crop, also in the winter. Short-day varieties would be needed in such climates and could possibly be generated transgenically.

13.3.2. Weeds

Weed competition causes about half the crop to be lost, but with hand weeding (even at suboptimal times), crop loss is only 8 percent. Even when nonselective herbicides are applied before plowing (which unfortunately do not control the perennial weeds), an additional weeding (hand or chemical) is required at early tillering to increase yield. However, if weed infestations are high, a second weeding is done at the stem-elongation stage. No selective graminicide has been reported to be effective for the selective control of grass weeds (or sedges) postemergence. The grass weeds *Digitaria scalarum* and *Phalaris* spp. are important alternate wild hosts of red tef worm, which is another reason weeds should be controlled. Thus, a need exists for a resistance to a postemergence graminicide for tef.

13.3.3. Small Seed Size

The thousand-seed weight of tef is only 265 mg, that is, 0.26 mg per seed, one-sixtieth the size of some maize seeds. This has implications at many levels:

1. The farmers' traditional practice is to hand broadcast tef at the rate of 40–50 kg/ha, yet about 15 kg/ha is considered sufficient in Africa, but it cannot be broadcast evenly by hand or most machines. A U.S. tef grower uses only 2.5 kg/ha. At that low rate, the farmers are planting nine million seeds per hectare, yet the stand is much less. Thus, stand establishment at all rates is a major problem. Indeed, tef fields in Africa appear very patchy.
2. Most combine harvesters and other mechanical threshing equipment cannot distinguish between tef seed and trash, and much seed is lost with the chaff.
3. Because much seed is lost with the chaff, tef can be a volunteer weed in the following crop.

4. Handling and transporting tef seed is a problem because the seeds tend to fall through any crack.

Thus, there is a dire need to increase seed size more than the small increment that has occurred in 35 years.[9]

13.3.4. Lodging

It is currently not advisable to use higher rates of fertilizer to increase yield because tef heavily lodges in response to fertilizer. Lodging also exacerbates tef rust.[256] Thus, there is a major need to decrease lodging. Breeding has not solved this problem, but it is being addressed genomically. Still, data cited in the next section suggest that linkage problems may exist that will have to be broken.

13.3.5. Harvest Index

Green revolution crops have been bred to have a harvest index in excess of one half, which far exceeds the maximum of 38 percent known within the variability of tef. Despite the purported excellent digestibility of tef straw, it is still worth enhancing yield through increasing the harvest index, which in the green revolution cereals went hand in hand with lodging resistance.

13.3.6. Shattering

Harvesting must be performed before the plants become too dry to prevent losses from shattering, yet drier harvesting is better for storage. Shatter resistance is necessary.

13.3.7. Insects

Welo bush-cricket (*Decticoides brevipennis*) is a major pest that exists only in Ethiopia with the early instars as flower feeders. Central shootfly (*Hylemya arambourgi*) and red tef worm (*Mentaxya ignicollis*) are also problematic. Insecticides do control these problems, but resistance to bush-cricket would be advisable to prevent its spread outside of Ethiopia, as well as within Ethiopia, to preclude the need to spray insecticides.

13.3.8. Diseases and Mycotoxins

Some authors state that diseases are not a serious problem; tef suffers less from diseases than most other cereal crops in the major production areas of

Ethiopia. Still, damping-off caused by *Drechslera poae* can be severe, especially with higher seeding rates and early sowing. Tef rust (*Uromyces eragrostidis*) and head smut (*Helminthosporium miyakei*) have been reported as the most important diseases on tef. Tef rust can cause yield losses between 10 and 40 percent of the crop.[256] None of seven thousand accessions and mutants tested was resistant to the disease, but twenty-two landraces had a modicum of tolerance to the disease.[256] The tolerant landraces were all early maturing,[256] a trait not desired by most large growers.[106]

Although mycotoxins are a major problem in other African grains[431] (Chapter 7), debilitating the population with cancers and malnutrition caused by liver damage, this is probably not the case in tef. The major vectors, stem borers and grain weevils, seem not to be pests in tef, and the fungi that produce mycotoxins, *Fusarium* and *Aspergillus* spp. are not listed among the pathogens of tef. Indeed, *Aspergillus* and aflatoxins were far more prevalent in samples of maize and sorghum in Ethiopia than in tef.[2] More current and direct evidence should be obtained on a large scale, as this could be an added reason to consider tef, which along with fonio[549] (Chapter 12) may be the sole African crop that does not need resistance to mycotoxin-producing pathogens.

13.4. Breeding Tef

Breeding tef is not easy because of the small size of the tef floret, its autogamous nature, and its unique pollination habit, and because pollination naturally occurs within an hour after 6:45 a.m. in Ethiopia. Techniques for delaying flower opening were developed to control pollination and to effect hybridization whenever required during the late hours of the day by putting potted plants overnight at a cold temperature or in darkness. The process of initial flower opening to anther dehiscence takes about thirty to forty minutes and once pollination is effected, the pollen grains take only three to four minutes to germinate on the stigmas. The female plants must be emasculated under a binocular microscope as soon as the flowers start to open and before the anthers dehisce. Various genetic markers are available to ascertain that crossing was affected.

The considerable variation within the species (Table 19) suggests considerable leeway for breeding for many traits. Yield per panicle and panicle weight are strongly correlated with grain yield per plant and have high heritability compared with grain yield per se. Grain yield per panicle and shoot biomass per plant negatively correlate with harvest index and positively correlate with most of the remaining traits in breeding experiments with three thousand lines representing sixty germplasm populations. Individual plant grain yield positively correlated with all the other traits except harvest index.[67] Thus, conventional breeding has not led to the vast improvement in harvest index that could be predicted from the green revolution dwarfing of other cereals. There seems to be no information on the genetic variation in seed size, which is a major constraint to mechanized cultivation of tef.

Direct selection within landraces, mutation breeding, and intraspecific hybridization were tried for developing lodging-resistant varieties, with no success. Lodging positively and significantly correlated with grain yield and shoot biomass in a recent study of thirty tef genotypes.[516] This suggests that means other than breeding are necessary to obtain high-yielding, lodging-resistant varieties, because breeding cannot separate yield from the propensity to lodge. While it is expected that dwarfing will reduce lodging, it is also expected that this will be hard to attain genetically, where dwarf is recessive. Recessive traits may be hard to obtain in an allotetraploid such as tef, as possibly all four alleles would have to be recessive.

Osmotic adjustment was significantly correlated across tef genotypes with delayed wilting and the maintenance of higher relative water content under conditions of soil moisture stress, but there was no association between root depth and osmotic adjustment among genotypes.[78]

No studies have identified specific economically important traits such as tolerance to diseases, drought, waterlogging, or low temperature in wild *Eragrostis* species, and thus no attempts have been made for large-scale interspecific hybridization programs.[577]

13.4.1. Genomics-Based Breeding

A genomics-based project has been proposed to decrease the lodging problem,[111] where the authors will first genotype the core collection of 320 diverse tef accessions for 50 expressed sequence tags/simple sequence repeat (EST/SSR) markers, 150 amplified fragment length polymorphism (AFLP) markers, and the homologs of wheat *rht*1 and rice *sd*1. The *rht*1 and *sd*1 homologs of the twenty most diverse accessions are to be sequenced for investigations of linkage disequilibrium and population structures. The most promising varieties, especially those with superior lodging resistance, will be appropriate for further breeding.

ESTs from rice and maize that are associated with the intermediates of the lignin biosynthetic pathway are being placed on the framework map of the tef chromosomes.[781] This is being done to assist in breeding for resistance to lodging under the assumption that stem toughness is a function of lignin content, an accepted assumption that has no evidence to back it up, only evidence to the contrary[1073] (Chapter 9). Even if enhanced lignification would prevent lodging, it will also strongly decrease the feed value of the straw. Thus, ways other than increased lignification should be found to reduce lodging. The best way to decrease lodging is probably by lowering plant height by shortening the internode length of stems.

13.5. Biotechnological Solutions

According to the text above and the summary in Table 21, tef has a complex of problems, some of which have not been amenable to breeding, and others where transgenics might provide a quicker fix than breeding. There is still considerable variability in tef that will facilitate breeding for traits where the genes exist.

13.5.1. Engineering Tef

As a prelude to any genetic engineering, it must be possible to regenerate tef from tissue culture. Three groups have reported a modicum of success.

1. Seed-initiated callus led to embryogenic tissue appearing on soft and amorphous callus, which developed into somatic embryos during a subsequent subculture. Most of the calli producing somatic embryos converted into plants. Regenerated plants were successfully transferred to soil. Neither chlorophyll-deficient plants nor morphological variants were found among regenerants, suggesting that somaclonal variations were not induced by the regeneration system. All regenerated plants were fertile.[568]

2. Another group induced callus on immature leaf bases explanted from one-week-old seedlings of four tef genotypes. They were able to obtain a high frequency of direct somatic embryogenesis using dicamba as an auxin. Plants from all four genotypes were grown to maturity.[718]

3. The third group achieved in vitro somatic embryogenesis and plant regeneration from root and leaf explants and from seeds from eight

Table 21. Constraints to the Cultivation of Tef

Factor	Constraint	Possible solution
Agronomic		
Photoperiod 12 h, LD	Need shorter photoperiod to grow in warm winters	Modify phytochrome genes
Seed size	Much too small	?
Yield	Low	Increase harvest index by dwarfing
Harvest index	Maximum 0.38 recorded is suboptimal for cereals	Increase harvest index by dwarfing
Lodging	Major problem, linked to yield	Dwarfing has helped with rice and wheat
Tillering	If seed size is enhanced and stand is increased, fewer tillers may be valuable	Consider when more information is available; auxin genes may assist in reducing tillers
Fertilizer response	Causes lodging	Reduce lodging (see above)
Mineral utilization	No problems discussed	
Mineral deficiency	Not known	
Plant protection		
Insect	Crickets	Bt, insect venoms
Disease	No major problem at present	Watch developments in other grains
Weed	Need postemergence graminicide	Glyphosate, glufosinate, or ALS[a] resistances should rectify
Bird	Unknown	Change morphology if a problem
Quality		
Mycotoxins	Seems not a problem	Follow further
Flavor	Seems not a problem	
Animal nutrition	Little information on animal feeding with grain	
Human nutrition	Seems not a problem	

Abbreviations: LD, long day; ALS, acetolactate synthase.

tef genotypes using (preferably) 3,6-dichloromethoxybenzoic acid or 2,4-dichloromethoxybenzoic acid as the auxinic hormones. Unlike leaf-induced callus, the root-induced callus was not significantly different among genotypes and treatments. Regeneration from leaf callus was better than from root callus.[105]

Reports of stable transformation of tef have yet to appear, but one could assume that the methods that have been so successful with rice and other small grains can be tweaked to work for tef, with one of the preceding regeneration systems.

The problems of linkage described in the previous sections suggest that a transgenic approach may well be easier than breeding to further domesticate tef. This is further accentuated by the polyploid nature of tef where many of the needed native genes would have to be in the recessive form; quite a feat to achieve in a polyploid. The foremost issue seems to be the lack of sufficient variability within the species, in photoperiodic requirement, in seed size, in harvest index, in insect resistance, and in herbicide resistance. In these days when many grasses can be transformed biolistically, and multiple traits can be introduced in tandem without splicing in advance,[599] it seems wise to simultaneously engineer as many traits as possible. It would probably be advisable to start with two groups of elite lines: a group of very short season lines for the arid areas and a series of long season elite lines for the wet waterlogged areas. Which of the following genes are chosen depends on the precise area where the final varieties are to be cultivated.

13.5.2. The Genes for Immediate Consideration

13.5.2.1. Herbicide Resistance as a Selectable Marker and Agronomic Necessity

The *bar* gene for glufosinate resistance, either of the genes for glyphosate resistance, and many of the *ahas* genes for imidazolinone or sulfonylurea resistance would allow postemergence control of all weeds. The choice of genes depends on the weed spectrum and the cost-effectiveness of the relevant herbicides in that area. The bromoxynil resistance gene would not be appropriate as the herbicide does not control grass weeds, the major problem weeds in tef. If parasitic *Striga* species continue to evolve to be a problem in tef, it will be necessary to use only target site resistances, together with gene flow fail-safes to prevent movement to other African *Eragrostis* species, to deal with these parasitic weeds. Other genes may be available in the future.

Probably the best gene for herbicide resistance would be EPSP synthase target site resistance to glyphosate, and not just because of its efficacy for killing weeds, and the low cost of generic glyphosate. Glyphosate proved to be somewhat effective in controlling both soybean and wheat rusts[43,340] on glyphosate-resistant crops; there is good reason to believe that it would do so with the tef rust pathogen as well. It is claimed to be efficacious for a month,[43] which is much better than topical fungicides and is equivalent to many systemic fungicides. The gene that confers metabolic resistance to glyphosate[194] would probably not allow this residual systemic effect, as glyphosate in the tissue would be degraded. This must be checked, though, as 80 percent of applied glyphosate remains on leaf surfaces and does not get in initially.[340] What happens to this herbicide over time is an open question. Perhaps it is the residual surface herbicide that controls rust.

13.5.2.2. Seed Size

First it must be asked, what might the implications be from increasing seed size. Is it the present minuscule seed size that precludes damage by grain weevils because each seed is too small to harbor a single larva? Is it the lack of vectors that keeps aflatoxin levels low[2]? Tef is already a tetraploid; will making it an octaploid increase seed size? The problem of mechanically harvesting the small seeds has been addressed by agricultural engineering. The U.S. grower has modified combine harvesters that do not pass the tef seed as weed seed. One variety DZ 01 99 had seeds a statistically significant 30 percent larger than the mean, but also yielded significantly less than the highest yielding variety DZ 01 974, which was also significantly the tallest and had the most days to heading.[1050] Basically, 35 years of breeding did not greatly increase seed size, and the only correlation with the increased yield was increased biomass.[1050]

13.5.2.3. Dwarfing

Presumably tef is like most grains and inhibiting the biosynthesis of gibberellic acid will cause dwarfing. The utility of choosing this group of genes can quickly be elucidated in pots; spray plants cultivated with high fertilizer levels at different ages with plant growth regulators that inhibit gibberellic acid biosynthesis (paclobutrazol, chlormequat, ancymidol, etc.) and see if they cause dwarfing while simultaneously increasing the harvest index and yield. Field trials with the best rates of application from the pot experiment that dwarfed the plants will be needed to ascertain if indeed there is a tolerance to lodging. We have seen no record of such experiments with tef. If the experiment above shows utility, the *GA20ox* gene from rice could be chosen,[811] which was found to be responsible for the green revolution dwarfing of rice, decades later. A consensus sequence from that gene and/or its tef orthologs could be introduced into tef in RNA$_i$ constructs.

13.5.2.4. Reducing Tiller Number

An overexpression of auxin (indole acetic acid) biosynthesis genes in the shoot apical meristem should lessen the number of tillers and provide a more stable plant with more uniform ripening. It is possible to use exogenous (bacterial pathway) genes *iaa*H and *iaa*M that are less likely to interfere with normal processes.

13.5.2.5. Insect Resistance

Two groups of genes are known to control Orthoptera, such as crickets. *cry* genes of *Bacillus thuringiensis* have been widely used to control many insects, including crickets. It is just because this gene group is so widely used, and because of the necessity to enhance biodiversity (and possibly delay the evolution of resistance by the insects), that it could be advisable to have genes encoding insect-specific (no mammalian toxicity) protein toxins from spiders and scorpions.[661,773,774] These genes have undergone considerable toxicological testing by independent academic groups, and are ripe for use for such purposes. They have the advantage of not being Bt, which is far too overused.

13.6 Biosafety Considerations

13.6.1. Outcrossing to Wild Eragrostis Species

Breeders had attempted to cross tef with wild *Eragrostis* species and were not successful in obtaining fertile hybrids with *E. curvula* (African lovegrass, the forage crop and also weed) or with tetraploid *E. cilianensis, E. pilosa,* or *E. minor,* suggesting barriers to gene exchange[1048] as cited in reference 577. Still, there are many other wild *Eragrostis* species where this has not been ascertained. These results are in a thesis that was not subjected to journal peer review, so it is necessary to consult further with local breeders and biologists on the risks involved, or experimentally demonstrate that a transgene is unlikely to establish in nearby weedy *Eragrostis* species.

13.6.2. Gene Flow Modulation

If dwarfing and preventing seed shatter are among the traits successfully engineered into tef, and they are engineered in tandem with other primary traits, they should serve as mitigation genes, preventing untoward risks of the outflow of genes and their establishment. Still, the fitness of offspring should be compared with wild type, as outlined in section 4.2.

CHAPTER FOURTEEN

Buckwheat

The Crop for Poor Cold Extremes

Common buckwheat (*Fagopyrum esculentum*) is a temperate crop that has persisted through centuries of civilization. It has a very short growing season and is thus often grown in areas with a short frost-free period, or as a second crop. It can be grown as an emergency crop after the first crop fails, producing satisfactory yields. *Fagopyrum tataricum*, or Tartary buckwheat, is also cultivated in many areas of the world but in general is consumed or traded locally.

The buckwheats are one of the few noncereal, nonlegume grain crops. They are broadleaf plants, members of the Polygonaceae, with an erect annual single main stem and a branching habit. The plants generally grow from 60 to 130 cm tall. The fruit is a triangular nut, sometimes winged. Both types of buckwheat have an indeterminate photoperiodic response. These species have many cultivars or landraces. Insects are attracted to the flowers of cross-pollinating buckwheat because of the nectar secreted by the glands at the base of the ovary, but they are also wind pollinated.

The main producers of buckwheat are China and Russia and its surrounding republics as a crop that grows on marginal, unproductive land. It is a subsistence crop in mountainous areas where at the higher altitudes Tartary buckwheat is preferred because of its frost tolerance. In many areas the trend is for replacement of common buckwheat with higher-yielding, frost-tolerant finger millet or other crops. Tartary buckwheat production in most areas appears to be remaining constant. This crop was extensively reviewed by Campbell,[180] who is used for pre-1997, unreferenced background information.

Buckwheat has a high level of aluminum tolerance due to external root excretions of oxalic acid[1195] and internal precipitation of aluminum by oxalic acid in the cell sap.[668] This is important for many of the marginal, acid soil areas where buckwheat is grown. At very high aluminum concentrations, induced higher levels of citrate also complex with this toxic cation.[972]

14.1. Major Uses

The seeds (achenes) are used as food for humans and as animal or poultry feed. The dehulled "groats" are cooked as a porridge, and the flour is used in the preparation of pancakes, biscuits, noodles, cereals, and so on. The protein of buckwheat is of excellent quality and is high in the essential amino acid lysine, with less glutamic acid and proline and more arginine, aspartic acid, and tryptophan than cereal proteins. The amylose content in buckwheat granules varies from 15 to 52 percent and its degree of polymerization varies from 12 to 45 glucose units, which is quite different from cereal starches. The digestibility of buckwheat protein is rather low, probably because of the high (18 percent) fiber content in buckwheat, which may be desirable in some parts of the world. Buckwheat fiber is free of phytic acid, a highly desirable character, and is partially soluble. Buckwheat is gluten free and is thus edible by people with celiac syndrome.[269,1190]

The buckwheats contain many secondary metabolites (flavonoids, especially rutin), flavones, phytosterols, vitamins, minerals, and proteins that place it in the category of a "functional food,"[649] which could make buckwheat more interesting in the high-end markets. Likewise, it has been proposed to market buckwheat sprouts that are claimed to have a soft and slightly crispy texture, and attractive fragrance.[584] They have an unusually high (more than 80 percent) unsaturated fatty acid composition. They have relatively high levels of the nutrients lysine, γ-amino-n-butyric acid (GABA), and sulfur-containing amino acids as well as rutin.[584]

Buckwheat straw contains allelochemicals, and when spread at one to two tons per hectare in rice paddies, there were "significantly" lower levels of some major weeds, with wild perennial types being more effective than Tartary, and Tartary more effective than common buckwheat.[1165] Despite the authors' optimism, this seems impractical, as straw from about one hectare of buckwheat would be required for one hectare of rice. The transportation alone would be more expensive than using herbicides, and herbicides would be more effective.

14.2. The Domestication of Buckwheats

F. esculentum subsp. *ancestralis*, a self-incompatible outbreeder that had been thought to originate from northwest Yunnan China, is the wild ancestor of common buckwheat. More recent studies have pinpointed the origin to the Sanjian region of eastern Tibet, not far from Yunnan.[606] Buckwheat was cultivated in China as early as the second to first centuries BCE, which is not considered to be very ancient. It spread to Europe via Russia in the Middle Ages, probably from Siberia, reaching Germany early in the fifteenth century. There are at least fifteen *Fagopyrum* species of buckwheat but only two are utilized as food or feed. Tetraploid wild buckwheat (*Fagopyrum cymosum*) is used on a sporadic basis as a green vegetable and cattle forage.

Truely wild Tartary buckwheat is found in Sichuan China, Tibet, Kashmir, and northern Pakistan. One wild group is identical with cultivated Tartary and is widely distributed. Another wild group, found only in Sichuan, is probably an older form. A weedy form of Tartary buckwheat with the morphology similar to cultivated Tartary buckwheat, but having characteristics of wild species including shattering ability and strong dormancy, was found in northern Pakistan.[810] Its properties suggest that it is a domestic form gone feral. The weedy form may also be a hybrid between wild Tartary buckwheat and the cultivated form.

14.3. The Constraints to Wider Use

There is considerable genetic variability in many key growth factors relating to buckwheat (Table 22). The constraints that do not appear to be easily amenable to standard breeding are discussed in section 14.5.

Table 22. The Genetic Variability Available to Breeders in Buckwheat[a]

Character	Range	Mean	CV	SE
Plant height (cm)	60–181	120	25	
	25–116	*70*		*0.96*
Number of branches	1–6	3.2	35	
	1–14	*4.5*		*0.13*
Number of internodes	6–28	13	23	
Number of leaves	10–45	20	35	
Number of leaves on main stem	*2–18*	*8*		*0.11*
Leaf length (cm)	2.8–8	5	23	
	1–8.2	*5.4*		*0.05*
Leaf width (cm)	2.1–8.9	5	25	
	3–11	*5*		*0.05*
Days to flower	24–78	43	37	
Days to 50% flowering	*26–45*	*28*		*0.16*
Number of flower clusters	*1–6*	*4*		*0.05*
Days to maturity	75–125	97	17	
Days to 95% maturity	*67–98*	*80*		*0.37*
Grain-filling days	*33–69*	*52*		*0.37*
Number of seeds per cyme	1–7	2.5	24	
	7–50	*20*		*0.4*
1000-seed weight (g)	12–50	30	28	
	10–32	*21*		*0.22*
Yield per plant (g)	2.3–20	9	47	
	0–9	*1.6*		*0.08*
Powdery mildew (0–9)[b]	*0–9*	*6.4*		*0.15*
Downy mildew (0–5)[b]	*0–5*	*0.4*		*0.05*

Source: Collated from data in Campbell[180] from data of Joshi and Paroda[554] and Baniya et al.[89] for Nepal.

[a]Data from more than three hundred accessions each in India and *Nepal (in italics)*.

[b]The scale is 0–9, where 0 = no disease and 9 = 100% disease.

14.3.1. Fertility

Buckwheat has two problems relating to fertility. Probably the most important problem is abscission of 50 to 90 percent of the flowers of common buckwheat. Many major crops, including those that are wind pollinated, are actually self-fertile (e.g., wheat and rice), and cross-pollinating buckwheat is an anachronism, as even some of its wild interbreeding relatives are self-fertile.

14.3.1.1. Floral Abortion

Except for the first flush of flowers, most of the flowers on buckwheat abort. Neither Tartary buckwheat nor *Fagopyrum homotropicum* abort in this manner and possibly interspecific breeding will lead to greater seed set. The factors causing floral abortion are not fully understood, despite being studied for more than thirty years, and no genetic variability has been found in this trait. An in-depth attempt was made to assess the reasons for lack of fertility,[1049] suggesting that if the situation can be overcome, yields can be greatly increased. Competition for nutritional resources on the same raceme was ruled out. An elegant recent study addressed the question of whether inadequate photosynthates are available to support later flushes of flowering, causing abortion. They used partial defoliation and partial removal of flowers.[450] They too concluded that the photosynthate source was fine, but the sink (the flowers) was defective.[450] Postfertilization abortion of the fertilized embryo was also ruled out. Careful anatomical studies provided evidence that the problem is a lack of fertilization because pollen tubes fail to penetrate the micropyle.[1049] Indeed, there is an early one-week period where most flowers become fertilized, and only a few later-forming ovules are fertilized.[1049] The floral abortion problem has not been rectified in a few thousand years of selection.

14.3.1.2. Self-incompatibility

It will probably be easiest to genetically correct the problem of self-sterility by bringing genes in from relatives by breeding. Hybridization of Tartary buckwheat with common buckwheat initially required embryo rescue and tissue culture regeneration from embryo-derived callus, yielding sterile plants. Later, fertile hybrids between these and other species were saved by embryo rescue.[809] Fertile interspecific hybrids of diploid buckwheat with *F. homotropicum* were obtained by controlled pollination and embryo rescue.[1163] The progeny have been backcrossed to *F. esculentum* to transfer the ability to self-pollinate and be frost tolerant, but unfortunately also carried severe seed shattering, which must be bred out. *F. homotropicum* has been reclassified as *F. esculentum* var. *homotropicum*.[210] Such hybrids are much easier to obtain at the tetraploid level than at the diploid level. Hybrids between diploid buckwheat and tetraploid *F. homotropicum* gave rise to a few $2n = 2x = 24$ progeny that then had their chromosomes doubled with colchicine. Many of the subsequent progeny were hexaploid, but some were diploid, with considerable variability introgressed from the wild that might be useful in future breeding.[1127] *F. tataricum* by *F. homotropicum* interspecific hybrids are being used as such a bridge for further hybridizations with other species.

14.3.2. Determinacy

Buckwheat seeds mature over a long period because of the indeterminate flowering habit. This wastes resources on producing flowers, most of which abort (see above) and many of the rest will never mature fruit by the end of the season, and a true determinate growth pattern might increase yields, as well as have other effects (see following). There are somewhat determinate lines of buckwheat, but their growth habit in experiments (with nonisogenic material) showed little effect on ripening.[366]

14.3.3. Photoperiod

The indeterminate nature of most varieties suggests that they flower and set fruit after a fixed time. This property is detrimental when buckwheat is planted as an emergency crop later in the season, as the crop may not be mature before frost. A fixed day-length requirement that guarantees flowering at a specific time in mid or late summer (depending on the climate) would be far more desirable. Some varieties have specific day-length requirements, which vary with growth stage,[726] suggesting that the variability needed for breeding this character may exist.

14.3.4. Shattering

Seeds mature over a long period due to indeterminate flowering. If harvest is delayed, early-formed seeds can shatter due to wind, even though no abscission layer is formed; the seeds are broken loose by winds. Determinate varieties might alleviate the problem.

14.3.5. Frost Tolerance

The lack of frost tolerance is often a major constraint. Little variability to frost damage is found among accessions. The frost tolerance found in *F. tataricum* and *F. homotropicum* brought in through interspecific hybrids is being ascertained.

14.3.6. Lodging

Lodging is considered to be an important constraint. This would probably be exacerbated under high fertilizer regimes or with fuller seed set.

14.3.7. Allergens

Buckwheat can be extremely allergenic to some individuals with a cross allergenicity to latex and many fruits,[95] and in some cases the seeds cause hypersensitive reactions, which in extreme cases can lead to anaphylaxis. About 5 percent of Koreans react positively in skin tests.[1002] Buckwheat husks are often used in pillows, and the small amounts of flour there can cause allergic reactions or asthma.[679] The development of hypoallergenic buckwheat would make this important pseudo-cereal available to allergic people.[1184] Animals that eat the flowers and seeds of buckwheat have been know since the beginning of the twentieth century to be exceedingly photosensitive, and the syndrome termed fagopyrism, is caused by a compound termed (can you guess?) fagopyrin.[1055] The syndrome was not reported in humans, suggesting that the active compound is heat labile and lost during cooking, and as few farm animals graze in buckwheat fields, the problem is not acute, except in production of the nutraceutical rutin or with cosmetic extracts from buckwheat. Extraction procedures have been elucidated that minimize the amount of fagopyrin, while maximizing rutin and other components.[503] Those who have proposed that people should eat fresh buckwheat sprouts[584] did not report the fagopyrin levels in the sprouts.

14.3.8. Protease Inhibitors

Low-molecular-weight protein inhibitors of serine proteases belonging to the potato protease inhibitor I family have been isolated from buckwheat seeds.[1078] They are relatively thermostable, which could prevent digestion of proteins from the seeds or flour. Still, the extended times of cooking of groats probably inactivates most proteases, although products with the flour should be checked. Their levels in the fresh buckwheat sprouts have not been reported,[584] and may indeed be a problem. No mention was made of postharvest storage insect problems in buckwheat. This could either be due to the temperate/cool climates where it is grown or due to the protease inhibitors, which are known in other species to be insect antifeedants. If the latter is the case, then it is possibly better to retain the protease inhibitors and then adequately remove them by extensive cooking.

14.3.9. Bitterness

Tartary buckwheat has a bitter taste that develops in the flour due to the activity of a flavonol 3-glucosidase, which degrades the desirable rutin to the bitter quercetin and other unknown compounds. Rapid gel electrophoresis methods were developed to discern the isozymes of these enzymes for screening deletion lines for where the glucosidase is lacking.[1033]

14.4. Breeding Buckwheat

Buckwheat is one of the few crops that are self-incompatible. Seed production in common buckwheat depends on cross-pollination between "pin" (long pistil, short stamen) and "thrum" (short pistil, long stamen) flowers. Flower forms with reduced style length were found, and they developed into self-fertile homomorphic lines, some of which were especially adapted to self-pollination, as the flowers have equal pistil and stamen heights. Unfortunately, the introduction of this character into other buckwheat lines resulted in severe inbreeding depression, probably due to deleterious recessive genes linked with the thrum gene, because this gene never occurs in the homozygous state. Breeders wish to develop self-pollinating buckwheat to ease in selection and to find spontaneous recessive mutations that are hidden by cross-pollination. Because so much of crop domestication is based on selection for recessive traits, this is not a wonder.

14.4.1. Interspecific Hybrids

Interspecific buckwheat hybrids have been obtained on occasion. *F. cymosum* × *F. tataricum* hybrids and *F. cymosum* × *F. esculentum* hybrids have been produced at the tetraploid level. *F. cymosum* and *F. tataricum* are closely related. *F. tataricum* is thought to have speciated from the perennial *F. cymosum* in the Tibet-Himalayan area.[1168] *F. esculentum* × *F. homotropicum* hybrids are fertile at the diploid level, as expected with their new classification of being conspecific.[210] *F. esculentum* × *F. tataricum* hybrids have been made at the tetraploid level. The barrier to hybridization between these two domesticated species was overcome by selecting for a line of common buckwheat that had an identical isozyme pattern for three alleles.[937] Even then, embryo rescue and callus regeneration were required, with less than 1 percent recovery.

It had been thought that the S supergene governs self-incompatibility, flower morphology, and pollen size in buckwheat. A series of crosses recently demonstrated that the short style length needed is controlled by multiple genes outside the S supergene.[699] Further crosses between the pin flowering type buckwheat and a homostylic (self-compatible) *F. homotropicum* whittled the number of genes down to two complimentary dominant genes that control self-compatibility in the wild species.[1129] The pin types are easier to cross, so if homostyly can be achieved in buckwheat it would be the preferred type. The breeders wish to bring in cold tolerance and better seed set from the wild, with a much a more rational breeding system than can be done with a self-incompatible species. From an evolutionary point of view it remains a conundrum why our ancestors chose this yield lowering, more complicated breeding system during domestication, while the progenitors had it so simple. There must be some linkage problem that the breeders may soon find that can only be overcome transgenically.

14.4.2. Dwarfing

Dwarfing that often leads to lodging resistance with a higher harvest index is usually a recessive trait in buckwheat, but a new gene seems to have been found that is semidominant.[731] Two of the youngest internodes are compressed (numbers 2 and 3), and dwarfing was unaffected by self-shading (planting density), nitrogen fertilizer regime, and photoperiod, unlike with tall cultivars. There have been no further publications about this or other dwarfing genes, suggesting that linkage problems may exist.

14.4.3. Antishattering

The two dominant complimentary *SHT*1 and *SHT*2 genes are responsible for the brittle pedicel that easily breaks, shattering the seed in buckwheat. These genes are linked on the same chromosome at a distance where 5 percent of the offspring segregate.[698] Most cultivars are recessive at *sht*1, which has been mapped by amplified fragment length polymorphism (AFLP), with two close nonrecombining primers. These were converted to markers according to sequence, for selection of nonbrittle plants.[700] Crosses with *F. homotropicum* increased the number of recessive genes that must be present to prevent shattering to three genes.[1128] This will further complicate attempts to bring in other genes from the wild, especially because a strong linkage exists between shattering and the so-desired homomorphic flower type in this species[1128]. This linkage will have to be broken to achieve progress with the genetic approach.

Genome analysis has been used more for relatedness studies within the genus (cf., reference 1176) than within common buckwheat. The use of marker-based breeding for antishattering has begun recently.

14.5. Possible Biotechnological Solutions

Buckwheat has been successfully transformed by *Agrobacterium* infection of excised needle-pricked meristems.[601] Some transgenic approaches to further domesticate buckwheat that might be considered are the following.

14.5.1. Fertility

Fertility is the most interesting nut to crack. Clearly more basic information is needed on the causes of the barrier to pollen penetration.[1049] It may be easier to ascertain what genes are expressed in the penetrating pollen of the progenitor of buckwheat, or its closely related *F. homotropicum = F. esculentum* var. *homotropicum* vs. the nonpenetrating *F. esculentum* pollen, and see if any do not hybridize back to *F. esculentum,* to ascertain whether the gene was somehow deleted early in domestication. This will be harder if the gene was truncated and silenced. If there are but a few differences in the expression profile between potent and impotent flowers, it will be relatively easy to see what must be augmented, harder if there are more differences. It is hard to predict how such a project will go, but it is both a fascinating basic as well as important applied area to follow up.

Perhaps it may not be possible to alter these late flowers and one should look in other directions. One possible alternative is to engineer buckwheat for determinacy, to stop flower initiation after the one week when most flowers are fertile. In this manner resources would not be directed to new growth and new flowers, and the resultant varieties could be close planted, increasing yield.

14.5.2. Determinacy

Various genes in floral development are being discovered through expression profiling.[292] Perhaps an expression profiling comparison between aborting buckwheat and nonaborting Tartary buckwheat, using mRNA extracted at the time of fertilization from the restricted areas near the micropyle, may provide information on the genes involved. This would allow either a gain-of-function transformation using the missing gene from Tartary buckwheat, or an antisense or RNA interference (RNAi) suppression of an overactive gene from buckwheat itself, depending on the findings. Before performing such complicated work it might be wise to ascertain whether buckwheat has an ortholog to the genes: *TERMINAL FOWER1* or (*TFL1* and *TLP2*) from *Arabidopsis*, which have orthologs among fruit trees,[324] and tomato where the ortholog is called *SELF-PRUNING*.[190] A single amino acid transversion of *TERMINAL FLOWER1* converts this repressor into *FLOWERING Locus* (*FT*), an activator of flowering.[461] From the study cited in section 14.3.1 demonstrating that photosynthate is not limiting,[450] buckwheat probably does not have genes mutated to the *FT* forms, producing more flowers than it can handle.

Another solution might be to generate parthenocarpic fruits, that is, obtain seed without fertilization. The *KNUCKLES* (*KNU*) mutant gene modifies the *Arabidopsis* flower such that it develops parthenocarpic seed.[838] The flowers are also male sterile, and this would limit gene outflow, a desirable trait from a transgenic biosafety point of view. Perhaps this gene or an ortholog from buckwheat might do the same. Overexpressing of auxin in the flowers of some species enhances parthenocarpy, and in other species gibberellins perform this function. Application of these hormones to the flowers might give an indication of which (if any) may work, suggesting which gene expressions to try. The genes required for both auxin and gibberellin production are listed in GenBank; they could be isolated and reengineered for higher expression with tissue-specific promoters.

14.5.3. Dwarfing/Lodging

Various dwarf lines have been described, but there seems to be a problem with them. If they had the expected higher yields and greater lodging resistance, it is clear that there would be more dwarf varieties cultivated and more publications about their success. Thus, there is good reason to consider using RNAi of gibberellic acid biosynthesis or reception, and/or brassinosteroid biosynthesis, to achieve higher-yielding varieties and overcome linkage problems with available lines.

14.5.4. Herbicide Resistance

Currently, no broadleaf herbicides are registered in the developed world for use in common buckwheat,[1121] although some should be active but have limited effectiveness. Buckwheat is a highly competitive crop where it is grown, and perhaps does not need herbicides under normal conditions. If buckwheat is dwarfed to increase the harvest index, the weed problem will surely be exacerbated. Thus, it imperative that the selectable marker for any such primary transformation for dwarfing be for resistance to a herbicide that is easy to register.

14.5.5. Frost Tolerance

Many disparate genes, with a wide variety of functions, have been isolated and provide a modicum of frost tolerance to other species (Table 23). Which ones, singly or stacked would be best for buckwheat is an open question. There is a valid question about which promoters to choose: in nature chilling tolerance is usually an induced trait and in most of the cases in Table 23 a constitutive promoter was used. Will this have a (Haldanian) fitness cost in nonselective conditions, and if the gene were to move to a wild relative, would this confer a (Darwinian) fitness advantage[532]? Jackson et al.[532] tried to address this, but used transgenic *Arabidopsis*, which is not easy, and measured productivity and not fitness (despite their claims). Some of the lines they used were less productive under nonselective conditions, but as competitive fitness was not measured, the results can only be considered ambiguous. Unless the cold-tolerance problem facing buckwheat is sudden drops in temperature, from warm to freezing without warning, it might be best to use cold-inducible promoters for the gene(s) of choice, especially if there is a yield drag with constitutive promoters.

14.5.6. Allergens

The cDNA coding for the major 22-kDa buckwheat Fag e 1 allergenic protein was isolated and its immunoglobulin E (IgE)-binding activity was confirmed using recombinant Fag e 1 and sera of allergic patients. The derived amino acid sequence from Fag e 1 cDNA was used to synthesize an overlapping peptide library for the determination of the Fag e 1 epitopes.[1184] Eight epitopes and the critical amino acids for IgE-binding within the epitopes were identified. This should aid in the development of hypoallergenic buckwheat using RNAi technology to suppress the expression of all the epitopes.

14.5.7. Bitterness

One of the reasons Tartary buckwheat is valued is due to the presence of large quantities of rutin, which is considered to have favorable functions for

Table 23. Examples of Genes that Might Confer Frost Tolerance on Buckwheat[a]

Gene ↑ enhanced ↓ suppressed	Function	Source	Demonstrated activity (often along with others; see reference)	Reference
↑DREB1	Transcription factor	Rice	General stress tolerance	530
WCOR15	Unknown / chloroplast targetted	Wheat	Tolerance in tobacco	976
Des C	Acyl-lipid Δ9 desaturase	Synechococcus	Tobacco cold tolerance	860
↑Osmyb4	myb transcription factor	Rice	Arabidopsis cold tolerance	701
↑TaLEA	Late embryogenesis abundant protein	Wheat	Yeast freezing tolerance	1186
↑ABF3	ABA-responsive transcription factor	Arabidopsis	Lettuce cold tolerance	1099
mtlD	Mannitol phosphate dehydrogenase	E. coli	Petunia cold tolerance	214
↑atRZ1a	Zinc finger glycine-rich binding protein	Arabidopsis	Arabidopsis freezing tolerant	585
ipt	Isopentenyl transferase (high cytokinin)	Agrobacterium	Green fescue grass at low temperature	513
P5CS	Pyrroline carboxylase synthase (high proline)	Vigna	Larch cold tolerance	394
↑mDHNs	Dehydrins (multiple constructs)	Arabidopsis	Arabidopsis cold tolerance	872
CaPF1	ERF/AP2-type transcription factor	Pepper	Arabidopsis cold tolerance	1179
	Antifreeze proteins	Many	Many	432
ARL	Aldose/aldehyde reductase	Alfalfa	Tobacco cold tolerance	486
PK1	Mitogen-activated kinase	Tobacco	Maize cold tolerance	978
betA	Glycine betaine synthase	E. coli	Maize cold tolerance	876
↓Lea Gal	α-Galactosidase	Tomato	Petunia freezing tolerance	844
inaA	Ice nucleation protein	Pseudomonas	Tobacco cold tolerance	82

[a]ABA, abscissic acid; ↓, antisense or RNAi down-regulation; ↑, up-regulation

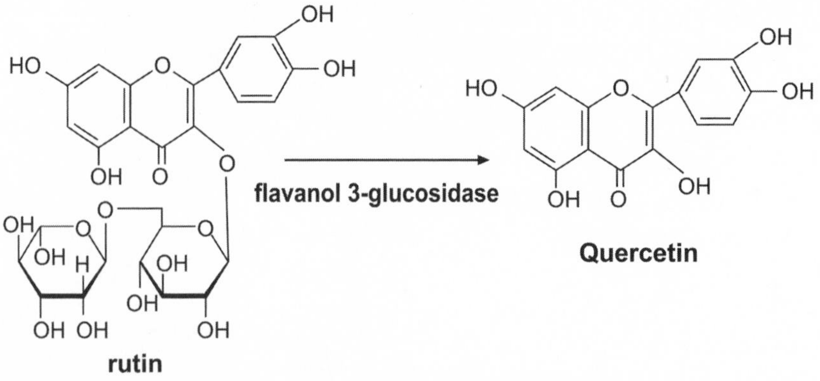

Figure 31. The undesirable conversion of rutin to quercetin in buckwheat. This reaction could be blocked transgenically using antisense/RNAi strategies.

human health (see reference 1033). Many lines of this crop are bitter because a well-characterized flavanol 3-glucosidase[1033] removes the disaccharide side chain on rutin, giving rise to bitter quercetin (Fig. 31). Quercetin is a well-known allelochemical inhibiting the growth of other plant species (e.g., reference 836) and is insecticidal (e.g., reference 684). The breeders approach is to use an enzyme activity assay to screen lines for low activity of the enzyme that synthesizes quercetin.[1033] Once such lines are found, they will have to be crossed and backcrossed with high-yielding lines that have other favorable properties. The resultant crop from such arduous breeding efforts may well possess less or no quercetin in the grain, but will also possess less in other organs, possibly lessening their ability to compete with weeds and withstand pests. The molecular approach would be to isolate the flavanol 3-glycosidase gene after microsequencing the enzyme to obtain the information needed to design primers and then antisense it back into Tartary buckwheat, but under the control of a seed-specific promoter so that this allelochemical will still be made in other tissues. The enzyme has been purified and partially microsequenced,[99] so the rest should be easy. This approach will provide the best of both worlds, allelopathy toward pests but not against humans eating the seed.

14.6. Biosafety Considerations

Tartary buckwheat has become a weed in Canada since its introduction.[467,799] This should not be confused with U.S. terminology where the distantly related pernicious weed *Polygonum convolvulus* is termed wild buck-wheat and another weed *Eriogonium longifolium,* in the same family, is termed longleaf buckwheat. Although Tartary buckwheat crosses with common buckwheat, there are no reports of these other species crossing with buck-wheat. Considering the problems of obtaining intrageneric crosses in *Fagopy-rum,* it is doubted that such crosses will occur.

Intervarietal pollen flow must be considered in this obligately outcrossing crop. Pollen flow was measured from a tall (dominant type) into a field of semidwarf (recessive) material. Substantial cross-pollination was limited to a few meters from the pollen source, but nearly 1 percent outcrossing occurred even at a hundred meters, the maximum measured.[8] This might be expected from an insect-pollinated species. Large isolation distances will clearly be needed for transgenics, if they bear traits whose movement is eschewed to other varieties, or containment or mitigation strategies (Chapter 4) will have to be employed.

CHAPTER FIFTEEN

Should Sorghum Be a Crop for the Birds and the Witches?

Sorghum (*Sorghum bicolor*) is a major world crop with a high degree of drought tolerance. Hybrid sorghum varieties have been successful in out-yielding maize in the dry lands of Africa and Asia as well as in the Western Hemisphere in places where aquifers have dried up and the amount of irrigation water has become limited. Sorghum cultivation has not replaced maize as much as might be expected in a supposed market economy (USA) due to illogical subsidies to both grow maize, as well as subsidies not to grow maize (set aside), which do not apply if sorghum is cultivated. Thus, maize and not sorghum is raised in such areas. Much of the important lore about this crop was summarized in an excellent book by Doggett[285] nearly two decades ago. The book is of value for more than its historical perspective. Many of the intractable problems described therein have yet to be solved, and Doggett does an excellent job of defining the problems and describing what has not worked; there is no reason to attempt to reinvent the wheel.

Sorghum is cultivated mainly for its seeds, but in the past more was cultivated for the sugars in its stalks (like sugar cane), or seeds (like sweet corn), for building material (like bamboo), but especially as forage for livestock, mainly as silage (in the Northern Hemisphere), and direct feeding. Recently much discussion has taken place about sorghum as a source of raw materials for bioethanol and other biofuel production, especially the sweet sorghums.

Sorghum originated in Africa and it has many relatives there, both in the wild, along with weedy relatives, as well as feral forms of the crop,[312] which are also major constraints to its production. The wild and weedy *Sorghum* species have also spread over many millennia to distant parts of the Old World from Africa, supplying quite a bit of diversity for the breeder and weed problems for the farmers. The species is beset with seemingly breeding-intractable pest problems such as stem borers and grain pests (discussed in Chapter 6) and mycotoxins (Chapter 7). It is also plagued by parasitic witchweeds and by birds, the two most important constraints to sorghum production in Africa (and very important elsewhere), which are the subject of this chapter.

The Gates Foundation is funding a major project to enhance quality traits ("biofortification") in sorghum (http://supersorghum.org). Will people benefit from it, or is this project just for the birds and witches? Solving the bird problem along with witchweed and other major problems enunciated by leading African scientists dealing with sorghum (see reference 431) such as parasitic weeds, stem borers, and storage pests would lead to a vast increase in sorghum production. Biofortification will only intensify the problems, especially the insect ones, as insects and rodents can distinguish between different traits and prefer a fortified diet when it is available.[752] Much of the increase in production of nonfortified sorghum could go to feeding animals, providing eggs, milk, and meat that would alleviate the same nutritional problems as breeding sorghum for better nutritional quality. The farmers and consumers prefer better nutrition from diverse sources than a single one. Higher sorghum yields would cause subsistence farmers in developing countries to divert land to lower-yielding legumes and other crops, further diversifying diets. One can sincerely question why the Gates Foundation has not chosen to attack the basic constraints to sorghum productivity, but have skipped ahead to biofortification.

15.1 Two of the Intractable Constraints to Sorghum Production

This chapter will deal with two issues with which sorghum has especially bad problems, with weeds and with birds. The weed problems are twofold: (a) the breeding-intractable issue of closely related interbreeding weeds, which are similar to those of rice (Chapter 12.2) and will not be discussed in detail again; and (b) the parasitic weed *Striga,* for which sorghum has the genes that could possibly be isolated and cloned to be used to deal with this scourge in other crops that lack the biodiversity and the evolutionary history of sorghum.

15.1.1. Sorghum Yield and Education

African lore has it that the higher the sorghum yield, the poorer children's education. When the children are out of school acting as human scarecrows, there is a crop to be harvested (Fig. 32). When they are in school, the birds and the rodents celebrate. The major problem comes from huge migrating flocks of the red-billed quelea also called the red-billed weaver (*Quelea quelea*), a colorful relative of the drab sparrow. Their numbers are so large and the situation is so bad that these birds have been called "feathered locusts." Flocks number in the millions, with hundreds of millions migrating from their nesting areas. One hectare of acacia scrub can have a roosting density of fifty thousand nests, producing one hundred thousand young and hungry birds that migrate with their hundred thousand parents, each eating twenty

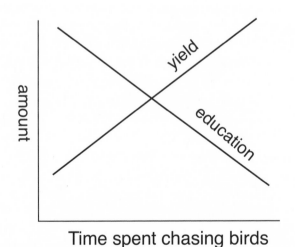

Time spent chasing birds

Figure 32. The purported inverse relationship between childrens' education and sorghum yield. Children stay home from school to act as human scarecrows to prevent predation by granivorous birds, especially the red-beaked quelea (feathered locusts) in Africa, when sorghum seeds are in the milk and dough stages of development.

to fifty grams of grain per day.[285] Some regions are luckier than others, because the sorghum harvest is finished before the migration, and the yield losses are lower. In areas others such as Somalia, losses of 80 percent have been reported. Doggett[285] considers the estimate (a few decades ago) of an annual yield loss of six hundred thousand tons of grain to be an underestimate.

A major symposium was devoted to the problems of controlling these devastating bird populations.[205] Two methods of control are being used: harvesting the birds for eating, and chemical control. In one nature conservancy in Zimbabwe alone three and a half tons of chicks were harvested from a large colony.[764] Crops within a thirty-kilometer radius of a roosting area are in danger of being decimated by roosting birds, and the grain short-fall of many countries is in part due to the roosting and later migrating bird flocks. Red-billed quelea are being chemically controlled by aerial or ground application of fenthion (an organophosphate pesticide) to roosting areas often using back-pack sprayers,[764, 1095] and an estimated thirteen million birds were sprayed in one year in Zimbabwe at a cost of thirty-eight thousand dollars, saving a crop worth many times more.[764] Even then, the cost of spraying pesticide is considered excessive (especially as the cost is not borne by the farmers but by the governments), with farmers abandoning sorghum for maize, cassava, or sugarcane. Because the birds are tasty, there is the worry that fenthion-killed birds will be served up, especially because the initial breakdown products of the insecticide are more poisonous than the pesticide itself. This worry has not materialized as actual occurrences of known poisonings of humans or other scavengers, but the possibility exists. Non-target-nesting birds that do not feed on the crop are also killed by the pesticide.

Although quelea is clearly the worst bird problem, other grain thieves exist: yellow weavers in Africa; sparrows, other weavers, and parakeets in India; and sparrows and blackbirds in the United States.[285]

15.1.2. Parasitic Witchweeds

The witchweeds (*Striga hermonthica* and *S. asiatica*) are serious weeds that decimate sorghum as well as maize, millet, upland rice, sugar cane, and napier grass throughout sub-Saharan Africa (see Table 24 for conservative estimates). A relative, *S. gesnerioides* attacks legumes in West Africa. *Striga* spp. are hemi-parasite plants, growing into the vascular system of the roots of crops and sucking out water and nutrients for their own growth. After they emerge from the soil they photosynthetically fix about 20 percent of their carbon, contin-

Table 24. Production Losses Due to Striga in Sub-Sahara Africa—Estimates

	Sorghum and Millets[a]	Maize	All crops
Area affected (million ha)	21.9	4.33	26.23
Estimated yield loss (%)	26	40	33
Estimated loss in production (million tons)	8.60	2.07	10.67

Source: Summarized by A. B. Obilana in reference 431, where a table with a country-by-country breakdown can be found.

[a]Includes cowpeas in West Africa

uing to steal from their hosts. *Striga* infests an estimated twenty to forty million hectares of farmland cultivated by poor farmers throughout sub-Saharan Africa, on average halving yields, but in bad seasons it can wipe out a crop. It is a major reason for partial land abandonment in many parts of Africa, with the men migrating to the cities, leaving the women to scratch the land. Thus, there is a relationship between this HIV (highly invasive vegetation) and the more common definition of the abbreviation. Migrating male populations have been prone to contacting sexually transmitted diseases throughout history. *Striga* control methods have been researched in Africa for more than fifty years and efforts have focused on agronomic practices, breeding host plant resistance, and herbicide applications. None of these methods have been widely adopted by farmers because the proposed solutions do not fit in with existing cropping systems, for example, they require rotating land out of their major food crop where their benefits will be seen only in the medium to long term. Present population pressures require intensification of land use for food production. Although a modicum of host plant resistance exists, the present lines are often ineffective under high levels of infestation or are region specific. Conventional herbicide applications are prohibitive, because the parasite damages the crop before *Striga* emerges from the soil when it can first be sprayed with a contact herbicide. Also many small-scale African farmers intercrop with herbicide-sensitive legumes, and herbicides would affect the intercrop.

One of the first signs of *Striga* infection, visible well before *Striga* appears above ground, is a "bewitching" of the whorl of crop leaves. Bewitching is manifested as a yellowing of the leaves and a cessation of growth. This appears as if it is a poisoning, but it is not known who produces the poison; the crop as a type of hypersensitive response, or the parasite to stunt the crop.

Striga can produce a large number of iridoids as well as other potentially toxic compounds,[887] including eight iridoid glycosides, two caffeoyl phenylethyl glycosides, as well as shikimic acid and trigonelline. This is not proof that *Striga* poisons the host, but it is indicative that it may have the capacity to do so. Another (non-African) *Striga* (*S. orobanchioides*) synthesizes flavones that act as contraceptive chemicals at 25 mg/kg body weight of rodents,[504] another bewitching trait.

15.2 Breeding Solutions

Sorghum was domesticated in Africa; thus it coevolved with the parasitic birds and weeds, and there is a modicum of variability in its germplasm to deal with these pests. The problem is that a biological balance between pest and host guarantees survival of the host, but the farmers want their crops to more than just survive and produce enough seeds to replenish themselves. The farmers want full yield without predation. The breeders have been trying to find a combination of genes that when stacked as QTL will fulfill the farmers' needs. It has yet to be demonstrated that sorghum has these genes in a manner that can be used vis-à-vis bird damage.[285] In many decades of breeding sorghum has not been demonstrated to have the genes to efficiently contend with stem borers, storage grain pests, and mycotoxins, subjects discussed in Chapters 6 and 7.

15.2.1. Breeding for Lack of Bird Predation

Breeders have tried to modify two aspects of sorghum to reduce bird damage: to modify the structure of the inflorescence to deter the birds and to modify the taste.

15.2.1.1. Modifying the Inflorescences

Doggett[285] states that various plant characters have been tested; awned sorghums, large glumes, close (compact) grain heads/panicles, and inverted grain heads (goose necks). Birds attack other varieties first when there is other food around but come back to the less accessible varieties when they have finished consuming the others.[285] A breeding program to combine these characters while retaining an acceptable yield proved difficult,[285] and the efforts seem to have been abandoned, while the problem gets worse.

15.2.1.2. Modifying Palatability

So far breeding for lack of palatability has been attempted by selecting for factors that modify flavor (both to birds and alas to humans); high tannins or phenols, which must disappear just before harvest, or must be removed by soaking, cooking, or polishing off the outer seed coat containing these antifeedants. The tannins act as antinutritional factors, and the food and feed value is less, and thus they have been bred out of many indigenous sorghum varieties to increase the acceptance and use of this crop, exacerbating the bird problem. Indeed, when there are varieties that are unpalatable to birds due to tannins, they only deter birds when grown next to tastier varieties in small plots (Table 25). When birds have no choice, that is, when large tracts of the supposed bird-resistant high-tannin varieties are planted, the flocks of birds attack and decimate the supposedly unpalatable varieties (Table 25).[103]

In bird taste trials a white, nontannin cv. Ark-3048 was found in the United States that was not attacked by birds.[1183] This variety may be an exception in the field as it was shown that the seeds have high contents of dhurrin, a natural cyanogenic glycoside in the milk and dough stages of seed development.[30, 1047] These are the stages when bird predation is highest. Dhurrin is often found in seedlings and stalks of sorghum and is toxic to animals.[389] Ark-3048 has a mutation that causes the expression of dhurrin in the seeds of cv. Ark 3048. Dhurrin is compartmented in the outer layers of the grain, and the

Table 25. Resistance to Bird Damage in Sorghum—
Effective in Interspersed Small Tracts

*Type/*Variety	Varieties interspersed	Cultivated alone
	% yield loss	
Putative resistant		
Savanna III	47	94
Susceptible		
X3101	67	
8D	88	
65D	92	

Source: Calculated from data in Beesley and Lee.[103]

Note: Comparisons were made between caged and uncaged rows of plants in a study in Botswana where the putative resistant variety was grown in small plots interspersed with the other varieties, versus a large plot of the resistant variety, with caged and uncaged rows.

β-glycosidase and hydroxynitrile lyase are in other tissues. When immature grain is macerated, cyanide is rapidly released due to the mixing in aqueous solution of dhurrin with these enzymes (Fig. 33). If this is in a pot, the cyanide boils off, as with cassava. If the macerate is made in the beak of a bird or mouth of a mammal, and the cyanide released in the gut, the effect can be lethal. Birds ingesting just Ark-3048 sorghum died, and when given a choice, birds have enough brain to avoid it.[30,1047] By the time the grain matures, dhurrin and cyanide have disappeared from the seed, so besides being an antifeedant, the cyanoglycoside may also be a storage compound. There may be a danger to villagers who harvest and roast immature seeds (especially when there is nothing else to eat); this may be worse than real junk food, as the consumers will be the first to macerate the tissues. Dhurrin appears early after germination, and thus is found in malted sorghum.[1088] Many attempts have been made to breed dhurrin out of sorghum stalks. The production of dhurrin in stalks and leaves is much higher when the plants are heavily fertilized with nitrogen, especially in high light,[171,1146] but there is no information on how this affects the dhurrin levels in the developing seeds of cv. Ark-3048.

Sorghum is not the only species producing cyanogenic glycosides. They are common in cassava (where breeding them out was better for people, but lowered the plants defenses), clover, and barley.[110]

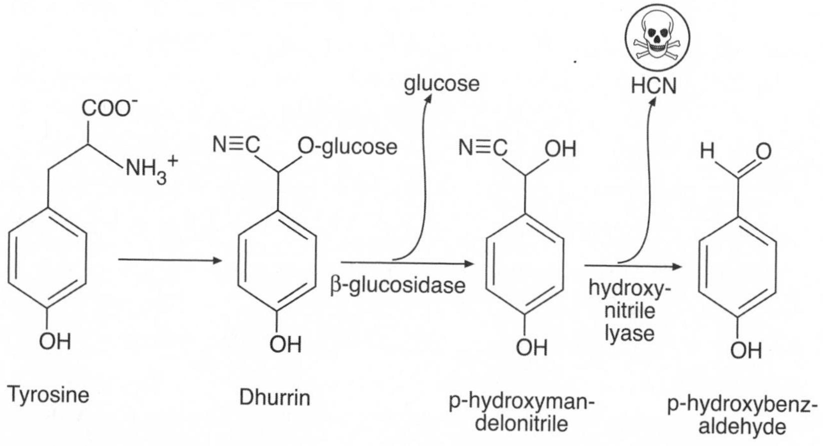

Figure 33. Dhurrin from sorghum is catabolized to toxic cyanide. Catabolism occurs when tissue is macerated in aqueous solution and dhurrin and the enzymes, each stored in different tissues, are mixed.

There are some who feel that all natural solutions are good, but many scientists may find the solution of dead birds killed by a crop to be unpalatable. A moral conundrum exists: is starving birds with tannins better than killing or deterring them naturally with cyanide? Or should we be leaving the birds to their own devices and starve people? The recent "farm scale evaluations of transgenic herbicide resistant crops" in the United Kingdom[346] did not focus on weed control per se. The important issue there was whether sufficient weeds were left in the fields to feed all the grain-eating birds. Well-meaning urbanites think they have a right to mandate what pests should be in farmers' fields stealing the crops. Can/should such developed world luxuries be exported to Africa? In Africa it is a matter of them versus us; human survival versus the feathered locusts. The chemical control of red-billed quelea with fenthion[764,1095] clearly has far greater off-target effects than would occur with a cyanogenic sorghum. Breeding efforts using Ark-3048 have not advanced because derived progenies appear to have yield penalties (G. Ejeta, pers. commun., 2006), suggesting linkage problems.

15.2.2. Breeding for Striga Resistance

Breeding sorghum for full resistance to *Striga* cannot be easy. No one has entered a heavily infested field and spotted a healthy sorghum plant among the half-dead crop. Breeders and pathologists would call having no *Striga* as being immunity; resistance is having some *Striga* in an infested crop but without yield loss; tolerance has some crop loss; versus devastated susceptible. Others refer to levels of resistance, and in such a discussion, definitions must be clarified, and resistance is used here for what breeders call immunity. Sorghum has coevolved with *Striga* for millennia and has modicums of tolerance in some cultivated lines and wild relatives. Conversely, the introduction of maize to Africa, and the subsequent recent evolution and spread of *S. hermonthica,* purportedly from *S. aspera*[17] may have wrought a much higher *Striga* seedbank as well as more virulent forms of *Striga* than sorghum had evolved to cope with.

Not in every season is the high level of *Striga* infestation apparent despite dense soil seedbanks of *Striga* seeds, confounding and confusing results and researchers. This is probably due to environmental factors that unpredictably control infestation levels. Because of the larger *Striga* seedbanks, sorghum may have become effectively less tolerant, because the larger numbers of *Striga* can gang up against the crop.

Based on the experience of breeding resistance of sunflowers to the other major parasitic weed group, the *Orobanche* spp. (broomrapes), single-gene resistance is not desirable, even when it can be found. Each new resistant sunflower variety provided the selection pressure for the evolution of new strains of the parasite, and no variety lasted for more than a few years. Wild sorghum that is tolerant to *Striga asiatica* is highly susceptible to the more recently evolved and more pernicious *S. hermonthica.*[442] A polygenic type of resistance would probably be the only type that should be bred, and if successfully bred, might have the necessary resilience to delay the evolution of resistance. Thus, conventional selection in segregating crop germplasm populations grown in heavily infested fields did not always yield promising material because of the complex host–parasite–environment interactions. When such material seemed resistant, it was not resistant in all environments or regions. *Striga* resistance breeding had only been for the bottom line, few emerging *Striga* stalks, without considering mechanisms.[911]

Enlightened breeders therefore defined a paradigm for *Striga* resistance breeding based on the beginnings of understanding the biological basis of the etiology of infection, while attempting to minimize environmental influences.[311,770,894] It has been wisely pointed out that, although the terminology of pathologists has been adopted vis-à-vis resistance, only superficial similarities exist between attack by a parasitic weed and by a fungal pathogen.[703] The underlying mechanisms are different, and only in penetration are analogous strategies used, yet the structures are very different.[703] Similarly, the strategies used by a plant for resistance to parasitic weeds are clearly different from those used by plants to combat fungi. The breeders characterized some of the essential signals exchanged between host and parasite to determine potential sites for breeding intervention. They developed useful bioassays to screen segregating genetic populations to determine which populations should be crossed based on the different modes of tolerance to *Striga:* low production of germination stimulants (lgs), low levels of a hypothetical haustorial factor (lhf), a hypersensitive response (HR) in the root tissues, and an incompatible response (IR) following infection (Fig. 34),[311] as described in detail below.

The lack of *Striga* germination near a mutant crop may be for a variety of reasons, because there are compounds that directly stimulate *Striga* germination and compounds that potentiate or facilitate it. Thus, this first step may eventually be divided into more than one genetic component. *Striga* germination is primarily controlled by compounds belonging to more than one group of chemicals, initially presumed to be synthesized via the sesquiterpene pathway; "strigolactones" including strigol, first isolated from cotton, which is not a *Striga* host.[139] Strigolactone is one of the germination stimulants exuded by sorghum[139,479] and maize roots. It is not clear which stimulants are used by other species.[139] Although the stimulants structurally appear to be sesquiterpenes, they are not synthesized via the classical sesquiterpenoid pathway, but by a novel offshoot of the carotenoid pathway.[702]

The germination stimulants stimulate ethylene production and trigger subsequent germination of *Striga* seeds.[81] This is in contrast to the parasitic *Orobanche* species, which plague vegetable crops around the Mediterranean and are ethylene insensitive. Cytokinins increase the capacity of *Striga* seeds to convert the natural ethylene precursor 1-aminocyclopropane-1-carboxylic acid (ACC) to ethylene and elicit germination of seeds.[81] All *Striga*-susceptible sorghum genotypes appear to be high stimulant producers.[314] Low germination stimulant (lgs) production (Fig. 34) in sorghum is inherited as a

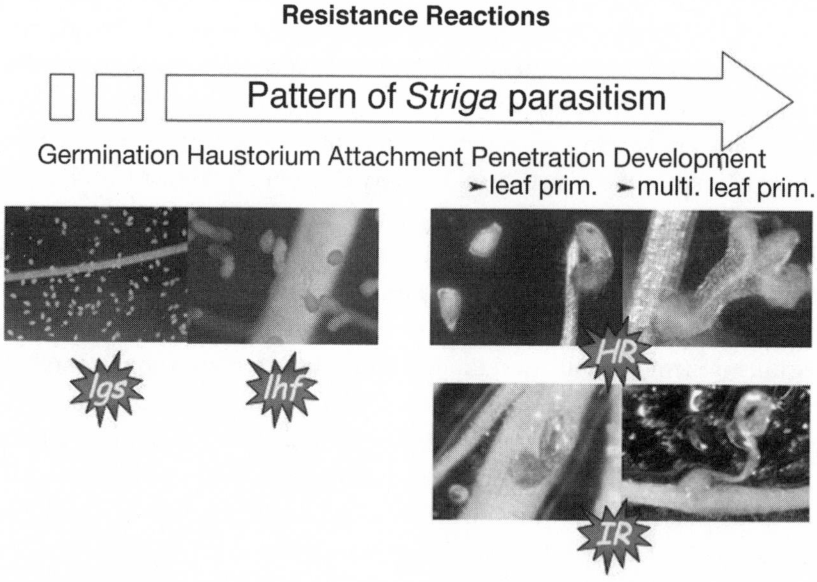

Resistance Reactions

Pattern of *Striga* parasitism

Germination Haustorium Attachment Penetration Development
➤leaf prim.   ➤multi. leaf prim.

Figure 34. Striga-resistant reactions of genetic variants of sorghum displayed at key developmental stages of germination (*lhs*), haustorial initiation (*lhf*), penetration and attachment (*HR, IR*). *Source:* Reproduced from Ejeta,[311] with permission of the publisher.

single recessive gene[1113] and the *lgs* gene mapped about 12 cM from a restriction fragment length polymorphism (RFLP) marker PIO200725. The *lgs* gene was transferred into high-yielding and widely adapted sorghum cultivars[311] and was highly effective in many African countries, with some exceptions. The lack of effect might be due to a *Striga* density in the soil that is so high that there is enough inoculum very close to roots that could cause infection, or it could be due to unclear environmental effects that confound the resistance mechanism.

The parasite attaches to sorghum and other hosts via a haustorium after the *Striga* germling is in proximity to the host root, triggered by an as yet unknown host signal. Kinetin, simple phenolic compounds, and 2,6-dimethoxy-1,4-benzoquinone (DMBQ) are exogenous haustorial initiators.[904] Parasitism is initiated after attachment to the host, facilitated by a secretion of a hemicellulose-based adhesive substance that fixes the parasite to the host root.[84] No *lhf* (low haustorium formation) lines were found among cultivated sorghum lines but some were discovered among wild sorghum lines (Fig. 34).[894] The *lhf* gene is simply inherited as a single dominant trait at about 19 cM distance from simple sequence repeat (SSR) marker TXp358, and is being stacked with other resistant genes.[311]

Further enzymatic activity after haustorial attachment degrades host cortical cells, leading to penetration of and direct connection with host xylem.[304] *Striga* does not establish direct connections with host phloem in contrast to *Orobanche*. Penetration may involve additional chemical or tactile signals from the host root. The lack of direct phloem connections does not seem to diminish the ability of *Striga* to steal photosynthate from its hosts. Further, *Striga* development can be stopped by a hypersensitive response (HR) where host tissue around the penetration site dies, with the necrotic tissue starving the *Striga* germling and preventing it from integrating with xylem. An incompatible response (IR) can also occur, where the *Striga* stops growth at an early stage, and the sorghum root continues on with its life without any tissue death (Fig. 34). Both cultivated and wild sorghum lines were found that have hypersensitive and incompatible responses.[314] The hypersensitive response is controlled by two interacting dominant genes 7.5 cM and 12.5 cM from SSR markers, XP96 and SBKAFGK1, respectively. Incompatible response lines were found in cultivated and wild sorghum germplasm,[311] as well as in rice.[443] So far stacking the different responses has brought low-level resistance that may be durable, but it has not brought "immunity." This is good enough to get an excellent yield, but will the longer-term goal of reducing the seedbank be achieved? Mathematical modelers have calculated that when the seedbank is high and resistance reduces seed input by 95 percent, it will still take many years to reduce the seedbank to half its size.[998] This is because *Striga* puts out such large numbers of seeds, that the remaining 5 percent gives rise to sufficient numbers of plants to produce more than enough seeds to replenish the seedbank.

15.3. Biotechnological Solutions

Although breeding or agronomic solutions may offer some local relief, no solutions in the field are really available over large areas. Many proposed "solutions" have not been practical in the African context. When truly good solutions are found in Africa, and are cost effective, they are rapidly adopted by farmers. Thus, there is reason to consider biotechnological solutions.

Plans are being made for the sequencing of the sorghum genome, and an international consortium has formed to press for this, while making immediate plans for the development of an anchored physical map to guide sequence assembly and annotation.[612] Considering the many uses of sorghum, and unlike sequenced rice, sorghum is a C_4 plant, and it has closely related weeds, there are many good reasons to have this crop sequenced, including the possibility to isolate and clone weediness genes. Sorghum, which underwent a whole-genome duplication seventy million years ago, should have fewer retained copies of duplicated genes than maize, which underwent the same process far more recently, "just" twelve million years ago.

15.3.1. Possible Solutions for the Birds

15.3.1.1. Cyanogenesis

If cyanogenic solutions are acceptable to the regulators, one could consider generating a designer sorghum that makes dhurrin only in the developing seeds so as not to poison livestock eating sorghum forage, and then have dhurrin completely self-destruct on maturity. Is this feasible? The pathway of dhurrin biosynthesis has been completely elucidated, originating at tyrosine (Fig. 33). A multigene pathway leads to dhurrin so the problem might seem daunting. It is composed of a multifunctional cytochrome P_{450} that converts the amino acid tyrosine in a three-step reaction to *p*-hydroxyphenylacetaldoxime.[558] The second enzyme is also a cytochrome P_{450} and it converts the oxime to *p*-hydroxymandelonitrile in two steps. These reactions are "expensive," costing four molecules of NADPH, as would be expected to form a single cyanide bond. The last step is the glycosylation of the latter compound to dhurrin.[558] The cytochrome P_{450} steps take place in microsomes, and the last step in the cytosol. All these genes have been isolated, cloned, and sequenced. Dhurrin was synthesized when the pathway was transformed into a heterologous *Arabidopsis* system achieving 4 percent of dry weight without seeming to cause any damage or morphological changes to the plants,[613] but it is not reported if anyone tasted the *Arabidopsis* sprouts or tested them for toxicity. On can still ask what yield drag the expensive synthesis of dhurrin may have on sorghum. Will the sorghum expend too much energy in defenses, when it could be diverting it to growth? The proverbial dilemma posed by many ecologists and evolutionists is to grow or to defend.[497] In economic terms it could and should be asked whether the pesticide being used in Africa is cheaper to spray, at the very low rates used, than the yield lost in producing dhurrin. Yield drag is always a question that must be asked, and it has to be posed against the benefits derived from the trait. Does one need to engineer this pathway into sorghum? Probably not, because it is already there. The missing link is the tissue-specific promoter found in Ark 3048 or elsewhere that turns on the pathway in the developing seed. Conversely, dhurrin production could be suppressed by RNA interference (RNAi) or antisensed under the control of stalk- and leaf-specific promoters in Ark 3048, leaving dhurrin production in the seed.

The genetics of high dhurrin production in stalks has been "variously reported as recessive, partially dominant or dominant."[285] Less than that is known about the genetics of high dhurrin production in seeds. The genes have yet to be mapped in sorghum, and it would be interesting to know whether they are clustered, which would make the genetic transfer of high dhurrin production easier. Pathways leading to secondary metabolites are often clustered, for reasons that are explained in evolutionary terms: each of the enzymes by itself would be useless to an organism; only when they can be inherited as a block can they be of use.

If these genes were put under a high expression, developing seed-specific promoter, the first need could be achieved. The cyanohydrin glycosyltransferase that degrades dhurrin has also been cloned from sorghum,[1059] and it could be engineered under a promoter that is expressed in the same compartments as the dhurrin, but late in seed maturation, just before harvest. This solution should be far more acceptable than spraying organophosphate pesticides. It may also be easier than crossing and backcrossing from Ark 3048, as we are dealing with a multigene pathway.

This biotechnological solution to the African bird problem is one that that would probably never be used or developed against avian predators in the West, as the outcry of well-fed westerners would be a deterrent. A brave government would have to make the decision to go ahead on a project like this, but clearly the will and needs of the people being starved by these feathered locusts would be behind such a decision. The problem would indeed be a nice academic exercise for ecology as well as ethics classes, but in Africa the issue is not academic.

15.3.1.2. Morphological Solutions

The gene jockey could also head for a morphological solution, for example, put the grains closer together on the stalk to make them harder to peck

Figure 35. Changes in flower structure wrought by a gene mutation in *Arabidopsis*. A wild-type inflorescence (*left*) and an inflorescence showing later produced flowers of a plant homozygous for the *Lapetala 2* mutation (*right*). The flower becomes determinate and carpelloid. *Source:* Photos and legend courtesy of Dr. John Alvarez.

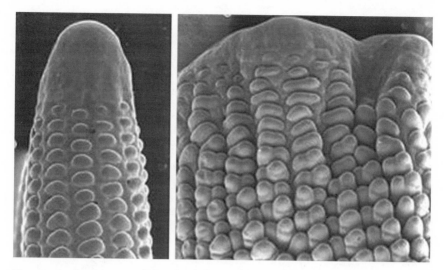

Figure 36. Changes in maize flower structure wrought by a single gene mutant *thick tasseled dwarf1* (*td1*). Normal (*left*) and mutant (*right*) ears twenty days after pollination are shown. Scanning electron micrograph by Peter Bommert; *Source:* reference 131.

apart, ensheath the grain heads to make them less conspicuous, and possibly put them somewhere less accessible, all as with maize. Can that be done? Surely yes in the future; carefully examine Fig. 1 to see the results of how Mexican Amerindians, without genetic engineering, selected the genes that turned puny teosinte with its open head into giant maize with its less accessible cob.[282] Multiple genes control the inheritance of the morphological traits that distinguish maize from teosinte. Fewer loci of large effect have been identified that control flower structure, the key innovations during maize domestication. A single gene is responsible for turning the teosinte stems with dispersed seeds into a tight maize cob, with seeds in orderly rows.[283] Some of the genes relating to flower structure are being elucidated, and some have immense major effects both in dicots (Fig. 35) and in monocots (Fig. 36). Very extensive and insightful reviews on the genetics and evolution of inflorescence development in maize and rice have recently appeared.[130,282,564] The maize *ramosa1* gene encodes a transcription factor controlling normal ear formation. The open inflorescences of sorghum are reiteratively branched like maize *ramosa* mutants.[1114] In sorghum, delayed production of spikelet pairs correlates with a protracted onset of *ra1* expression.[1114] One might ask what

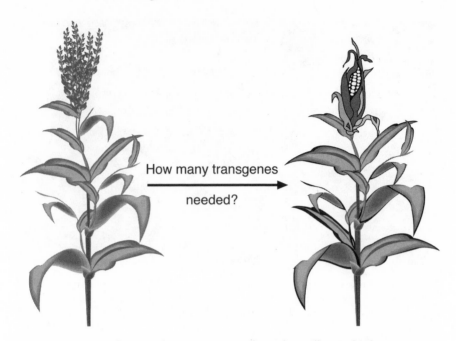

How many transgenes
needed?

Figure 37. An artist's view of a transgenic sorghum that will resist birds. How many transgenes will be needed to attain this protection? By Ms. Amit Mishali, Weizmann Institute of Science graphics department.

would appear on a sorghum plant if the native *ramosa* were to be transgenically replaced by the maize one?

It is up to the readers' imagination and skill to use this small number of major genes to reshape sorghum to deter birds (Fig. 37). It will have to be well beyond what was performed by breeding. When more of the appropriate orthologs, or genes analogous to those described for *Arabidopsis*, maize, and rice are known in sorghum, it should be possible to reengineer sorghum to prohibit access to birds and send children to school more of the time, and have added sorghum yield to pay for their tuition, books, and clothes.

15.3.2. Witchcraft-Biotechnologies for Witchweed

15.3.2.1. Herbicide-Resistant Crops

The possibility that the parasitic weeds might be controlled by systemic herbicides, on herbicide-resistant crops was suggested more than a decade ago,[407] but it took a few years to obtain transgenic crops to first validate the concept with *Orobanche* using model crops, tobacco, and oilseed rape.[550] Since then, potatoes[1031] and carrots[75] were specifically transformed with herbicide resistance genes, and the concept was demonstrated to work with transgenic glyphosate-resistant tomatoes (E. Kotoula-Syka, pers. commun.). The concept should work equally well with target-site glyphosate-resistant maize, but such material has not been made available to researchers who wished to test it. The technology would not be effective with the newer metabolically glyphosate-resistant crops,[194] because the herbicide would be metabolized before getting to the parasite.

One of the transgenes used, encoding a modified acetolactate synthase conferring resistance to a wide range of herbicides affecting that target, does not require genetic engineering to achieve parasite control. The gene is highly mutable and maize resistant to this group of herbicides was first obtained by tissue culture selection and regeneration[786] and later, far more simply, by pollen mutagenesis.[404] The mutant gene was introgressed by breeding from U.S. maize by CIMMYT (the international maize and wheat development center for developing countries). They bred the gene into African open pollinated varieties as well as in inbreds for hybrid seed production.[561] Methods were developed to apply small amounts of herbicide to seed (more than tenfold less than would normally be sprayed on a field), precluding the need for expenditures on herbicides or spray equipment.[3,559] The seed coating treatment is "appropriate" to African farming regimes, as intercrops of legumes are not affected by the treatment.[560]

Large-scale experiment station and farmers' field trials have been ongoing in four East African countries, with excellent *Striga* control with maize yields nearly tripled on average.[561] A major advantage is that the herbicide provides season-long control in short season, double-cropped maize in western Kenya, at even the highest *Striga* infestation levels; yields are normal on treated maize, where untreated maize is a total loss.[561] Even when there are some late-season attachments by the parasites, they never set seed before harvest, preventing replenishment of the *Striga* seedbank. This would be termed immunity by the breeders, albeit chemically facilitated immunity. CIMMYT has facilitated varietal registration and released material to local seed companies for bulking up and successful commercialization in western Kenya.[561] The herbicides used are perforce water soluble, the only type that are systemic in the crop plants. The same property leads to soil leaching and the herbicide would not stay long enough in the root zone of long-season varieties of the crop. Slow-release formulations of some herbicides have been developed based on high-capacity ion exchangers to keep the herbicide where needed for a longer duration, which will be applied to the crop seed.[169]

Similarly, sunflowers with the same type of mutation (e.g., cv. MAS93.IR) are being released in Europe for *Orobanche* as well as general weed control.[680] Whole-field spraying will be used there to control other weeds as well.

Because of the high natural frequency of resistance to this group of herbicides, modeling suggested that there is a great chance that resistance could rapidly evolve[426] and monitoring should be instituted. The model overstated the risk; it predicted that there would be five new resistant *Striga* plants emerging per hectare per year, each developing into an expanding clump. The models were predicated on the heterozygote mutation frequency for this trait that is typically dominant at the field doses used as a spray in western agriculture. Despite the use of low dose per hectare, the dose in the immediate vicinity of the treated maize seed is very high, requiring that resistance be homozygous, as was found while backcrossing the gene into maize. Thus, resistance would be much longer in coming; as with a recessive frequency, there may be as few as five new resistant plants per million hectares per year.[416] It is very easy to be off by a factor of a million with models when key assumptions are inaccurate. Alternatively, it may be advisable to introgress the gene into putatively resistant or tolerant sorghum having a direct mode of *Striga* tolerance to attempt to delay the inevitable genetic evolution of herbicide resistance in *Striga*. The levels of herbicide that will be sprayed on sunflower fields may be too low to require two copies of the resistant gene, and resistant heterozygotes could appear. So the modeled five resistant parasite plants appearing per year per hectare[426] may approximate what will happen under widespread use of herbicides in sunflower as low-dose spray applications to fields to control *Orobanche*, or late in the season with long-season maize and sorghum with *Striga*.

15.3.2.2. Engineering Sorghum for Direct Resistance to Parasitic Weeds

Some crops are immune to parasitic weeds yet cause the parasite to germinate. They might have the right genes for conferring resistance to other crops. To the best of our knowledge there is no ongoing research to find these genes, a rather complicated approach possibly being somewhat simplified with new "chip" approaches of differentially displaying the functioning genes.

A simpler approach is to have the crop root emit a toxic allelochemical. Because there is a metabolic cost to constitutively producing toxins, as well as the possibility of autotoxicity, the approach has been taken to put such toxin under promoters that are activated by parasite attack. Such a promoter has been isolated and the attacking parasite strongly activated a reporter gene.[1145] This also works with the antibiotic sarcotoxin as well; plants expressing sarcotoxin are more resistant to *Orobanche aegyptiaca*. Parasite development was abnormal and more parasites died after attachment than for nontransformed plants.[455] Sorghum also emits the allelochemical sorgoleone (not to be confused with the strigolactone stimulator of *Striga* germination), a potent photosynthesis inhibitor. This allelochemical is synthesized as droplets on root hairs and inhibits various other processes in higher plants as well.[258] Sorgoleone was once thought to be a *Striga* germination stimulant, but *Striga*-resistant sorghum lines that are poor germination stimulators synthesize as much sorgoleone as the wild type.[979] It might be possible to attain resistance by engineering the excretion of a hemicellulase under the control of a wound-inducible promoter to dissolve the hemicellulose that glues the *Striga* haustorium to the sorghum root (see section 15.2.2).

There has been a theoretical proposal to engineer *S. hermonthica* with a multiple-copy transposon containing a lethal gene under the control of an inducible promoter.[421] This is based on a concept proposed and partially tested for the control of insect populations.[434] The transformed *Striga* would be released in the field, and all crossed progeny would bear the gene construct (instead of half the progeny, as with Mendelian inheritance). After the construct spreads through the population, the gene could be turned on, either by a systemic chemical applied to the crop, or by a new crop variety secreting the chemical inducer. *S. hermonthica* is the only one of the major parasitic weeds that is an obligate outcrosser, and thus the only one where this concept might work.

15.3.2.3. Breeding versus Engineering Host Resistance

The breeding project described in section 15.2.2 has successfully brought the release of many *Striga*-resistant sorghum varieties in Africa, but also some early failures where the low-stimulant production varieties were ineffective. The first varieties released depended on a single recessive gene conferring a single mechanism of resistance, and carried the risk of resistance rapidly breaking down. The stacking of different resistance traits has clearly helped, as demonstrated with the most recent releases in Ethiopia (G. Ejeta, pers. commun.). If that is so good, why want to do the same by engineering? There are a variety of reasons to want to isolate and characterize the genes, way beyond a desire for basic knowledge. First, there is a good chance that the genes could be used to engineer resistance into other *Striga*-sensitive crops that do not possess the genetic diversity of sorghum, for example, maize, which evolved in *Striga*-free Meso-America. There are even reasons within sorghum to consider engineering. The low-germination-stimulant gene is recessive, complicating backcrossing into each inbred for hybrid production. When the gene is cloned, it can be transformed as RNAi, where it would then be dominant. Some of the other genes come from wild sorghum lines, and the wildness must be backcrossed out of them. Even with careful breeding the associated linkage drag may carry undesirable genes that may limit crop productivity or poor grain quality.

There is always the chance that the *Striga*-incompatible response gene from rice [443] or other genes from other sources will be complementary or better than the endogenous sorghum genes. They should be stacked together to enhance resilience and decrease the likelihood that *Striga* will evolve resistance to the host mechanisms, but sorghum does not cross with rice. Indeed it is conceivable that other genes that will help fend off *Striga* will soon be isolated and could be transgenically stacked with the genes being bred. Efforts are underway to elucidate the pathways of synthesis of the isoflavone allelochemicals that are secreted by *Desmodium* that keep *Striga* from attacking grain crops in intercropping.[581,1077] After the enzymes are found, isolating the genes for use in other species is relatively mundane. Similarly, it is also possible that genes will be found encoding enzymes that degrade the poisons *Striga* seems to be secreting to bewitch the crops,[887] which would decrease the damage to the crop, and let sorghum shade the *Striga* when it emerges from the soil and is photosynthetic.

It clearly will not be simple to stack the myriad of genes needed for resilient resistance by breeding into a wide spectrum of varieties to obtain and retain extensive varietal diversity. It might be far easier biolistically to cotransform a mixture of genes, or to transform a tandem construct of a group of these genes into the varieties than to backcross the traits, even from a single variety containing them. Backcrossing is unnecessary with transformation, the only task is to isolate stable, nonsegregating, healthy F_2 families and bulk up the best ones.

15.3.2.4. Biological Control

Insects and fungi have been isolated that attack parasitic weeds (for a review, see reference 42). Biocontrol can also be accomplished by using allelopathic plants,[581] a form of biocontrol. No insects have been reported that kill the underground portions of *Striga*. There are insects that mainly attack the seed pods, eating most, but never all of the seeds, or even enough seeds to matter. Thus, the replenishment of the seedbank is sufficient to sustain the weed population[998] while having little (if any) increase in yield, and thus biocontrol by insects will not be further discussed.

Fungi have been tested both for *Striga*[225] and *Orobanche* control,[42] but not yet in wide-scale field testing. Regulatory authorization may be a problem for even indigenous strains, as the best pathogens are parasite-specific formae speciales of *Fusarium oxysporum*. DNA fingerprinting[41] suggests that these strains diverged from crop-pathogenic strains of this species over one hundred thousand years ago (L. Hornok, pers. commun.), yet regulators are fearful of all *F. oxysporum* formae specialis.

The strains that attack *Orobanche* have not been successful in providing near the level of control desired by farmers when tested in the field, or even in large-pot experiments in the greenhouse.[41] Transgenes encoding auxin production were introduced into an *Orobanche*-attacking fungal species, significantly doubling the virulence,[229] which is still far from what the farmer needs. More potent toxic genes are needed to enhance virulence, and the *NEP1* (necrosis-eliciting protein) gene, used to enhance a different mycoherbicide,[40] also was active in enhancing the virulence of a *Fusarium* sp. that is specific to *Orobanche* spp. The biosafety aspects of using transgenically hypervirulent biocontrol agents are specifically addressed in reference 414, where possible ways to contain such organisms by using asporogenic mutants to limit spread are discussed along with methods of gene flow mitigation. Another set of genes that might be added specifically to biocontrol agents controlling *Striga* would be the genes of the ethylene biosynthesis pathway, as ethylene causes false germination, independently of the host, as previously discussed. Bacteria and fungi that excrete ethylene also cause *Striga* germination,[14,15,115] but are not necessarily pathogenic to the parasite. Ethylene-producing genes in a pathogen should cause more *Striga* to germinate, providing more food for the pathogen. The fungi could additionally be engineered with the *ipt* gene to stimulate cytokinin production, as this group of hormones potentiates germination.

While intercropping of parasite-susceptible and parasite-resistant crop species is not considered to be a "biocontrol" measure, an interesting allelochemical compound has been described from research using *Desmodium* as an intercrop.[581] If parasite-susceptible crops could be engineered to produce this compound (when the genes responsible are found, and are amenable) resistance might be possible, without "intermediates," that is, herbicides, biocontrol agents, or intercrops.

Most of the strategies outlined above are multigenic and produce "expensive" products. The genetic engineer and the breeder must always remember the conundrum posed to plants: to grow or to defend.[497] If the yield loss is more than the cost of alternative strategies, for example, chemicals that control birds or witchweeds, the farmers will choose the more cost-effective solution.

15.4 Gene Flow Biosafety Constraints

15.4.1. Sorghum Transformation

Various groups have transformed and regenerated sorghum. Most recently, conditions for microparticle bombardment were optimized for four types of sorghum by using transient expression of the *uid*A gene as a reporter. Fertile transgenic plants were regenerated from immature embryos and from shoot tips. Stable integration and Mendelian inheritance of the selectable marker gene was demonstrated in all transgenic plants.[1041]

15.4.2. Biosafety Considerations—Introgressing Weeds

Sorghum is quite capable of crossing with its progenitors and various related wild and weedy relatives (Fig. 38).[312] Surprisingly, very little is known, and no publications could be found, about natural gene flow among the various *Sorghum* species in the center of origin. Thus it is hard to analyze risk vis-à-vis any of the possible transgenic solutions to the bird and parasitic weed species problems. Two related species that interbreed with sorghum are intractable major weed problems almost everywhere sorghum is cultivated, especially in monocultures or limited rotations:

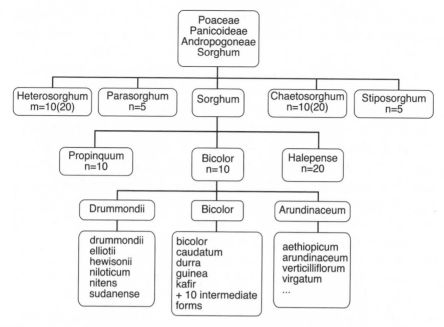

Figure 38. Evolutionary interrelatedness of sorghum species. The sorghum taxa as per de Wet[262]: the *Sorghum* genus is a member of five major subgenera in the Poaceae family. Species *bicolor* is the cogenitor of the crop-wild-weed complex in the sorghum genus.

15.4.2.1. *Sorghum halepense* (johnsongrass)

Sorghum halepense is one of the world's most noxious weeds.[509] Indeed, a selective herbicide that could distinguish between the crop and the weed is one of the new traits most desired by farmers in the Western Hemisphere, but of course only if the gene responsible for selectivity would not appear in this and other sorghum-related weeds. Sorghum itself is considered to be one of the progenitors of this amphitetraploid hybrid species, which propagates vegetatively and also produces some seed. Of course *S. halepense* still crosses with its crop progenitor. A few groups in the United States have studied gene flow between the crop and this weed. Hybrids between the weed and the crop did not show any significant increase or decrease in time to flowering, panicle production, seed production, pollen viability, tiller production, or biomass.[63] This let the researchers conclude "that a transgene that is either neutral or beneficial to *S. halepense* would probably persist in populations growing in agricultural conditions under continued gene flow from the crop."[63] The long-term persistence of sorghum genes in *S. halepense* populations was elegantly studied by another group.[750] They surveyed two hundred eighty-three loci on all ten sorghum linkage groups and identified seventy-seven alleles at sixty-nine loci that are found in U.S. sorghum cultivars but are absent from a worldwide sampling of *S. halepense* genotypes. These putatively cultivar-specific alleles were present in nearly a third of *S. halepense* individuals in populations growing next to production fields where sorghum has been repeatedly cultivated, but at far lower frequencies where there has been no recent sorghum cultivation. Their genetic data are inconsistent with alternatives to the hypothesis of continual gene flow, such as evolutionary convergence, or joint retention of ancestral polymorphisms,[750] leaving gene flow as the best explanation. Although most people worry about the crop pollinating the weed, another group has found that genes from *S. halepense* can just as easily find their way into the crop. They used a recently evolved herbicide-resistant biotype of the weed and readily found the dominant gene in the offspring of male-sterile sorghum inbreds.[996]

15.4.2.2. Shattercane

Shattercane with its weedy syndrome properties such as seed shatter is a feral form of domesticated sorghum. It is hard to control shattercane in many other crops, and the weed has evolved many herbicide resistances,[483] and of course it is obviously impossible to chemically control shattercane within a conspecific sorghum crop. This is another reason why farmers desire a transgenic herbicide resistance that would allow shattercane control. Of course the crop and its feral form are fully genetically compatible,[312] so containment and mitigation are imperative.

15.4.3. Risks from Transgene Flow

Clearly transgenes could flow from sorghum to its relatives just as other genes have moved around. The question then is the proverbial "so what"? The implications of transgene flow are almost the opposite in different areas of sorghum cultivation. Indeed, the discussion below demonstrates how hard it may be to decide what gene flow may not be an issue, and what is contraindicated.

For example, herbicide resistance transgenes would be highly undesirable if the genes moved in North America into *S. halepense* or to shattercane, because it would preclude control of these weeds in sorghum as well as in other crops. At present, herbicides are rarely used in Africa, so the genes would not have an impact on the fitness of these and other relatives of sorghum. Indeed, the developers of mutant herbicide resistance in sorghum for *Striga* control use this as an excuse for not worrying about gene flow, a worry that is not a regulatory issue as they are not under the same constraints as developers of transgenic resistance. When Africa overcomes its institutional problems, begins job-generating industrialization, and subsistence agriculture subsides and larger-scale production agriculture produces, herbicides will become a part of agriculture. Then there may be a reevaluation of whether transgenic resistances with transgene flow containment and mitigation would not have been better than using mutants.

The flow of transgenes from sorghum carrying direct *Striga* resistance might be a trait that enhances the fitness of the related wild and weedy species. Even if this is the case, and resistance does spread through wild populations, one could ask if this is not desirable. Some would argue that all biodiversity is good (but why then is the biodiversity generated by transgenics bad in their eyes?). Most people are more balanced, and are all for eradicating that addition to urban biodiversity generated by the existence and world spread of *Rattus* spp., using the excuse that rats are recent intruders that vector diseases (but so are parrots). *S. hermonthica* is also a recently evolved species; it is a rat of the plant kingdom and may be equally undesirable, and if a wild sorghum is a garbage can harboring *Striga* as a secondary host, then putting poison in the garbage can may be what is needed.

The flow of genes from an avian-poisonous cyanogenic-sorghum would also be a highly undesirable side effect in the north, but perhaps would be desirable in Africa, where severely limiting all quelea populations is a strategy. This would have to be balanced against the effects it would have on other bird populations, as well as on domestic and wild animals grazing on wild species. Perhaps in Africa a seed-specific promoter for poison production and an RNAi in the stalks against it would be desired to protect animals. This might be better than the natural varietal poison producers, which probably overproduce in stalks as well as seeds. (No data seem to be available about dhurrin production in stalks of Ark. 3048 discussed above.)

The flow of genes from nutritionally enhanced sorghums would probably have little impact. The rare hybrid offspring from such individuals would probably be less fit, due to a greater nutritional value to herbivores. As weeds are high propagule producers, these less fit hybrids would be outcompeted by their sibs and would disappear from the populations.

15.4.4. Dealing with Transgene Flow

The relatedness of the crop to other *Sorghum* species (Fig. 38) and the existence of weedy feral forms of the crop[312] present a transgenic biosafety hazard, vis-à-vis developing transgenic forms. These hazards are similar to the problems described for rice. Indeed, one can substitute "sorghum" for rice in much of Chapter 12; thus, the reader is referred to that discussion, and to Chapter 4, where gene flow risk, containment, and mitigation strategies are discussed.

Male sterility has been tested as a specific mechanism to contain gene outflow in sorghum. The ability of A(3) cytoplasmic male sterility to control transgene flow through pollen (using nontransgenic pollinators) to decrease the risk of viable pollen flow was tested under field conditions. The normal percent seed set of F_2 individuals averaged 74 percent, and on A(3) F_2 individuals averaged four per ten thousand.[839] PCR analysis confirmed that the four male fertile individuals from a population of a thousand contained the A(3) cytoplasm.[839] This severe restriction of gene flow through pollen would be helpful in slowing gene flow but not in staunching it. A trait that increases fitness of shattercane or johnsongrass would slowly appear because of the slowing of gene flow by male sterility, but would quickly spread once it appeared in sexually crossing material. Male sterility only prevents the outflow of genes. The male sterile variety could be pollinated by a related weed, and the same F_1 forms as with pollen flow. This is not just "theoretical" because it has been demonstrated that *S. halepense* can transfer "useful" genes to sorghum.[996] Thus, the use of A3 cytoplasmic male sterility would have to be stacked with other containment or mitigation strategies, outlined in Chapter 4, especially those that would prevent pollination from the wild, or would mitigate the effects of a transgene.

A series of genes has recently been elucidated that specifically encodes processes leading to storage of metabolites in the *S. halepense* rhizomes.[537] The ability to store such metabolites confers "phoenix" resistance to herbicides on this species; herbicides appear to kill the shoots, but there is phoenix-like regrowth from the rhizomes. The more active and rhizome specific of these genes in antisense or as RNAi could act as excellent mitigator genes, transforming *S. halepense* into an annual instead of a perennial. Used together with mitigators preventing shattering and conferring dwarfing should turn hybrids with this pernicious pest into wimps without affecting the crop. Gene constructs with these rhizome-specific genes in antisense or RNAi under control of a rhizome-specific promoter could be transformed into disarmed viruses, and the *S. halepense* plants could be infected as outlined in Chapter 8. If indeed this works, the plants could now be killed by the herbicides that had resulted in phoenix-type resistance. Such constructs should have no effect on standing crops that do not possess rhizomes.

CHAPTER SIXTEEN

Oilseed Rape

Unfinished Domestication

Oilseed rape (*Brassica napus*) (also called rapeseed or canola) is a relatively recent crop (in evolutionary terms) and has in the past decades become a major source of edible oil in temperate/cool climates. Elegant work by U[1081] demonstrated that *B. napus* (genomes AACC, 2n = 38) is probably a hybrid between *B. rapa* = *B. campestris* (AA, 2n = 20) and *B. oleracea* (CC, 2n = 18). *Brassica rapa*, the pernicious weed, wild turnip, has also been domesticated (confusingly called turnip rape or Polish rape or just rape by many authors) as an oilseed crop, as well as being domesticated as turnips. *Brassica oleracea* is the progenitor of cabbage and relatives.

Oilseed rape has clearly been domesticated compared with its progenitors. Recent synthetic hybrids between the two progenitors are very weedy.[18,747,820] The differences between domesticated oilseed rape and its ancient hybrid progenitor include the following: it lacks of seed secondary dormancy, it is self-pollinated (but retains about 20 percent outcrossing), it does not shatter as much as the wild type, and it has far less genetic diversity.[451] Its further domestication in the past four decades has been biochemical, changing it from its status as a producer of a low-value industrial oil with feeding deterrents in the resulting residue, to a much higher-value crop with edible oil and a meal that could be served to livestock. The high level of long-chain erucic fatty acid in the oil rendered it inedible for human consumption, and the high glucosinolate content in the meal affected livestock. These were bred out in Germany and Canada, and the crop renamed in Canada (but not trademarked) as Canola (*CAN*adian *O*il *L*ow *A*cid), for the 00, or double null varieties lacking these two components.

Oilseed rape is grown mainly in cooler northern (or southern, as in Australian) climes. It is considered a good crop, because it is one of the few broadleaf crops that can be rotated with wheat, in a much-needed rotation in areas too cold for soybeans. In Europe its culture is politically motivated, in part, to reduce dependency on imported (much cheaper) soybean oil, necessitating substantial European Union (EU) subsidies to compete. Additionally, yellow flowering oilseed rape is a beautiful sight to see in a gloomy European spring, and landscape in agriculture has gained in importance in Europe. The last time fields in Europe looked so beautiful was before World War II, prior to phenoxy herbicides, when related brassicas gloriously bloomed in the wheat fields (heavily lowering crop yields). China is the largest producer of oilseed rape (thirteen million metric tons), followed by Canada (7.7 Mt), India (6.8 Mt), Germany (5.3 Mt), France (4.0 Mt), and Australia (1.5 Mt).[329]

Oilseed rape can be transformed,[739] and transgenic herbicide-resistant oilseed rape is the predominant type cultivated in western Canada, the largest exporting region in the world. Various biotech companies have elegantly transformed oilseed rape to obtain virtually any oil composition that can be desired. Calgene, now part of Monsanto, generated many different lines of oilseed rape with different compositions, including a high laurate line (resembling cocoa butter) for chocolate.[930]

The rotation of wheat with oilseed rape is important for the wheat, reducing weed and disease problems in the wheat. Rotation is presently imperative for oilseed rape, because rape monoculture is almost impossible due to a buildup of its own insect and disease problems. It is also important because oilseed rape is still insufficiently domesticated, and monoculture could lead to shattered crop seed dedomesticating and becoming feral.

The cultivation of oilseed rape can be questioned on environmental grounds. Despite the ban on the use of the soil fumigant methyl bromide, the reduction in ozone-layer-depleting alkane halides has hardly decreased. This is because of the release of such compounds into the atmosphere from mainly natural sources, from algae and fungi through to higher plants. Most of these natural sources cannot be controlled by humans, except the release from crops.[473] Among the crops *Brassica* species emit orders of magnitude more methyl bromide than all others.[932] The 1998 estimate is an emission of seven thousand tons of methyl bromide per year from oilseed rape,[373] which translates to nine thousand tons for 2005 due to the expansion of cultivation of this crop.[329] It is telling that the so-called environmental groups that so vehemently campaigned against methyl bromide are mute as to the environmental impact of this major anthropomorphic source of pollution. Could it be because the "fix" to this problem may be transgenic?

16.1 The Problems of Underdomestication

There have been eight thousand years of domestication of allopolyploid wheat, probably a third as much for oilseed rape. Wheat has two traits quite different from oilseed rape because of the many more millennia and greater areas, with more farmers acting as domesticators and selecting the most desirable individuals: wheat is more than 99 percent self-pollinated in the field, whereas oilseed rape is only 80 percent self-pollinated. Wheat hardly shatters, oilseed rape under good conditions shatters about 5 percent of its seed, enough to replant itself,[747] rendering oilseed rape a major volunteer weed problem. When poor weather delays harvest, even more rapeseed shatters to the ground.

16.1.1. The Problems of Cross-Pollination

The consequences of weedy-type cross-pollination in oilseed rape have been quick to appear. Wheat breeders can grow families for testing in adjacent blocks, the level of cross-pollination is so low, but not oilseed rape.[452] The advent of mutant imidazolinone herbicide-resistant oilseed rape and transgenic glyphosate- and glufosinate-resistant oilseed rapes and the concomitant selection pressure facilitated easy visualization of pollen flow that occurs between fields. Pollen flow has allowed the genes to stack naturally, and double- and triple-resistant oilseed rape is common.[100] Much of this might be the result of cross-pollination between volunteer oilseed rape from a previous season bearing one transgene, with a standing crop bearing a different transgene. The problem is an expensive nuisance, but not insurmountable at present.[101] Either preplant cultivation must be used in wheat (yet wheat farmers prefer more energy-efficient, cheaper, minimum or no tillage that prevents erosion), or a return to phenoxy herbicides in wheat. There would be nothing to stop a biotech company from developing and registering a transgenic phenoxy-resistant oilseed rape with the present regulatory rules. If this were done, very limited technologies would be available to control volunteer oilseed rape in wheat. Additionally, pressure groups are demanding the deregistration of phenoxy herbicides, which would have the same effect.

Thus, the next generations of transgenic oilseed rapes should clearly have built-in methods for limiting gene flow to other varieties, as well as the previously described fail-safe mechanisms (Chapter 4) to prevent pollen flow to related weeds and wild species.

16.1.2 Shattering Rape

Too many seeds of oilseed rape shatter prior to harvest.[232, 440, 834, 869] Under normal conditions in Canada, seed loss is 100 kg/ha, or about 6 percent of the potential yield.[440] This means that there is a seed rain of three thousand seeds per square meter, while a farmer typically plants more than an order of magnitude less. Luckily, most seeds do not overwinter (in western Canada); many are devoured by insects and diseases, and others succumb to freezing after premature spring germination. Still, not many viable seeds are needed to cause a volunteer weed problem, and what farmer wants to lose 6 percent of a crop? If harvest is delayed because of weather or other constraints, 30 percent or more of the crop can shatter. Even 70 percent has been reported.[232] Breeding has been exceedingly successful in bringing down the percent shatter. Just a few decades ago 15 percent shatter was considered the norm. Still, it appears that shatter resistance has reached the glass ceiling and little more can be easily done within the genome. One might hazard a guess that the genes that must be modified to enhance nonshattering have other functions, and their further modulation by genetic means has a yield drag. Two genes are known that together control nonshattering when recessive *sh1* and *sh2*, yet the dominant shattering can be found with a single RAPD marker, suggesting that a gene duplication was needed to obtain shatter resistance.[741] Do we need another duplication to lower shattering even more? Thus the ability to modify tissue-specific promoters by genetic engineering may be the key to dealing with shattering. As will be seen below, this may not be as easy as it might seem.[741]

16.2. Potential Biotech Solutions/Biosafety Issues

If the key problem in oilseed rape, shattering can be solved transgenically, a key biosafety issue is precluded, as antishattering is one of the best mitigator genes to prevent unwanted establishment following gene flow (Chapter 4). Likewise, if the overpropensity for outcrossing can be prevented through biotechnology, the gene movement would be even more contained. Thus, biosafety is not discussed in a separate section. The two issues described above needing further domestication can be solved in the next generations of oilseed rape by transgenic containment and mitigation (Chapter 4), or by solutions that contain elements of both. Because oilseed rape is one of the few crops adapted to very cold climes, many valuable transgenes will be further introduced into the crop. It would benefit all if these transgenes are introduced in tandem constructs with mitigating genes that preclude gene flow, keeping the valuable transgene from leaving proprietary material, while preventing seed shatter, raising yields, and rendering the cultivation of the rotational crop easier. These genes should probably be introduced with dwarfing genes, which have the potential of raising yield,[19, 21, 758] while mitigating volunteer weed establishment[21] and establishment in crosses with weedy relatives.[19]

16.2.1. Can Volunteer Oilseed Rape Go Feral?

Oilseed rape seemed to be the crop most likely to dedomesticate and go feral from volunteers, both because of the large number of volunteers produced enhancing the likelihood, and because the crop is not completely domesticated to begin with. The major hindrance to going feral was the fact that oilseed rape is rarely grown as a monoculture. Good agronomic practices with a rotational crop controls incipient feral individuals. Thus, the greatest chance from ferality would come from ruderal areas, disturbed lands near fields where seed spillage might occur. Volunteer populations of oilseed rape are usually generated annually from the spillage of seeds, but some populations self-perpetuate over several generations. Volunteer populations usually gradually decline until the arrival of new seed and any claim for the evolution of ferality was countered by the possibility of renewed spillage. If there had been evolution of feral populations, the population would be expected to genetically diverge from the original cultivar.

This possibility of the evolution of feral biotypes is checked by comparing a persistent feral population with many different winter and spring varieties by use of multidimensional scaling.[132] The feral population investigated was located within a field and along the field margin, having persisted for at least six years. The positioning of the feral plants within the multidimensional scaling plot (Fig. 39) reveals that most feral plants cluster separately from both winter and spring cultivars,[132] although five feral plants are in or near the cluster of winter cultivars. These individuals were probably established from recent seed spillage. Most of the feral plants are clearly separable from the cultivars tested (Fig. 39). The authors suggest (but alas did not perform the experiments to demonstrate) that strong selection pressure for winter hardiness and seed survival may have caused the feral population to diverge from

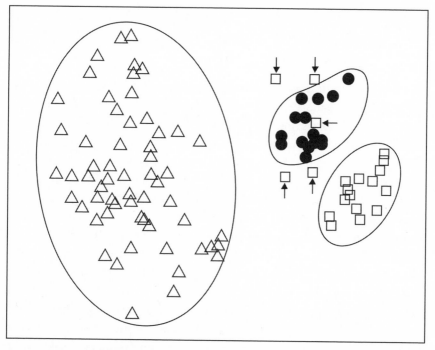

Figure 39. Feral oilseed rape probably evolved from winter rape varieties. Most feral rape populations are genetically dissimilar to winter and spring cultivars, suggesting evolutionary differentiation, although some are similar to winter rape varieties, suggesting recent origins as shattered volunteers. The multidimensional scale cluster data are replotted from an analysis of microsatellite information in Bond et al.[132] using individuals from eight spring cultivars (open triangles), four winter cultivars (filled circles), and separate feral plants (open squares). Individuals marked with arrows may be volunteers into the feral populations from the winter varieties, or hybrids between feral individuals and the winter varieties.

the agricultural cultivars.[132] There is a possibility that this long-standing feral population is related to an untested winter cultivar. The genetic dissimilarity is greater between the spring cultivars and the feral plants than between the winter cultivars and the feral plants.

This suggests that transgenic spring cultivars are less likely to persist in a feral population than a trait introduced into a winter cultivar, at least in the climatic conditions of Scotland where the experiments were performed.[132] One looks forward to seeing further studies with this population, whether they have a higher degree of shattering and whether the seeds have secondary dormancy that lasts beyond one year. These characters would be the best proof that indeed a feral population has been found and that ferality is more than a theoretical worry.

16.2.2. Preventing Gene Flow to Other Varieties of Oilseed Rape and to Other Brassica Species

There are various reasons to worry about gene flow in oilseed rape. When the original glucosinolate/erucic acid–free varieties were bred, gene flow from volunteer populations and older varieties to the new varieties was an issue. Intervarietal gene flow in seed production areas was always an issue for seed producers, and distances between fields were mandated. The stacking of both transgenic and mutant commercial herbicide resistances became a worry to the farmers who had to know what herbicide would work to control volunteers in wheat. Even though none of the transgenes used has any properties that are even remotely considered dangerous to the public, mandated admixtures of less than 0.9 percent are part of the current EU regulations. This has to be measured in the seeds before crushing, as the oil is free of DNA.

There are two ways to estimate the gene flow. One is to develop simulation models at a computer console in the lab, and then write about the results in language that suggests that gene flow was actually measured: "Gene flow increased with the area of the pollen and seed producing field."[231] The authors then concluded that nearly half a kilometer is needed between fields. The mathematician this author has collaborated with in many a modeling exercise continuously muttered the mantra: "garbage in, garbage out"; if assumptions are inaccurate, it is so easy to be wrong in modeling by factors of a million, as described in an analysis of three other models (including one of the author's).[416]

Instead of modeling gene flow in an office with a computer, another group performed field experiments. Unlike with other gene flow measurements in cereals and conifers, there was no effect of wind direction. The vectors carrying the pollen (bees) are far less predictable than wind.[367] When acceptor plots were just three meters from the donor plots the level of gene flow was well beneath the EU-mandated levels, and their conclusion was that to stay below the mandated limits separation was the only requirement, but no distances are required because of plot dilution.[367] These results imply that most gene flow problems reported must have come from hybridizing within the fields; between the crop and volunteers from previous years. The results demonstrate how easily models can be misused.

Despite the apparent lack of consequential gene flow to meet the EU requirements, the other issues of transgene establishment, and oil contamination are issues that need addressing. It is easy to prepare specifications of what is needed to deal with gene flow, because the mechanisms already exist. That does not mean that the genes have been isolated and can be taken off the shelf. One could introduce into each variety a copy of the pollen recognition and rejection systems from a distantly related species. The numbers of such species-specific systems is huge, that is, one variety should contain the system that allows *Arabidopsis thaliana* to recognize only *A. thaliana* and reject all other species. Another oilseed rape variety should have the analogous gene(s) from rice, and so on. This would mean that each variety would only allow fertilization by pollen of the same variety, and reject all others (they would possibly recognize *Arabidopsis* and rice pollen, neither of which are major species near oilseed rape). Pollen grains from these original gene donor species may rarely find their way to an oilseed rape stigma, but postfertilization chromosome problems would cause abortion, or at worst F_1 infertility. Such a solution will also take care of the problems of crossing with related weeds and wild species. No recognition, no fertilization, no problem. This proposed solution sounds easy, until it is realized how little we know and how much we have to learn about fertilization systems.

16.2.3. Engineering the Shatter out of Rape

The breeders have probably reached the genetic glass ceiling for genetic resistance to shattering in oilseed rape, except by obtaining it from crosses with other *Brassica* species.[748] The siliques (seed pods of *Brassica* species) have two valves, which at maturity separate along lines of weakness, allowing the seeds to shatter (Fig. 40). These lines of weakness that make up the zones of dehiscence are composed of simple unthickened cells.[200,653,714]

One genetic approach was to resynthesize *B. napus* from its progenitors to increase the diversity.[747] This was done with shatter-resistant lines in the progenitors. The hybrids had more vascular tissue within the dehiscence zone. Similarly, less shattering could be obtained by nonhomologous introgression of nonshattering genes from *B. juncea*[865] or in hybrids with *Sinapis alba* by using embryo rescue.[158] It is fascinating that wild species have shatter resistance genes to contribute to a domesticated species that has been selected for shatter resistance. This further suggests that selection for other agronomic traits allowed a regression toward shattering. In early agronomy of oilseed rape, the crop had been cut greener and wind-rowed to dry, and shattering was reduced. Perhaps if the combine harvester is used for a few centuries, there will be natural selection away from shattering.

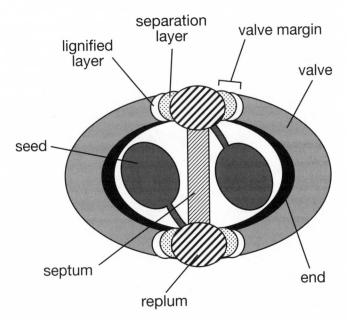

Figure 40. Schematic cross section of a mature silique prior to shattering. *Source:* Modified from Dinneny and Yanofsky.[278]

The shatter resistance cannot be too strong or perhaps it will be impossible to adjust the combine harvester to thresh the siliques to remove the seed. The synthetic line derived from the progenitors described above required a readjustment of the combine to have a 33 percent increase in combine rotor speed than is used for normal varieties, and the amount of damaged seeds was then doubled. Immature seed of the synthetic line required a doubling of rotor speed to be threshed, resulting in yet higher seed damage.[161] Thus it is clear that increasing shatter resistance will have a cost in seed damage, or that not just genetic engineering will be needed, but a reengineering of harvesting equipment. Engineering has been used to find a way to assess shatter resistance so that efforts to reduce shatter can be assayed with just a few siliques. A previous method to measure pod strength was based on cantilever bending until fracture occurred.[557] This method suffered from a low (albeit significant) correlation with field results. A random-impact assay was developed that measures the breaking point of twenty siliques until half are opened, which varied from three seconds to two minutes among lines.[162] A simplification uses a standard seventeen seconds, where a quarter of the seeds of a commercial cultivar shatter.[162]

The biochemical and molecular changes correlated with the dehiscence of siliques has been reviewed at length,[278, 907] and it is necessary to understand how these factors interact to possibly prevent shatter. The enzymes involved in opening the fracture lines of the siliques include cellulases (but which of this multigene family are involved?), pectinases, expansins, and pathogenesis-related proteins (what function?). A layer of lignin is just outside the dehiscence zone (Fig. 40), which increases the tension during maturation, allowing splitting. A *Brassica juncea* that is less lignified does not shatter.[1005] Additionally, shattering is less when the main vascular bundle of the silique valve that traverses into the pedicel is 60 percent larger, as in the synthetic *B. napus* described above.[215]

The genes that mutate to prevent shatter are actually controlling elements. The genes that associate with the dehiscence zone itself are mundane genes that control cell wall integrity, and their control must be subject to an element of timing. The timed disappearance of hormones from the siliques may be related; treating pods with the low rates of the auxin analog herbicide 2-methyl-4-chlorophenoxyacetic acid (MCPA) delayed natural dehiscence four days.[204]

Thus, it seems apparent that if one wants to use RNA interference (RNAi) or antisense to suppress senescence by using one of the cell wall-affecting genes, the controlling elements on the construct will be critical unless the enzyme involved is a member of a family that is only turned on during senescence. Indeed there is one such case; it is claimed that a specific pectinase is expressed only during dehiscence and at no other time in the plant.[846] If a general promoter is used with most cell wall-degrading genes, then other repercussions to the plants are likely, so tissue-specific promoters will be required.

Early patents[527] jumped on the possibility that preventing the expression of the cell wall-degrading genes would prevent seed shatter. Later papers and patents homed in on controlling element genes from *Arabidopsis* such as *SHATTERPROOF* and *FRUITFULL*,[652,653] *AGL*,[1173] *indehiscent1*[1174] genes, but the patents all were devoid of actual data showing an effect in oilseed rape, just predicting such an effect (Table 26). Whether antisensing individual or groups of such genes will work is an open question.

Attention was focused on MADS box transcription factors of *Arabidopsis*, a fellow member of the Brassicaceae with anatomically similar siliques to oilseed rape. These MADS box factors control many elements in reproductive development, from flowering initiation through seed shatter. Their role in flowering will not be discussed (see reference 200). Mutants in the *FRUITFULL* gene resulted in short siliques with crowded seeds. *FRUITFULL* negatively regulates *SHATTERPROOF* 1 and 2 transcription factors, which results in siliques that do not open.[342, 652] In one case the ortholog of *FRUITFULL* from *Sinapis alba MADSB* was overexpressed under the control of a 35S promoter both in winter and spring rape.[200] Silique anatomy was modified (Fig. 41), and shatter was prevented. The level of shatter was a function of expression in different lines. As the same construct changed the time of flowering, the construct might not be appropriate, requiring a different promoter; one that is only activated some time after anthesis. The authors claim, without presenting data, that the spring rape siliques with modified anatomy (Fig. 41) will easily be opened by threshing equipment. Winter rape, with similar expression levels, will remain tightly closed. They did not use any of the physical tests developed, so it is not clear whether the spring rape will shatter less than wild type.

Another group ectopically expressed the *Arabidopsis FRUITFULL* gene under the control of the 35S promoter, not in oilseed rape, but for an unclear

Table 26. *Many Cloned Genes are Involved in Shattering (Abscission/Dehiscence) in Brassicaceae*

Gene	Encodes	Function	Species	Reference
SHATTERPROOF I & II	MADS box transcription factors		*Arabidopsis*	342, 652
FRUITFULL	MADS box transcription factor; close homolog to MADSB	Silique valve cells do not elongate; negative regulator of SHATTERPROOF	*Arabidopsis*	437
INDEHISCENT	Atypical basic helix-loop-helix transcription factor	Mutants have eliminated lignin layer in silique	*Arabidopsis*	653, 1174
ALCATRAZ	Basic helix-loop-helix transcription factor	Mutants have no separation layer in silique	*Arabidopsis*	883
REPLUMLESS	BEL subfamily of homeodomain transcription factor	Mutant converts replum into a valve margin	*Arabidopsis*	914
MADSB	MADS box transcription factor; close homolog to MADSB	Ectopic expression in oilseed rape prevents shattering (see Fig. 41)	*Sinapis alba*	200, 723
Cellulases		Degrade enough cellulose to loosen wall		
Pectinase(s) (polygalacturonidases)		Degrade pectin between cells		
Expansins		Loosen cell walls allowing slippage		

Note: For a more complete discussion of these gene effects and the interactions that occur among them, see references 200, 278, 907.

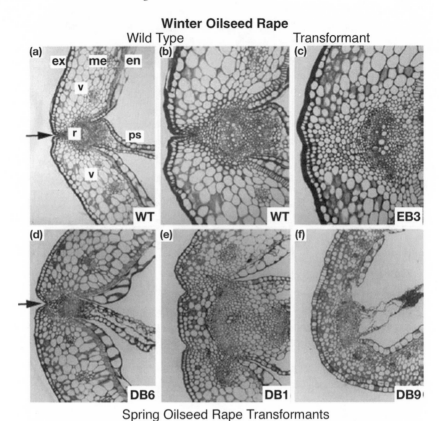

Winter Oilseed Rape

Figure 41. Modification of silique structure, preventing shattering, by engineering *Sinapis alba 35S:MADSB* into oilseed rape. The arrow points to the zone of dehiscence. Different spring oilseed rape transformants show the effects of different levels of expression, and it is unknown how much the transformants are affected vis-à-vis shattering. The winter rape variety is shattering resistant. *Source:* Modified from Chandler et al.[200]

reason in *Brassica juncea* (Indian mustard).[829] The results are similar to those described above with winter oilseed rape, siliques that are so tightly closed that they cannot be threshed. Fine tuning is clearly needed, but shatter resistance should be forthcoming.

16.2.4. Shattering and Shedding in Other Species

Assuming that the genes discussed above will provide the necessary anti-shattering in oilseed rape, will the same genes work with grain amaranth (a dicot), and will they work with grains such as tef and other underdomesticated millets that shatter? In the grain species shatter is due to separation of the seed from the branch that bears it, and not an internal opening of the fruit holding the seed. A dominant spike fragility leads to shedding the seed in the progenitors of wheat and domestic wheat, but in a hexaploid wheat that evolved ferality (dedomesticated), shattering is dominantly inherited as an abscission caused by a break in the rachis that causes the seed to shatter.[1029] Nonshattering in sesame is controlled by a recessive gene that induces more cell layers in the region of separation, making it harder to split.[257] In each case the anatomy of the dehiscence zone or the abscission zone is quite different. It has recently been claimed that even within cereals it is "to each his own" vis-à-vis shattering, with little similarity in morphology of shatter layers, mechanisms, and control genes among the various species studied.[650] Still the hydrolytic enzymes responsible for shatter are the same or very similar. When the comparative annotations of the rice genome and the *Arabidopsis* genome are better, and our understanding is better, we may know more. At the time of this writing, no cereal orthologs of *SHATTERPROOF* and *FRUITFULL* are known.

It is not always desirable to prevent an abscission zone from forming. A case in point is olives, where premature fruit drop is not desirable, but hand picking (usually by stripping with a stick) is time consuming and injures the fruit. Mechanical picking by shaking the tree or branches does not work well, because the abscission zone forms at different times in different fruits. Here is a case where plant growth regulators are useful to induce the genes that cause a rapid formation of an abscission layer. An olive grove is sprayed a day before picking with such a regulator, and crews come through mechanically shaking the trees with the fruit falling lightly on canvas placed under each tree.

16.2.5. Other Genes for Further Domestication of Oilseed Rape

Various groups have discerned needs to further engineer oilseed rape to have other properties. Successful efforts to modify its oil composition to meet various consumer needs are described in Table 17. A group that thought to manufacture pharmaceuticals in oilseed rape abandoned this crop because of the gene flow issues (Chapter 17). As noted earlier, oilseed rape is a more valuable crop than its typical rotational partner, wheat, yet rotations are a must because of insect and disease problems. These problems are being addressed, although it is hoped that even if successful, the farmers will stick with rotation so that the current crop is protected and the insects and diseases do not evolve resistance to the transgenes.

Disease resistance was addressed in an elegant manner to deal with two problems simultaneously. The group wanted to attain disease resistance by engineering in the gene stilbene synthase, a single gene that transforms a primary metabolite in a single step to the pathogen-killing phytoalexin resveratrol[517] (better known as the magic component of red wine that purportedly prevents heart attacks in France). The problem was that much of the primary pathway was being siphoned off by a competing secondary pathway that makes antinutritive sinapate and sinapine. Thus, they transformed oilseed rape with a construct that contained stilbene synthase under a strong seed-specific promoter, along with an RNAi for a gene encoding a key step in the sinapate pathway, considerably elevating the stilbene levels in the seeds.[517] It does not seem probable that this will protect leaves from disease, and there is no mention of resveratrol in the oil, but the lack of the antifeedants should enhance the value of the meal. A database search could not find information on whether the antiherbivory effects of sinapate/sinapine carried over from livestock to insects, but sinapate has also been implicated as an ultraviolet protectant.[971]

Resistance to insects has been addressed by engineering Bt Cry 1 Ac at effective levels,[1138] as well as by using a mustard trypsin inhibitor MTI-2.[344] The latter gene is useful because it does not affect the natural predator of the diamondback moth, while controlling the moth itself. As the diamondback moth has evolved resistance to the low rates of Bt used as sprays in organic agriculture (but not to the higher levels in transgenic crops), it might be wise to stack the two genes to guarantee a delay in evolution of insect resistance to either of the insecticidal genes.

Experimental field releases in the United States also include oilseed rape containing an alanine aminotransferase from barley to increase nitrogen utilization efficiency and unstated genes to increase seed size and yield, to reduce environmental stress, and to confer male sterility.[56]

16.2.6. Lessening the Environmental Impact of Brassicaceae

It is clearly necessary to lower the levels of the alkane halides (methyl bromide and its analogs) emitted by oilseed rape and its relatives. *Brassica oleracea*, a progenitor of oilseed rape, was shown to possess a bifunctional methyltransferase using S-adenosyl-L-methionine (SAM) that methylates halides to methyl halides as well as bisulfides to methanethiol.[73] Methanethiol is transformed to sulfuric acid and comes down as acid rain. The enzyme was purified to homogeneity, characterized,[73] and later cloned.[74] This was taken further in *Arabidopsis*, where an ortholog was found and euphemistically named *HOL* (*HARMLESS TO OZONE LAYER*).[892] A TDNA disruptive insertion into the gene resulted in plants that produced less than 1 percent the methyl halides as the wild type.[892] Many of the plants emitting methyl halides are salt tolerant. Cotton is among the crops that are salt tolerant and it possesses an ortholog,[892] but it has not been tested for methyl halide production. If one were a believer in conspiratory theories, or if one accepts some recent historical analyses about governmental interference in environmental research,[743] one would wonder if the there are not pressures not to find out more about crops producing methyl halides. The group that first ascertained that plants emit such large quantities of methyl halides in 1998[373] has subsequently published many articles on agricultural replacements for synthetic methyl bromide as a fumigant, but nary another paper on the natural production by crops. The results with the *Arabidopsis* mutant that does not produce these gasses suggest that there is little cost to not producing methyl halides, if at all, but this remains to be seen at a field scale with a real crop. It clearly should not be hard to RNAi the gene in oilseed rape and in other methyl halogenic crops to ascertain whether this can be easily done without detriment to the crop, but who has the incentive? The so-called environmentalists are activists that hate transgenics more than they love the environment, and no environmental regulatory authority has taken up the challenge. Not until authorities demand that, within a certain period, methyl halide emission of crops must be under a certain threshold will there be an incentive to deal with this issue.

CHAPTER SEVENTEEN

Reinventing Safflower

Safflower (*Carthamus tinctorius*) is a composite related to sunflowers that is cultivated mainly for seed, which is used to produce an edible oil as well as marketed as birdseed. Safflower is a highly branched, herbaceous, annual thistle with many long, sharp spines on the leaves. Two or three rows of safflower planted around a cereal field can fence out cattle. Needless to say, thistles are uncomfortable to cultivate and harvest. The crop has also been grown for its flowers (and still is grown only for flowers in China), is used for cosmetics, and was used as a dyestuff for foods and fabrics before cheaper aniline dyes became available, as well as for medicines. In ancient times it was also used as a ceremonial ointment to anoint mummies and as an oil for lighting. In the past few centuries additional documented uses have been as a purgative and for alexipharmic (antidote) effects, as well as in a medicated oil, to promote sweating and to cure fevers. Safflower was dropped from the European pharmacopoeia about half a century ago, but the Japanese pharmacopoeia detailed use of safflower more recently. Safflower is the subject of a lengthy monograph,[244] which is the source of much of the background information in this section (where not stated otherwise). Another source of information is the review by McPherson et al.[712] *The Sesame and Safflower Newsletter*, published since 1985, offers a considerable amount of information as well.

Safflowers are one of humanity's oldest crops. Its progenitor is thought to be *Carthamus palaestinus*, a self-compatible wild species restricted to the deserts from southern Israel to western Iraq. It evolved into the weedy species *C. oxyacanthus*, a mixture of self-compatible and self-incompatible types, and *C. persicus*, a self-incompatible species. These in turn are considered the parental species of the cultivated species *C. tinctorius*.[64] The four species have the BB genome formula and $2n = 24$ chromosomes. Intercrosses of the above, in all combinations, produce fertile hybrids. Pairing of chromosomes is essentially complete in hybrids between these species; this is not the case when the parents are *Carthamus* species having different chromosome numbers.[64] Gene introgression between the weedy and cultivated species may still take place.[712] The weedy progenitors of cultivated safflower are widely distributed in some of the areas where safflower is grown. Only a limited number of chemical herbicides are registered for use on safflower, mainly because of the high cost of testing required in many countries, which makes it too expensive for this minor crop.

17.1. Uses and Decline

Safflower is mainly cultivated for its edible oil, which has the highest polyunsaturated/saturated ratios of any oil available (Table 16). The linkage between health and diet has increased the demand for the oil in affluent countries. High oleic safflower oils are very stable on heating and do not give off smoke or smell during frying; it was used in Japan for tempura, but this quality is now mimicked by many other oils. Most safflower oil in Japan is sold in gift packs for special occasions, which reportedly are rarely opened and used, just circulated.

The known genetic variability in safflower allows the breeder to achieve edible oils with varying contents (Table 27). Three major recessive genes appearing at different loci control production of oleic, linoleic, and stearic acids (*ol ol, li li,* and *st st,* respectively). Increases in stearic acid are accompanied by decreases in the percentage of oleic or linoleic acid or both. Cooler growing temperatures reduce stearic and oleic acids while increasing linoleic acid in certain genotypes.

Table 27. Genetypic Variability in Fatty Acid Content Safflower

Oil type	Genotype	Fatty acid content in oil (%, range)			
		Palmitic C16:0	Stearic C18:0	Oleic C18:1	Linoleic C18:2
Very high linoleic	OlOlliliStSt	3–5	1–2	5–7	87–89
High linoleic	OlOlLiLiStSt	6–8	2–3	16–20	71–75
High oleic	ololLiLiStSt	5–6	1–2	75–80	14–18
Intermediate oleic	ol'ol'LiLiStSt	5–6	1–2	41–53	39–52
High stearic	OlOlLiLiListst	5–6	4–11	13–15	69–72

Source: Information in reference 244

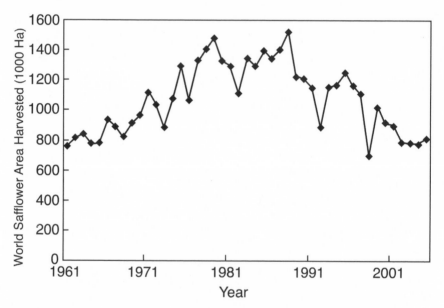

Figure 42. Rise and fall in world safflower production for oil. *Source:* Data in FAOStat.[329]

In recent years the area in safflower production worldwide peaked at one and a half million hectares in 1988 and is now down to half as much (Fig. 42). This can be compared with more than twelve million hectares in oil palm, which yields fourteen times as much oil per hectare, at far lower production costs. Chinese safflower production for its florets on about forty thousand hectares per year is not included in most crop estimates.

Soybean, oilseed rape, and probably palm can now be engineered to produce oils with any fatty acid composition desired, including that of safflower oil (see Chapter 11). The economy of scale and mechanization presently practiced in olive production now even render olive oil competitive with safflower oil as the healthy oil of the cognoscenti.

17.2. Constraints to Safflower Production

Besides the economic constraints above confronting the farmer wishing to grow safflower, the following constraints to safflower production have been described.[244]

17.2.1. Susceptibility to Disease and Insect Pests

Safflower is very susceptible to foliar diseases in moist atmospheres, in particular, leaf blight caused by *Alternaria carthami* but also *Botrytis cinerea, Cercospora carthami, Pseudomonas syringae, Puccinia carthami,* and *Ramularia carthami;* root-rotting organisms (*Phytophthora* as well as *Fusarium oxysporum* f. sp. *carthami* and *Verticillium dahliae*), especially under irrigation; and numerous insects, in particular, the safflower fly (*Acanthiophilus helianthi*) and aphids. These problems are most acute in the center of origin of safflower and its related species.

17.2.2 Developmental Pattern

Earlier maturity would make safflower more competitive with wheat and permit double cropping. Too long a duration at the rosette stage exacerbates weed competition; so early bolting types could have an advantage. The lack of primary dormancy at maturity, that is, premature germination of mature seed in the heads of standing plants (ovivivipary) after rains can be a major problem.

17.2.3 Morphological Ideotype

A changed angle of branching to achieve denser stands with more heads per hectare would facilitate harvesting. Varieties with shorter or no spines but having high yields are needed. Reductions in seed hull thickness increases oil to more than half the seed weight, but such changes may incur losses to birds, insects, and seed breakage during mechanical harvest.

17.2.4. Resistance to Stress

Further increased resistance to drought, greater resistance to salinity, and greater resistance to cold are on the wish lists of safflower agronomists and breeders.

17.3. Increase Yield through Breeding

The change of ideotype (discussed above), heterosis with sterility systems, and possibly interspecific hybridization with and without induction of polyploidy and changes in chromosome structures might enhance yields. One can still question whether dealing with the above, along with better agronomy, can raise the genetic ceiling of this crop in competition with other oilseeds. Thus, the area dedicated to this ancient species is probably relegated to further reductions, with its products expected to gather dust in health food stores and in the limited market of herbal medicine, despite initiatives to call attention to safflower and revive it in its classical sense.[244] The gene jockeys are reviving it as a new ultra-high-value crop for the production of pharmaceuticals and industrial products. To them, the very limited area in production for oil as a food product is a distinct advantage.

17.4 The Need for a Pharmaceutical Crop

Safflower is being recreated to produce pharmaceuticals and industrial products. To understand why, one must first discuss why plants are being targeted for such uses, due to problems with the other proposed crops for transgenically producing pharmaceuticals and industrial chemicals.

17.4.1. Production of Pharmaceuticals and Industrial Products in Plants

Many protein pharmaceuticals have been produced in animal systems. Insulin is a classic example; initially it was produced from the slaughterhouse-derived pancreas tissue of pigs. Too many people were allergic to the porcine protein so human tissue cultures were initially tested, as humans are not similarly available. Producing human insulin was exceedingly expensive as human tissue cultures routinely use calf serum albumin as a vital substrate, which can be contaminated with viruses and prions. Human insulin is now produced inexpensively after genetic engineering of the human insulin gene into bacteria. This works fine for insulin, because humans are not allergic to human insulin. The approach was not successful for other human products needed as pharmaceuticals (e.g., human growth hormones, glucocerebrosidase, the missing enzyme in people with Gaucher's disease, and so on) because of a basic difference between humans and microorganisms. After translation on the ribosomal machinery various sugar groups are added on the surface of many eukaryotic proteins (glycosylation), which are needed for activity. Insulin is not "posttranscriptionally glycosylated' in humans, so there is no problem with production in bacteria, which keep their proteins sugar free. Plants glycosylate proteins, not always in the same manner, but this seems not to matter, and where it does matter, the plants can be modified.

17.4.2. Beat the Micro-Middleman on Industrial Products

Many organic chemicals are needed as feedstocks in industry that are now being made in microorganisms in large, stainless steel fermenters. The bacteria start with sugar and synthesize the desired product. The sugar comes from plants. Why not engineer the plants to use their own sugar to directly synthesize the same organic feedstocks without the need for microbial middlemen who take a cut (in the form of sugar for their own growth) and live in expensive stainless steel houses? Plants live in much cheaper abodes, with dirt floors and no roofs, with natural (although occasionally stressful) air-conditioning and lighting.

The rush was on to make pharmaceuticals and feedstocks in plants by a variety of companies. The very expensive pharmaceuticals (e.g., the drug for Gaucher's disease) are being made in plant cell cultures in fermenters[870] and should be released at a price ten times lower than the equivalent from animal cell cultures. This poses no biosafety risk of gene flow, because no flowering plants are involved.

The world supply of other expensive orphan drugs can also be made in enclosed greenhouses. Tobacco was a plant of choice because of the ease of transformation and the backing of the tobacco industry, looking for alternate uses for this poisonous crop. It does not take that much space to make enough drugs for the affluent ill, and greenhouses can also be contained to prevent pollen flow to cultivated varieties and to related species. Biosafety is an issue with the need for less expensive, large-scale production of enzymes needed for food or feed processing, antibodies or antigens needed for large-scale immunization programs, or for industrial feedstocks.

The ideal crop seemed to be maize and field trials started, with at what seemed to be adequate distances to prevent pollen flow, but no efforts were made to install some of the transgenic fail-safe mechanisms (Chapter 4) to prevent volunteers in the field. The following year volunteers were not adequately removed, and it was also claimed that volunteer maize plants bearing the transgene were in neighboring fields. The ensuing fiasco of finding traces of transgenic DNA from volunteer maize in the leaf trash dust on soybeans, because of human error, resulted in the company going bankrupt. The uproar was based on principles, as no claims were made that anything dangerous to the public was emitted. Biotech industry spokespersons who typically rant that regulation should be product safety driven called for stopping the use of maize as a production vehicle, that is, called for process regulation instead of product regulation. They did not call for "no residue maize," which would fit the product regulation philosophy that the same spokespersons claim to believe in. A company using rice is having trouble obtaining permission for field development, even though their paddies are more than 300 km from the nearest rice production areas. Knowing how oilseed rape pollen flows from field to field, and drops volunteer seed,[452] a company developing oilseed rape as a vehicle quickly looked for other crop species. In this hysterical climate, the race was on to find a crop that could be fully segregated from common food crops. A production vehicle was sought among crops not widely grown, and safflower became an ideal target, especially for Canada.

17.5. A New Life for Safflower; an Engine for Pharmaceuticals

Safflower has proven to be amenable to biotechnological manipulations. *Agrobacterium tumefaciens*-mediated transformation and regeneration of transgenic safflower was reported more than a decade ago with stable integration of transgenes,[825,1181] and again as a novel finding more recently, when whole embryos were transformed using *Agrobacterium*.[915.]

Safflower is especially amenable, as an oilseed crop, to the production of recombinant proteins, although this may appear to be contradictory. One of the most expensive steps in commercial production of recombinant proteins is product recovery. With safflower, genes encoding a fusion protein between oleosin and the chosen gene product are transformed into the plant. As the plant grows and the seeds develop, the oleosin/gene-of-choice fusion is expressed, producing a recombinant protein that forms a recombinant oil body (due to the oleosin in the construct).[6] Seed is produced from the genetically engineered plants in the field. The harvested seed is then processed using oil body extraction. The company developing this technology claims to have achieved recombinant protein expression levels of more than 5 percent of total seed protein for some proteins. The oil bodies are then removed from extracted seeds via centrifugation. In some cases the oil body/recombinant fusion protein is the finished product. In others an enzyme[627] or chemical that recognizes a cleavage site between the target protein and the oleosin is added to purified oil bodies, cleaving the recombinant protein from the oleosin.

The company claims to have produced a wide array of proteins in this manner, ranging from peptides of nine amino acids to more than one hundred kilodaltons in size, retaining complex secondary and tertiary structure, including those structures requiring the formation of multiple disulfide bridges. They do not report whether the proteins are glycosylated and, if so, whether they are allergenic to people.

17.6. Biosafety Considerations

Anecdotal reports suggest that safflower does not become established outside of agroecosystems.[712] However, volunteer safflower has been documented in the United States, suggesting that safflower has the potential to become established in some locations. If a pharmaceutical-synthesizing safflower is grown in the vicinity of existing feral populations, the pharmaceutical traits may move via pollen. Cross-pollination of safflower plants is facilitated predominantly by insects, but wind can also move pollen short distances (up to less than a meter and a half) between plants grown close together.[65] Bees and other flying insects contribute to gene flow among safflower and its wild relatives over relatively large distances,[553] with the frequency of outcrossing, in part, under genetic control. Thus, safflower cultivars considered for the production of plant-made pharmaceuticals and chemical feedstocks should be evaluated to determine the potential outcrossing rate.

MacPherson et al.[712] suggest that prior to the release and/or growing of broad-scale pharmaceutical-synthesizing safflower under confined release conditions, considerable information is required to assess environmental biosafety impact, including:

— the likelihood of volunteers surviving and perpetuating in the natural environment.
— a quantification of gene flow from pharmaceutical-synthesizing safflower between feral populations and conventional varieties.
— the risk of introgression of the transgenes to wild/weedy populations.
— the potential for persistence in the environment of pharmaceutical-synthesizing safflower × weed hybrids.

They state that two new world species, *Carthamus oxyacanthus* and *C. creticus* are cross-compatible with cultivated safflower. Hybrids between safflower and these wild relatives could serve as sinks for pharmaceutical traits and sources of feral traits. Thus, hybrids could facilitate introgression of pharmaceutical or feedstock genes into conventional safflower or weedy relatives. Alternatively, hybrids could transfer feral traits to transgenic safflower, which may enhance the possibility of escape to ferality. For this reason, such transgenic safflower should not be grown in India or the Mediterranean area where many cross-compatible species are known and where extensive cropping of safflower for human and animal consumption has occurred. Further, regions in America where wild relatives or conventional safflower are grown should be avoided, or appropriate isolation distances determined and maintained. These wild hybridizing species occur in Argentine, Chile, the southwestern United States and Oregon.[713]

Luckily for the Canadian company concerned, no wild *Carthamus* spp. seem to be in Canada.[713] In 2003 safflower production was limited to 2,000 ha in Canada, none of which was dedicated to oil (FAOStat, 2003), so that separation distances are maintained. This company has been field testing their transgenic pharmaceutical-containing safflowers in geographic containment for the past few years in the desert state of Arizona, as well as the northern state of Washington,[57] where safflower is not commercially cultivated for oil. Thus, it seems that the paradigm of cultivating this species for pharmaceuticals in Canada does not pose a biosafety risk.

CHAPTER EIGHTEEN

Swollen Necks from Fonio Millet and Pearl Millet

Millets are a group of small-seeded cereals from a variety of genera in two separate tribes of Graminae.[804] They are cultivated on dry lands, especially in sub-Saharan Africa and India, with a total world production of about twenty-eight million tons per annum.[804] One of the rarer ones, tef (9 percent of world millet production), is discussed in Chapter 13. This chapter will mainly deal with one aspect of two other millets: fonio (4 percent of world millet production) and pearl millet (76 percent of world millet production). The chapter will highlight total disconnects between the agronomy/breeding community, the medical community, and the molecular biology community, and their lack of communication and knowledge of each others' results and problems. This disconnect occurred even though the information resides on the same databases available to all these groups, demonstrating that many scientists are so narrow that they are unaware of pertinent literature dealing with their own area. The same common databases have facilitated the present author's efforts to write about crops he hardly knows, to discover their problems and bridge this gap, and while writing this chapter, to connect all three groups, which may lead to a solution to the problem delineated below. In this respect the author is not a good reporter, as good reporters are not supposed to become involved in their stories.

18.1. Fonio as a Crop

Fonio millet is actually annual two species; *Digitaria exilis* (called acha in West Africa) and *Digitaria iburua* (black fonio). They appear to be very similar to perennial weedy *Digitaria* species (e.g., crabgrass, *D. sanguinalis*) such that one wonders how farmers manage to harvest the tiny seed, and then how they find what to eat. Yet, fonio is called the "grain of life" in West Africa, as it is the first grain to ripen in the farming season, well before others. There are short-season varieties that can be harvested forty days after sowing, and others that are harvested after a hundred days.[502] Fonio is also called "hungry rice" because of its low yields and the tedious threshing to get some food.[439] The grain is less than half the length the grain of pearl millet and a tenth the weight. Fonio is used as whole grain to make porridges, steamed products (similar to couscous), and alcoholic beverages.[5,49] This native West African crop, cultivated for thousands of years, grows in poor soils, is drought tolerant, and requires a minimum of inputs. This "Lost Crop of Africa"[1109] is still grown on a few hundred thousand hectares, mainly in Guinea and Nigeria. Various groups have been breeding fonio for easy threshing, as well as developing simple machinery for dehusking (threshing),[524] and production has gone up.[620]

Fonio is also cultivated in the Dominican Republic, having been brought there with slaves in the sixteenth century. It is considered a gourmet food with aphrodisiac properties because of its popularity with ex-dictator Trujillo and playboy Rubiroso, who both consumed it to enhance virility.[745]

As with tef, fonio suffers the classic problems of low yield, poor response to fertilizer, lodging, seed shatter, birds, insects (especially in storage[439]), diseases,[620,1109] and *Striga* and other weeds. Considering that the area under cultivation is low, it would be expected that there is limited genetic variability within the species, which is borne out by a morphometric analysis of the sixty-two extant accessions, which clustered narrowly into the acha and black fonio types.[524] DNA marker techniques have yet to be applied to fonio.[294,620] As these constraints and solutions are similar to those of tef (Chapter 13), there is no need for duplication in their discussion. It was naively proposed to increase variability by generating somaclonal variation,[620] a technique forsaken by most of the earlier proponents.

Pearl millet (*Pennisetum glaucum,* with many synonyms) is a much more widely cultivated crop than fonio, both for grain and for forage, and is the subject of considerable breeding work, but the cultivated gene pool is narrow.[167] Breeding is being performed at ICRISAT, the international pearl millet breeding organization in India, as well as in public and private institutions, yet pearl millet too has most of the same agronomic problems as fonio, which might be easier to address by genetic engineering than by breeding. Fonio and pearl millet are being discussed because of the presence of related specific secondary metabolites that are marketed as health foods in the developed world. They are not healthy when these grains compose most of the diet.

18.2. The Disconnect between Scientific Communities—Goiter and Cretinism

An extensive medical literature correlates the high consumption of fonio and pearl millet with goiter, a fact not to be found in any of the agronomic breeding literature, nor is it known to leading breeders contacted who deal with these crops. Goiter is clinically observed as a swelling of the thyroid gland of the neck and is due to a deficiency in the iodine-containing thyroid hormone thyroxin. Many inland areas of the world, which were not ancient seas, have a deficiency in iodine in the soil. Those iodine-poor areas are where goiter is or had been prevalent. Most western countries now only allow the sale of iodized salt to prevent goiter. At its worse, endemic goiter leads to endemic cretinism, a severe form of mental retardation in children. In such an interior area of Guinea, 70 percent of the inhabitants had goiter.[602] This area had suboptimal levels of soil iodine that are not low enough to cause goiter to that extent. Those with goiter were people who almost exclusively consumed fonio.[941] Those that consumed other local foods were free of the syndrome. Similarly, enlarged thyroid glands were assessed in a large proportion of rural children in the state of Gujarat in western India,[145,146] and there the syndrome was correlated with consumption of pearl millet. Goiter was also found in a fifth of children tested in the southern Blue Nile region of Sudan (85 percent of children in the infamous Darfur area), an area with sufficient iodine, and the syndrome was correlated with the consumption of pearl millet.[319] Why had the medical scientists not seen fit to alert the agronomic/breeding community about their discoveries of the cause of goiter?

Initially it was presumed that thioglucosides or cyanoglucosides yielding thiocyanates in these millets were the nutritional agents causing goiter, by binding iodine and exacerbating an already low dietary iodine intake. The Sudanese studies where iodine was sufficient suggested that this could not be, and other components were sought. Various studies homed in on the flavonoids of these grains. An in vitro assay was utilized on the key enzyme of thyroid hormone biosynthesis, thyroid peroxidase, a cytochrome P_{450} type enzyme that attaches iodine to the enzyme precursor of the hormone. The culprit in fonio was discovered to be the flavonoid apigenin,[941] and in pearl millet vitexin.[368] which is apigenin with an added sugar group attached to the molecule (Fig. 43). Apigenin and vitexin were potent inhibitors of thyroid peroxidase (Fig. 43) at low concentrations and they clearly are the cause of the thyroid deficiency causing goiter. There was no mention of goiter in a recent workshop devoted to enhancing the nutritional and functional properties of sorghum and pearl millet,[107] nor in one dedicated to rejuvenating fonio for West African food security,[1110] accentuating the depth of the knowledge disconnect.

And now we come to another disconnect: north–south. Apigenin and vitexin are both marketed in the developed world as dietary supplements or "nutraceuticals." One supplier recommends 100 mg of apigenin per day, pur-

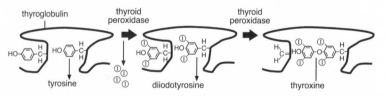

Apigenin and vitexin inhibit thyroid peroxidase

Figure 43. Apigenin and vitexin from fonio and pearl millet (respectively) inhibit thyroid peroxidase, the last enzyme in thyroxin production. This causes hypothyroidism, leading to goiter in adults and to cretinism in children. An RNAi of flavone synthase (FNS) engineered into these crops should preclude this problem.

ported to "protect brain and cardiovascular metabolism as MAP kinase inhibitors" (www.affordable-jarrow.com). Indeed there is scientific evidence for the inhibition of a mitogen-activated protein (MAP) kinase, inhibiting breast cancer cells, albeit only demonstrated in vitro,[1180] yet very few in vitro promising drugs are active in vivo, and no ethical pharmaceutical company would make claims based only on such tests. A hawthorn berry supplement containing 2 to 3 percent vitexin is claimed to be "cardiotonic and dilates blood vessels" (www.orcas-intl.com/products.html). Another purveyor standardizes their passion flower extract for anxiety treatments as containing 4 percent vitexin (www.seredyn.com/zzherbasupplementsforanxiety.html). No mention is made of the goiterogenic properties of either apigenin or vitexin in any of the advertisements, nor is there any evidence that regulatory authorities have considered the goiterogenicity of these "nutraceutical" "health foods" associated with cretinism. Even if apigenin and vitexin have these claimed properties, and many plant products do have a multitude of properties, one might wonder if the goiterogenicity does not make their use contraindicated, or if people with hypothyroidism should be warned against using them.

This disconnect continued at the scientific level to other scientists. A group has been working for many years on apigenin and vitexin biosynthesis in various members of the carrot/dill/celery family, the Apiaceae (Umbelliferae).[378,695] They have cloned the key flavonone synthase genes for apigenin biosynthesis along the pathway from narigenin in various Apiaceae, as well as closely related genes that encode enzymes of narigenin conversion to other flavonoids.[378,695] Many species express a microsomal-type flavonone synthase FNS II, the soluble FNS I apparently is confined to a few species of the Apiaceae. FNS I belongs to the Fe-II/2-oxoglutarate-dependent dioxygenases characterized by short conserved sequence elements for cofactor binding. The group thought that FNS I had evolved uniquely in Apiaceae, especially as no similar genes were to be found in *Arabidopsis*.[694,695] They were unaware that apigenin and vitexin are found in Graminae, and additionally that they were not cognizant that their research objects are goiterogenic, despite working in a department of pharmaceutical chemistry (U. Matern, pers. commun., 2005).

18.3. Biotechnological Solutions

The molecular biologists dealing with apigenin have been bridged to the groups dealing with fonio and pearl millet during the writing of this chapter. If they can clone the specific flavonone synthase genes of fonio and pearl millet responsible for apigenin biosynthesis using consensus primers from Apiaceae, then it would be possible to reduce apigenin levels in both fonio and pearl millet by using RNA interference (RNAi) technology (Fig. 43). Another *Digitaria* species has already been transformed,[212] so it should be possible to adapt the procedures to fonio. Pearl millet has also readily been transformed.[396]

A different biotech approach could be utilized. If the pearl millet flavone synthase gene is known, its primers could be used in breeding projects to assay for deletion or low-expression mutants. Is this the better approach? This may be daunting even though a randomly amplified polymorphic DNA (RAPD) analysis shows that the species is exceedingly diverse.[502,621] One might logically guess that the synthesis of apigenin is dominant and the lack of synthesis would be recessive. Fonio is a hexaploid,[502,1109] and if the gene for flavone synthase is expressed on more than one set of chromosomes, the breeding approach will not be simple. It might be easier to select for non-producing strains in pearl millet, a diploid, but the existence of a continually crossing weedy type within pearl millet[736] would negate the breeding effort if production of apigenin is dominant. Thus, RNAi will be simpler (in theory) if there is a consensus of sequences between the FNS genes of Apiaceae and Graminae. The same RNAi construct could probably be used with both crops (Fig. 43). One might postulate that apigenin and its derivatives are present as secondary products for a reason, possibly to allelochemically ward off insects or diseases, or other plant species in their ecosystem. A mutant producing less apigenin and derivatives would produce less in all tissues, and the positive aspects would be lost. An RNAi construct can be made with a seed-specific promoter, turning off apigenin production only in the seeds, protecting the children of India and Africa from cretinism and the adults from goiter, while protecting the rest of the plant from pests. It is not known whether livestock eating hay or straw of grain of fonio and pearl millet suffer from hypothyroidism; but goats eating pearl millet developed goiter in three months.[367a]

18.4. Biosafety Considerations

RAPD analyses have shown that fonio *D. exilis* clusters into groups,[502,621] these groups are related to *D. ternate* and *D. fuscescens* within the same distances, and *D. longiflora* falls within one of the fonio clusters.[502] Still, no one seems to have ascertained whether the two fonio species cross with other *Digitaria* species. As some of the *Digitaria* species are pernicious weeds of agriculture, engineering agronomic traits into fonio might require some scrutiny, as gene flow could be undesirable and containment and/or mitigation would be needed. Engineering nonproduction of apigenin or vitexin should have few consequences to rare gene flow recipients, which at worst would be unfit and disappear.

Highly outcrossing pearl millet is conspecific to both a weedy type that predominantly grows in or near it, as well as a truly morphologically distinct wild type that grows in nearby West African semideserts. These are often given other species names but are easily crossed and in nature produce hybrid swarms.[736] Indeed the weedy type differs from the crop type by a single allele of a supergene, which is present in the weedy type as a heterozygote, with the homozygote being lethal,[736] a rather unique case. Clearly this weedy type is a serious enough problem not to want agronomic traits to transfer, and containment and/or mitigation would be needed. Containment would not be possible if pearl millet can mutate to the weedy form, but mitigation could provide an unfit weed. Luckily, this weed problem only occurs in the center of origin of pearl millet, the possibility of recurrent mutation is almost impossible. Weedy pearl millet has not been reported on the Indian subcontinent where huge areas are planted to this crop, which is convincing epidemiological evidence against back mutation as being the source of the feral weedy type. As the weedy phenotype is controlled by a supergene made up of a cluster of several genes, a back mutation at one site cannot reproduce the weediness haplotype.

It is hard to envisage any adverse ecological effects from an RNAi construct that would prevent apigenin synthesis in either crop if expression is only in the seed. Indeed, the human need should clearly override any worries about whether there could be such an effect on the wild or weedy relatives.

CHAPTER NINETEEN

Grass Pea

Take This Poison

Grass pea (*Lathyrus sativus*) has been receiving interest as a legume adapted to arid conditions. It contains high levels of protein, a nutritional component that is increasingly hard for people to acquire in many developing countries. The advent of the green revolution cereals converted many arid areas into new irrigation districts, resulting in a constriction of areas devoted to dryland legumes. Because of this, coupled with a price structure that penalizes legume growers, and the presence of parasitic weeds, diseases, and insects that accompany large-scale legume production, legume cultivation has been scaled down. This has resulted in a caloric sufficiency from cereals coupled with protein deficiency due to lack of legumes in the developing world. Farmers cannot be forced to cultivate legumes when they achieve a higher profit per hectare from wheat. Still, many arid areas are undercultivated, and many sources of water are drying up, so there is a need for arid legumes. Chickpea is widely cultivated in such areas, as are lentils. Grass pea is considered to be the poor peoples' lentils, especially as grass pea contains β-N-oxalyl-L-α,β-diaminopropionic acid (ODAP or sometimes called BOAA), a neurotoxic poison that causes a syndrome obviously termed lathyrism, as well as another toxin and some anti-nutritional compounds, which severely limit its suitability for human consumption. Many open questions about ODAP exist that have caused the crop to be banned in certain locales, which may have been a bureaucratic overreaction. Still, the ability of this crop to provide a yield under stressful conditions has stimulated interest, as summarized in a lengthy monograph,[179] which is the source of much of the background information in this section (where not stated otherwise). Another source of information is the annual publication *Lathyrus* and *Lathyrism Newsletter* available online at: http://www.clima.uwa.edu.au/lathyrus/.

19.1. The Crop

Grass pea is an annual legume commonly grown for its grain but also used for fodder or green manure. The less-used vegetative types are utilized in the production of fodder or forage for animals, being more widely tested in the developed world as a source of hay in arid regions.

The emphasis of this section will be on grass pea grown for seed for human consumption. Mixed use is common, for example, in Bangladesh the young plants are used as a fodder for cattle or for strip-grazing by cattle, and the crop regrows, producing seed. Grass pea is produced mainly in the Indian subcontinent (India, Pakistan, Bangladesh, and Nepal) and Ethiopia (on 8 percent of arable area). It had been cultivated in small areas in Europe, the Middle East, and northern Africa.

There are two main ways of growing grass pea as a dry season crop.

1. The grass pea is utilized in many areas of in Southeast Asia as a "utera" crop, that is, the seeds are broadcast into a standing rice crop near the end of the season. The seeds germinate after draining for the rice harvest and the crop utilizes the remaining moisture for growth in the dry season.

2. Planting after normal plowing as a winter season crop, on whatever water is available after rice. It is relatively tolerant to frost and withstands some freezing temperatures, as in Pakistan. Grass pea is often intercropped in Southeast Asia as insurance against adverse weather conditions that might affect growth or yield of the other crop in the mixture.

Utera underseeding in rice in India typically yields half a ton per hectare, monocropping yields three times as much. One wonders whether it would be better to wait until after the rice has been harvested to plant, as is done with irrigated wheat after rice.

19.1.1. Major Uses

The grains are sometimes boiled whole in the countries of the Indian subcontinent, but are most often split with a dal mill to obtain dal, and then added to rice to make a porridge or soup-like dish. The unleavened bread and pancakes made out of grass pea flour is a staple for landless peasants. Grass pea dal and flour have been used to adulterate more expensive lentil or pigeonpea dal and chickpea flour. Tender young shoots are cooked as a green winter vegetable. Grass pea flour is used in Ethiopia/Eritrea to adulterate tef flour to make injera (Chapter 13). The young pods are also boiled and then salted as a snack food. A concoction of grass pea has been purported to have been used as a homeopathic prophylactic against polio in Argentina,[1159] probably without more than a placebo effect.

19.1.2. Grass Pea and Its Relatives

This annual nitrogen-fixing species, with its heavily branched, straggling, or climbing stalks is attractive for drought-stricken, rain-fed areas where soil quality is poor and extreme environmental conditions prevail, yet it is not affected by flooding. It has a very hardy and well-developed, penetrating taproot system, with rootlets covered with small, cylindrical, branched nodules clustered in dense groups. Therefore, grass pea can be grown on a wide range of soil types, including very poor soil and heavy clays. Unlike other legumes, the grass pea is resistant to many pests including storage insects, although the reason for this resistance is ominous; constituents that poison insects may well poison other herbivores, including *Homo sapiens.*

Of nearly two hundred *Lathyrus* species and subspecies only *Lathyrus sativus* is widely cultivated as a food crop. *Lathyrus cicera* and *L. tingitanus* have been cultivated for grain, and *L. ochrus, L. latifolius,* and *L. sylvestris* as forage species. *Lathyrus amphicarpus* has potential as a self-seeding forage species in the Middle East. The inedible ornamental sweet pea (*L. odoratus*) is a common garden species. About 10 to 28 percent of grass pea seeds result from intraspecific cross-pollination. The morphological similarity of grass pea to *L. cicera* and *L. gorgoni* may be a result of hybridization or common ancestry, although the fruit of grass pea is closer to that of *L. amphicarpus, L. blepharicarpos,* and *L. marmaoratus.* High interspecific incompatibility exists among *Lathyrus* spp., but grass pea has been artificially hybridized with *L. cicera* and *L. amphicarpus,* which accentuates the close relatedness among these species. *L. sativus* and *L. cicera* are sympatric in Turkey where they are often weeds in other crops. Certain cross-fertilizations result in embryo abortion at different stages of development according to the cross-combination.[179]

Cytological studies of the F_1 hybrids between *L. amphicarpus* × *L. sativus, L. amphicarpus* × *L. cicera,* and *L. odoratus* × *L. chloranthus* showed 50 to 70 percent chromosome homology and pollen fertility in conformity with the meiotic pairing. Breeding strategies involving alien genetic transfer for the improvement of grass pea are possible through the readily crossable species *L. amphicarpus.* There also appears to be a high probability of success in using embryo rescue techniques to obtain interspecific crosses between some species that do not viably cross because of embryo abortion after fertilization.[179]

19.1.3. Variability within Grass Pea to Deal with Constraints

The variability in various grass pea traits is summarized in Table 28. There is a correlation between small seededness and earlier flowering. There is also a converse correlation between large seeds, pods per plant, and plant height, which is normally associated with later maturity and increased plant biomass. The smallest-seeded lines originate from southeastern Asia with the largest ones being from Europe and northern Africa. This wide variation in seed size should be of value to the breeder because the correlation between seed size and yield, as well as seed size and vigor. The small-seeded lines of the Indian subcontinent usually have a tendency to shatter seed because of an early splitting of the pod down its ventral rib before the pod matures, a problem lacking in the European lines. Shattering is desirable for self-seeding grazing lines, a rare case where shattering is actually valuable.

The genotypes that are devoid of condensed tannins also contain lower levels of other phenolic compounds. Low amounts of low-molecular-weight phenolic compounds, unlike condensed tannins, do not directly cause any harmful effects to swine and cattle, except to impart a bitter taste. Condensed tannin levels are positively correlated with darker seed coat pigmentation and lower nutritional value of the seed. Lines that were almost devoid of tannins were all characterized by a white to creamy yellow seed coat. Moderate to high

336

Table 28. Genetic Variability of Grass Pea in **India** (**in bold**) and Elsewhere

Factor country—except India	Mean	Minimum	Maximum	COV (%)
Days to 50% flowering	**62**	**47**	**94**	**12**
Days to maturity	**107**	**86**	**127**	**5**
Canada	110	97	121	
(Hanbury)		76	123	
(Sarwar)		43	88	
Plant height (cm)	**34**	**15**	**68**	**23**
Canada	108	24	172	
Branches/plant	**9**	**2**	**28**	**34**
Pods/plant	**19**	**2**	**59**	**46**
Nepal	36	13	59	
Pod length (mm)	**30**	**19**	**52**	**12**
Canada	32	17	56	
Pod width (mm)	**9**	**3**	**13**	**11**
Seeds/pod	**3**	**2**	**5**	**13**
Canada	3	1	4	
Seeds/plant	**55**	**6**	**200**	**50**
Seed index (g)	**6**	**2**	**19**	**30**
1000 Seed weight (g)				
Canada	145	56	288	
Australia		190	220	
Bangladesh		30	68	
(Robertson)	87	34	226	
Seed density (g/liter)				
Canada	761	612	829	
Biological yield (g)	**8**	**0.4**	**52**	**56**
Yield/plant (g)	**4**	**0.6**	**20**	**60**
ODAP (%)	**0.4**	**0.1**	**8.7**	**58**
Condensed tannins (g/kg) (Canada)		0	4.3	
Trypsin inhibitor	"little variation where studied"			
Total phenolics (mg/kg) (Canada)		39	999	
Thrips	**"little resistance where studied"**			

Source: Collated from various tables and text citations in Campbell.[179]
COV - coefficient of variation

tannin-containing lines were all characterized by a dark brown to black pigmentation of the seed coats. Flower color was highly correlated with the seed color: the blue, pink, or red flowers usually produce speckled, colored seeds, whereas the white flowers are associated with white to creamy yellow seeds, which is an aid to the breeder.

A large range of variability in neurotoxic ODAP content exists in grass pea germplasm collections. Actual ODAP levels are also influenced by the environment, growing conditions, and locality. Lines totally lacking in ODAP have not yet been identified in present breeding programs, but there are several reports of levels as low as a hundred parts per million.

All the grass pea genotypes studied are characterized by very high levels of activity of another anti-nutritional component, trypsin inhibitors, with little variation that could be exploited. There appears to be little known resistance to thrips, despite their being a serious pest of grass pea.

19.1.4. The Domestication of Grass Pea

Grass pea has been cultivated for so long that it is unclear where the centers of origin and diversities are located.[179] Archaeobotanical and phytogeographical evidence suggests that grass pea cultivation began in the Balkans, in the early Neolithic period (sixth millennium BCE[589]), from where it expanded to the rest of Europe.

19.2. Nutritional Constraints to Wider Use

The various constraints to the production of grass pea are summarized in Table 29. Some are discussed in detail in the following sections.

19.2.1. Neurolathyrism

Grass pea, owing to its drought resistance and high protein content in seeds, has been cultivated in arid northwest China, reaching more than twenty thousand hectares in the early 1970s. A strong epidemiological basis exists between the consumption of grass pea and lathyrism, a motor neuron disease resulting in paralysis of the lower limbs. A 1973 epidemic outbreak of lathyrism attributed to the consumption of grass pea in China resulted in the prohibition of cultivation of what would otherwise be a valuable crop. Similar outbreaks have occurred in the Indian subcontinent and in Ethiopia, result-

Table 29. *Breeding-Intractable Constraints to Grass Pea Cultivation*

Factor	Constraint	Possible solution
Agronomic		
Shattering	In small-seeded varieties	
Mineral deficiency	Possibly zinc	ODAP[a] is a zinc carrier, higher when deficient
Plant protection		
Insect	Thrips	Engineer resistance
Disease		
Weed	Parasitic weeds	Engineer herbicide resistance Engineer parasite resistance
Quality		
Neurotoxins	ODAP	Breed out most Engineer out rest if needed
Flavor	Tannins/phenolics	Breed out
Nutrition	Trypsin inhibitors	Leave in—are responsible for resistance to grain weevils—boil seeds, or remove and replace with other weevil resistances

[a]ODAP (also called BOAA) is β-N-oxalyl-L-α, β-diaminopropionic acid.

ing in similar bans in some Indian states. The causative neurotoxin is ODAP (or BOAA).

It is fascinating how indigenous knowledge can almost be correct in understanding cause and effect. Lathyrism is common in Ethiopia; a recent medical survey showed that more than half of community respondents associated the disorder with walking or lying on the straw and the stalks of grass pea. Health workers commonly considered contact with vapor or steam of grass pea foods as the cause of lathyrism.[387]

Whereas the disease can now be demonstrated in animal models, it is not too prevalent in human populations, even poor populations subsisting heavily on grass pea. The incidence of neurolathyrism in humans is much lower than that in the other mammals ingesting grass pea.[925] This was traced to the ability of most (but not all) people to metabolize ODAP, lowering the incidence.[925] In areas with populations subsisting heavily on grass pea the incidence of neurolathyrism was 0.14 percent in Bangladesh[463] and a hundred to seven hundred fifty cases per million population in Ethiopia.[449] The incidence was highest among young males, only 13 percent of the patients were female in Bangladesh,[463] and 38 percent in Ethiopia, where they found that females exhibited a milder form of the disease.[449] The average family size was nearly five members and of seven hundred thirty families with lathyrism evaluated, six hundred twenty-two had only one affected member, and a hundred eight families had two to eight affected members.[463] This suggests a genetic component, with the disease being recessive, possibly on the X chromosome. Most of the human population would have a dominant gene conferring (probably) metabolic resistance, and that sensitivity is due to a recessive loss of function.

If the incidence of lathyrism is really low in human populations because of a genetic defect, not overconsumption by a subpopulation of mainly males in some families, the use of this species should be compared with other species commonly consumed. Another legume, *Vicia faba* (broadbeans) induces a syndrome of blood hemolysis in humans (favism) with a defective (mutant) glucose-6-phosphate dehydrogenase in some ethnic populations at higher-frequency levels.[634,1037] This is caused by two components, vicine and convicine.[535] Peanut allergies can be toxic, but the incidence is lower.[586] Thirty percent of infants are highly allergic to allergens in soybeans.[1158] Is the outright banning of such a crop in China and parts of India justified on two levels? (1) Should such bans be on a crop species, or on varieties with toxin levels above a regulatory defined threshold? Having regulatory thresholds for varieties would stimulate breeding activity instead of suppressing it, especially conventional and/or molecular breeding to achieve lower levels. (2) When a crop is as useful as grass pea, should genetic testing for susceptibility among the human population be called for? Should there be an educational system that can teach certain early warning symptoms to warn susceptible people against eating grass pea? Both approaches are taken with other allergenic foods, where a small proportion of the populace is susceptible. One wonders whether before media hyperbola (including that used by scientists), the incidence of lathyrism would have been called "epidemics" or whether they just would have been called "genetic clusters."

An anti-nutritional role in food digestion was seen in chicks fed 40 percent of their diet of low- and medium-ODAP lines or 60 percent of low ODAP line.[920] This aspect has not been critically examined, however. As a forage crop for livestock, it is only toxic when it makes up more than 30 percent of the diet for a three- or four-month period.

Zinc deficiency and oversupply of iron to the roots of grass pea increases in the content of ODAP. As the transport of zinc to the shoots is enhanced by the addition of ODAP it is proposed to function as a carrier molecule for zinc ions. Soils depleted in micronutrients from flooding by monsoon rains (Indian subcontinent), or otherwise poor in available zinc and with high iron content (Ethiopian vertisols), may be responsible for higher incidence of human lathyrism in those areas.[630] No experimental animal model for neurolathyrism could be produced by direct feeding either the seeds or the toxin. Experimental neurolathyrism could be produced in guinea pigs by making them subclinically deficient in ascorbic acid and feeding them the grass pea seeds or extracts.[533] Lathyrism is another case where it is hard to extrapolate studies from one animal to another, demonstrating why human studies are also necessary.

The biosynthesis of ODAP from its precursor β-(isoxazolin-5-on-2-yl)-alanine (BIA) was demonstrated in young seedlings.[520,521,618] ODAP was found in all tissues of grass pea plants, irrespective of age or variety, but maximum content was observed in the leaves of the vegetative stage and in the embryos at the reproductive stage.[866]

No correlations are discussed between high neurotoxin levels and insect resistance, suggesting that the pests may have evolved resistance to ODAP over the millennia.

19.2.2. Osteolathyrism

A related syndrome of skeletal deformities, often found among the same patients, is called osteolathyrism. It is associated with eating the green parts of grass pea, which contain 2-cyanoethyl-isoxazolin-5-one, a compound that chemically and metabolically can produce the osteolathyrogen β-aminopropionitrile (BAPN).[464]

19.2.3. Tannins/Phenolics

Tannins are strongly astringent (owing to their protein-binding properties), leading to a depression of feed intake, which lowers animal productivity.

19.2.4. Inhibitors of Mammalian Metabolism of Grass Pea Protein

There appear to be fairly high levels of a trypsin inhibitor in grass pea compared with many of the food legumes, except soybeans. These trypsin inhibitors, which are enzymes themselves, prevent the activity of the digestive tract enzyme trypsin, which is responsible for metabolizing ingested proteins to amino acids. Their amino acid composition and molecular weights suggest that grass pea contains the Kunitz class of trypsin inhibitors, a type generally absent from other agriculturally important legumes. The ingested proteins are digested at different sites by chymotrypsins as well, and a synergy exists between the two enzymes, such that both are needed for optimal dietary protein utilization. Although trypsin inhibitor activity was detected in all the genotypes studied, three genotypes were devoid of any chymotrypsin inhibitor activity. The reduction of trypsin inhibitor would make the crop much more desirable as animal feed.

Heating during food preparation almost completely eliminates this problem, so it does not appear to be a major concern for most human consumption of cooked grass pea. Uses are reported in Bangladesh, however, where the seeds are ground, mixed with water, and eaten as paste balls. In cases such as these the amount of trypsin inhibitor would affect the nutritional value of grass pea.

19.3. Pest and Agronomic Constraints to Grass Pea Cultivation

19.3.1. *Weeds*

In general, weeds are controlled by hand in most growing areas of developing countries where grass pea is grown, although this is changing as countries industrialize and agricultural labor becomes scarce. Weeding may or may not be practiced, as this crop often is considered a low-input crop with lower returns.

Although not discussed in the section on weeds by Campbell[179], in the Middle East where the parasitic *Orobanche* species are indigenous, they can cause considerable yield losses.[657] Being the center of origin of both crop and weed, there is a modicum of tolerance in this crop, unlike the situation in introduced legumes such as *V. faba*. Still, the losses are unacceptably high for the farmer. One must be exceedingly careful in introducing breeding material from the Middle East elsewhere, as the minute *Orobanche* seeds can easily stick to crop seeds.

Grass pea yields decreased with increasing density of volunteer wheat or barley in Canada. Grass pea is not competitive with weeds, in particular, when moisture is a limiting factor to plant growth. Herbicides are available that could selectively control most annual grass weeds, but none are currently registered for use in this crop. No herbicides are available that could selectively control parasitic weeds at a single normal application rate. Some legume-selective herbicides are degraded in legumes, where multiple applications control parasitic weeds in other legumes,[375] but this could never be economical in grass pea.

19.3.2. Arthropods and Diseases

Aphids (e.g., *Aphis craccivora*) and thrips are major pests in India, Bangladesh, and Ethiopia. Powdery mildew (*Erysiphe polygoni* and *E. pisi*) and downy mildew (*Peronospora lathyri-palustris*) are the two major diseases that infect grass pea. Losses due to these organisms and varietal reaction have not been studied sufficiently.

19.3.3. Shattering

The small-seeded lines often have a very severe shattering problem, with the splitting of the ventral vein of the pod as the seeds enlarge and before they are mature. This necessitates early harvest if seed shattering and loss are to be avoided.

19.4. Breeding Grass Pea

The floral structure of grass pea favors self-pollination. It is not known if wind or insects are the major vector of the pollen in the field that facilitates the approximately 20 percent outcrossing. Artificial crosses are performed in controlled conditions in the greenhouse or under netting by emasculating the anthers. The styles are fertilized the following morning with pollen following dehiscence of the anthers. Several seeds typically develop from each pollination.

Male sterility has been observed, controlled by both cytoplasmic and genetic factors, suggesting both single- and double-restorer genes, but this has not been used to develop commercial hybrids. Mutation breeding was attempted in the 1970s, with gamma rays being more efficient than ethyl methanesulfonate (EMS) and *N*-nitroso-*N*-methylurea (NMU) for induction of mutations, but it is not clear whether useful mutants were discovered.

A lengthy recent review describes advances in grass-pea-breeding efforts against biotic and abiotic stresses, including marker-assisted selection.[1107]

19.4.1. Breeding out Nutritional Problems

19.4.1.1. Lathyrism

Breeding has been successful in lowering the content of ODAP, the toxin, but not in eliminating it. As so little is known about this syndrome in humans, a few questions must be answered.

1. If indeed susceptibility to ODAP is genetic in humans, as it appears to be, it is necessary to know the threshold levels of ODAP toxicosis in the susceptible human genotypes to ascertain whether the small amount of toxin in the varieties bred to contain low amounts will affect susceptible individuals. If they are unaffected by eating grass pea having low ODAP levels, breeding may be sufficient to lower the levels, unless low toxin level is linked to low yield.

2. If less ODAP is required than has been achieved by breeding, a transgenic approach may be needed.

3. If ODAP is really a zinc transporter, one must look at the zinc nutrition of the low ODAP mutant lines. Zinc deficiency is easily dealt with in other crops by inexpensive seed dressing with small amounts of zinc sulfate. Also, if low soil zinc is responsible for plants synthesizing high levels of ODAP, then zinc seed dressing may lower the ODAP levels.

It must be ascertained whether high ODAP has high insecticidal or fungicidal activity, and if so whether it must be replaced by pesticide use or by transformation with pesticidal genes encoding products that lack toxicity to humans.

19.4.1.2. Other Nutritional Problems

It seems that there is sufficient variability to deal with phenolics and tannins by breeding for light flower color and light colored seed. It was earlier stated that it might be ominous that among all the tropical grains, grass pea is not infested with grain weevils. It must be determined whether this is due to genetically controllable ODAP, to phenolics/tannins, or to invariably high trypsin inhibitor levels, or none of the above. Grain weevils will have to be dealt with, if a nutritional deterrent common to humans is suppressed, allowing grain weevils to celebrate on a new crop.

19.4.2. Breeding for Disease Resistance—Mildew

Too little is known about the variability for resistance of the two types of mildew that attack grass pea. Lines showing moderate resistance to powdery mildew have been identified in India and Syria and efforts are underway in India to transfer the powdery mildew resistance to higher yielding, more adapted lines. Lines have recently been reported in Spain and Portugal that have reduced severity of disease due to powdery mildew, despite being easily infected. Conversely, germplasm has been screened for resistance against downy mildew but resistance genes have not been found in the available germplasm.[1107] Lack of severe disease in the growth chamber correlated well with field experiments, but at the other extreme, the intensity of severity of bad infections did not correlate as well. If indeed further germplasm screening for downy mildew resistance comes to naught, it may be necessary to consider molecular approaches. More information should be garnered before an approach other than breeding is proposed.

Lines of grass pea that were highly resistant or immune to *Cercospora pisi sativae* f. sp. *lathyri* should be a good source for breeding against that pathogen.

19.4.3. Breeding for Resistance to Arthropods and Nematodes

No genetic basis has been found for thrip resistance in more than a thousand lines screened, suggesting that standard breeding may be of little avail. This justifies a biotechnological approach. There seem to be no breeding solutions to the aphid problems. Resistant genotypes for cyst nematode (*Heterodera ciceri*) and root knot nematode (*Meloidogyne artiella*) have been identified, but the severity of infection and occurrence of these pests are not presently well documented.

19.4.4. Dealing with Orobanche by Breeding

The center of origin of grass pea is close to that of *Orobanche* spp., the weeds parasitizing them. The damage is not devastating, as it is for species introduced into regions with indigenous *Orobanche,* suggesting that the species is in a balance that would be acceptable for a wild species, although not for a crop. Thus, it is also a question of how much more tolerance can be achieved, because the farmers/selectors have obviously been selecting for *Orobanche* tolerance for millennia.

A breeding effort has begun to deal with the *Orobanche* parasitic weed problem through interspecific crosses because insufficient variability in resistance to *Orobanche* exists within the grass pea species, or in *Lathyrus annus, L. aphaca, L. cicera, L. gorgoni, L. inconspicuus, L. szowitsii,* and *L. tingitanus,* which were all highly susceptible.[982,983] Conversely, none of the accessions of *L. clymenum* and *L. ochrus* tested allowed *Orobanche* development in two seasons in the field, nor under controlled conditions, suggesting that they might be used in interspecific crosses.[983] Considering the known incompatibility within the genus *Lathyrus,* it can be wondered how easy the interspecific transfer will be, especially if resistance is polygenic, yet polygenic resistances are usually more resilient.

Breeding efforts were capable of temporarily delaying *Orobanche* in sunflowers, with new races of resistant *Orobanche* strains quickly evolving resistance to each of the resistant sunflower hybrids bred shortly after release. Thus, such a mutation breeding approach has limited sustainability. Recently, wild sunflowers evolved resistance to imidazolinone herbicides[23] and this resistance was bred into cultivated sunflowers, and is being used for *Orobanche* control. Such herbicide resistance can be achieved by pollen mutagenesis,[404] which was used in maize to achieve imidazolinone resistance. Similar herbicide resistances could be achieved with pollen mutagenesis in grass pea.

19.4.5. Shattering

There is clearly enough genetic variability to deal with shattering resistance at the subsistence farming level. Whether the inherent antishattering is sufficient for mechanical harvesting is an open point.

19.5. Biotechnological Solutions

19.5.1. Solutions to Anti-Nutritional Factors

As discussed in the section on breeding, removing the anti-nutritional factors may require engineering further insect resistance.

19.5.1.1. Trypsin Inhibitors

As there was no variability in the very high levels of trypsin inhibitor activity, reducing the levels by genetic engineering is called for to reduce this trait. The grass pea trypsin inhibitor has an apparent molecular mass 22 000[921,922] and comprises five protein "isoinhibitors" of identical isoelectric points, which each contain 203 to 212 amino acid residues.

Based on sequence homology between the very similar trypsin inhibitors, it should be simple to design an RNA interference (RNAi) construct to a major consensus sequence, and thus suppress the biosynthesis of the trypsin inhibitors. Tissue-specific promoters may be chosen based on the proposed use of the grass pea. If the hay is to be used as forage, then leaf/stem promoters should be used; if the seeds are to be used as fodder, or where they are un-cooked before human consumption, there should be RNAi expression in the seeds.

Recent evidence suggests that this would be an exercise in futility. Trypsin inhibitor isolated from seeds of *Vigna*, a related legume, inhibits weevil larvae and adults.[359] When barley trypsin inhibitor was transformed into rice, grain weevils were inhibited.[28] Thus a very good chance exists that the native grass pea trypsin inhibitors are responsible for the resistance to grain weevils, and should be left alone. Grass pea seeds should be boiled before consumption by humans and should never be more than a small part of animal diets.

19.5.1.2. ODAP

If further suppressing ODAP levels below those achievable by genetics is needed, or if unwarranted linkage losses occur in yield, then a transgenic approach would be worthwhile.

No correlations have been reported between high neurotoxin levels and insect resistance, suggesting that the few pests attacking grass pea may have evolved resistance over the millennia. Still, reducing ODAP to zero might allow susceptible insect species to attack the crop. There seem to be no reports about whether the low toxin lines are more susceptible to insect attack than the high toxin lines. This must be elucidated, when it is decided what insect toxins are necessary for engineering, beyond the genes needed to deal with thrips.

There are two ways to deal with ODAP: prevent its biosynthesis or degrade it before it gets to humans.

The pathway of ODAP biosynthesis is outlined in Fig. 44. Two nonprotein amino acids of grass pea, BIA and its higher homolog α-amino-γ-(isox-

Synthesis of ODAP

O-acetyl-L-serine (OAS) + isoxazolin-5-one $\Rightarrow$

beta-(isoxazolin-5-on-2-yl)-L-alanine (BIA) $\Rightarrow$

2,3-diaminopropionic acid (DAPA), $\Rightarrow$ beta-ODAP

Figure 44. Synthesis of ODAP in grass peas, as per references 520 and 521

azoline-5-on-2-yl)-alanine (ACI) were tested for excitotoxic potential. BIA (0.5–2.0 mM) but not ACI (2.0 mM) produced a concentration-dependent neurodegeneration in mouse cortical explants. These results suggest that grass pea plants engineered to block the synthesis of ODAP may accumulate a neurotoxic precursor, and therefore must be tested for the presence of both BIA and ODAP.[899]

The incidence of neurolathyrism in humans is low in sharp contrast to other monogastric mammals, which is paralleled by a low excretion of dietary ingested ODAP in humans.[925] This indicates a metabolism or detoxification of ODAP, which may be unique to humans. Considering the chips available with the complete human genome, it might be possible to ascertain quickly which mRNAs are being expressed by immune humans versus those expressing the disease who are ingesting the toxin versus humans not ingesting the toxin. This could lead to isolation of genes that could be used to express degrade enzymes in the seed. Given that the syndrome is eight times more prevalent in males than female human beings,[463] priority might be given to considering genes on the X chromosome.

19.5.2. Solutions to Orobanche and Thrips

19.5.2.1. Herbicide Resistance as a Selectable Marker for All Grass Pea Transformations

Herbicide resistance genes should clearly be the selectable marker for use in this crop. The herbicide of choice should be readily available in the target countries for this crop, or the manufacturer should be committed to its registration. Weeds are a problem, and cheap herbicides will replace vanishing cheap agricultural labor. The various available genes are discussed in great detail in reference 412.

If *Orobanche* is presently a weed problem in grass pea, or if there is a good chance it will be so in the future, then the herbicide resistance used as a selectable marker must confer target site resistance, that is, not a gene that encodes a protein that degrades the herbicide. This would eliminate the *bar* gene-encoding resistance to glufosinate. The newer and seemingly better version of the *GAT* gene for glyphosate resistance achieved by gene shuffling[194] would be inappropriate, as the gene encodes an enzyme-degrading glyphosate, albeit effectively. It will be necessary to use the CP4 5-enolpyruvylshikimate-3-phosphate (EPSP) synthase gene for glyphosate resistance from *Agro-*

bacterium,[1197,] which, while widely used in many crops, is inherently inefficient.[412] Transgenic herbicide resistance allowing *Orobanche* control has been
demonstrated with glyphosate[550], asulam[1031], imidazolinone or other acetolactate synthase inhibiting herbicides[75]. Methods have been developed to apply both the imidazolinone herbicides[559] and glyphosate[420] as seed dressings
to prevent parasitic weeds, using far less herbicides and precluding the need
for spray equipment.

19.5.2.2. Orobanche—Nonherbicidal Approaches

Although herbicides are a useful quick fix for the parasitic weed problem, especially when making transgenics for dealing with other problems
and using herbicide resistance as a selectable marker, other longer-term
strategies are also called for. This is especially true with parasitic weeds,
which put out nearly one hundred thousand seeds from each flowering
stalk, and the rapidity of mutations appearing in the field is a question of
numbers.[416,426p]

No one has yet isolated specific genes that confer resistance to parasitic
weeds, but a good possibility exists of finding the genes in *Lathyrus*. As noted
above *L. clymenum* and *L. ochrus* appear to be immune to *Orobanche*.[983,984]
Molecular techniques, either with chip technology or subtractive hybridization or even proteomics, and/or metabolomics, could be considered
to see what functions get turned on, or ascertain which proteins and/or
metabolites are synthesized in these immune species after attack by
Orobanche versus what happens in grass pea. Genomics can be considered,
especially as the genome of the closely related legume, *Medicago truncatula*
is being sequenced and gene chips are being made.[362] *L. clymenum* and *L.
ochrus* failed to induce *Orobanche* germination in recent in vitro studies,
suggesting that they do not make the germination stimulants. This suggests
that a metabolomics approach may also prove to be worthwhile, if coupled
with the genomics and/or proteomics.

19.5.2.3. Thrips and Aphids

Various crops have been engineered to withstand these pests. A multidomain cysteine protease inhibitor provided resistance against western flower
thrips.[831] Aphids were controlled on transgenic plants bearing genes encoding an oryzacystatin,[893] the snowdrop lectin gene *gna*,[1131] and the *Arisaema
heterophyllum* agglutinin gene.[1175] These genes would have to be tested in

grass pea against its specific aphids and thrips. It is easiest to first test the gene products, isolated from producing organisms, before embarking on genetic engineering.

19.6. Transformation and Regeneration

Regeneration techniques must be developed before transgenic biotechnology can be developed for grass pea. Two decades ago, legumes were considered totally recalcitrant to transformation by leading lights in the area. They were convinced that a biological barrier exists that cannot be overcome, to explain their lack of success. Slowly others succeeded with soybeans, and other legumes followed. Today, legume regeneration is considered hard, but doable, and so with some effort, it should be possible to crack the barriers for grass pea, as there is already a start.

Grass pea plants have been successfully regenerated from explants of stem, leaf, and root tissue. The resulting plants showed a high amount of somaclonal variation.[923] This technique may be successfully exploited in the production of agronomically desirable types in low ODAP lines, faster than conventional crossing and backcrossing methods. Regeneration techniques with less somaclonal variation are desirable for transformation.

Pea and grass pea protoplasts were fused and more than 10 percent of viable heterokaryons were obtained. Heterokaryon cells divided and formed small calli,[296] but no more. This is presently viewed as a dead end approach. More recently the same group demonstrated that, although they could not regenerate grass pea plants from protoplasts, they were able to get true-to-type regenerants from stem explants of some cultivars,[805] suggesting that transformation by biolistic bombardment may work.

19.7. Biosafety Considerations

Superficially, there seem to be no great biosafety concerns with this crop and the genes involved. Whereas grass pea does cross with other *Lathyrus* species, it does so reluctantly and at low frequency. Herbicide resistance should be of little consequence because (1) grass pea is not a volunteer problem in other crops, and (2) without herbicide use in the other crops (and herbicide use, in any case, is low in most places) resistance would pose no selective advantage.

The other genes described should offer no real huge selective advantage to the wild but not weedy *Lathyrus* species, except possibly for thrip resistance. It is doubted that thrip resistance alone would much change the balance of a wild species in the wild. The other genes described above should be neutral or slightly unfit in the wild. A closer examination of these factors is called for, using the precepts in Chapter 4, before the first transformation is contemplated.

CHAPTER 20

Limits to Domestication

Dioscorea deltoidea

"I yam what I yam"

(POPEYE, THE UNDOMESTICATED SAILOR MAN)

The yam *Dioscorea deltoidea* is probably the plant species that has effected the greatest change in human customs and practices in the past half-century. This species resides in the jungles of Mexico and is hunted down, dug up, and then extracted for diosgenin, a plant steroid. In the 1930s and 1940s, Russell Marker collected ten tons of this wild, edible yam, dried and reduced them to a syrup, which he brought back to his university laboratory. He developed a relatively simple, high-yield conversion of the diosgenin present in *Dioscorea* into progesterone, converting the syrup into 3 kg of progesterone, with a 1943 market value of about a quarter million dollars. He proposed that progesterone could be derivatized to make estrogens that could act as oral contraceptives. Marker became disillusioned when U.S. pharmaceutical companies refused to consider the industrial applications of his work; they were making small amounts of female steroid hormones from human female and mare urine as well as pig ovaries for research purposes. Birth control was not an issue publicly discussed at that time, even within the medical/pharmaceutical community. Marker moved to Mexico, where he became one of the founders of Syntex, and produced over 30 kilos of progesterone in his first year, before leaving the company.[281] Nearly all steroids in clinical use in the 1970s were synthesized from diosgenin, including the first oral contraceptive pills, which had been introduced in the late 1950s.[1151]

When the number of the earth's inhabitants reaches eight billion people, thirty-two tons of progesterone per year will be needed to protect the half-billion women of child-bearing age from unwanted pregnancies.[1151] This would require about twenty thousand tons of yams per year at the current dose of hormones in the pills. Just a small fraction of that amount of yams can be sustainably collected from the wild. Countless agronomists have tried to transplant this highly valuable yam and cultivate it as an agronomic crop. Despite the great human interest in its natural history and use, *D. deltoidea* is not even found in botanical gardens, even though botanical gardens put forth great efforts to display species that have impacts on civilization. It's not that the curators are prudish; they have been unable to cultivate the species. This yam species has been intractable to domestication, and its value suggests that it could become extinct. As negative findings are rarely published, the reasons for its intractability to cultivation are not adequately recorded. We know all too well that the vast majority of bacterial species cannot be cultivated on artificial media ("domestication"). There are animals that humans have tried to domesticate without success (e.g., the zebra and the African ass), and this yam seems similarly intractable. Two temperate *Dioscorea* spp., *D. villosa* and the weedy *D. batatas,* are cultivatable and their tubers contain about 150 milligrams diosgenin per kilogram fresh weight,[308] much less than in *D. deltoidea.*

20.1. Biotechnological Solutions

One cannot attempt at this juncture to propose genes that might be used to transform this species into a standard crop without knowing the nature of the intractable agronomic problems that could not be overcome. That does not mean that biotechnologists gave up, and three approaches have been or can be considered.

20.1.1. *Cell Cultures of the Species*

Cell cultures in bioreactors seemed to be the right direction for producing this high-value product, even though undifferentiated cells typically produce fewer secondary metabolites than the parent species. The challenge to get such

Table 30. Elicitation of In Vitro *Diosgenin Biosynthesis in* Dioscorea *spp.*

Species	Elicitor	Effect	Reference
D. deltoides	Norflurazon	7-fold increase in half the time in cell cultures due to inhibition of carotenoid biosynthesis, shunting metabolites to steroids	1044
	Entrapment		528, 908
	Mycelia	72% increase in diosgenin in fermentor to 230 mg/liter	917
	Abiotic	Light and limiting phosphate increase 40-fold to 750 mg/liter	888
D. galeottiana	Abiotic	Light and limiting phosphate increase 40-fold to 0.4%	916
D. floribunda	Ethylene	72-fold increase	259
Trigonella		Doubling to 7 mg/g dry weight	398
D. bulbifera	$CuSO_4$	10-fold increase to 1% dry weight, but 4-fold growth inhibition	777

cells to produce high levels of diosgenin was taken on by several groups and continues to this day. Cell cultures of *Dioscorea* were produced and grew well (e.g., references 172, 562, 1043; Table 30). The content of diosgenin in the plant cells is far too low for commercialization, nowhere near the 6 percent in yams. The cell cultures are bright orange because of high carotene content. Because carotenes are formed on a branch of the same pathway as steroids, it was hypothesized and found that suppressing a key enzyme (phytoene desaturase) of carotenoid biosynthesis with the antimetabolite norflurazon shifts the metabolism to the steroids, causing diosgenin to accumulate earlier and to a much greater extent.[1044] Other less specific, but effective stresses also effectively enhance the levels of diosgenin in cell cultures (Table 30), but still not near the levels that are competitive with the wild yam.

20.1.2. Move the Key Steroid Biosynthesis Genes from D. deltoidea to Currently Cultivated Yams

If the pathway is not long and/or if the genes are clustered it may be possible to clone the key genes and transform them with appropriate expression vectors and achieve high yields of diosgenin in related species. The early genes in the triterpene saponin biosynthesis pathway leading to steroids through to β-amyrin are being elucidated in *Medicago truncatula*, the model legume that is being fully sequenced.[1032] If this is indeed the same pathway in monocot *Dioscorea,* and some of these genes are limiting, there may be some utility to this genetic knowledge from a legume dicot. Because it is possible to induce diosgenin production in *D. deltoidea* cell cultures with antimetabolites (Table 30), it may be possible with microarrays or differential display techniques to easily fish out the genes controlling production. The technologies to transform *Dioscorea* species have not been fully worked out. Initial attempts were stymied by getting false positives to the gus (β-glucuronidase) marker reaction because of cryptic contamination of the yam cells by endophytic bacteria that naturally expressed the *gus* gene.[1067] It was later possible to obtain a stable biolistic transformation of one *Dioscorea* species[1068] and transient protoplast transformation of a few species,[1069] but plant regeneration has not been reported. If it could be done, the technologies are in place for cryopreservation of shoots.[279] One can still transform *Dioscorea* cells; one does not need to regenerate plants if the *Dioscorea* is to be grown in culture. Conversely, if the transformation is into a yam that can be propagated vegetatively, one can transform the bacterial endophyte that is already there,[1068] if it is not causing a yield drag, or one could attenuate one of the systemic viruses attacking yams, for example, DaBV (a Badnavirus that has been fully sequenced)[153] and YMV (a Potyvirus),[727] transform one of them, infect the yams, and get expression.

20.2. Do We Need These *Dioscorea* Biotechnologies to Prevent Overpopulation?

It is telling that some plant biotechnologists clearly do not fully read the literature before setting off on their endeavors, as also discussed in other chapters. They have set out with idealistic goals of saving a plant species from extinction, and mammalian populations from overpopulation (mammalian is used advisedly, as there are also similar veterinary uses of the derived hormones with pets, not only with the pets' owners), and monetary goals of technology replacement.

More than three decades ago chemical and microbiological methods were already replacing the yams, not just from a "prevention of extinction" point of view, but as a hedge against government monopolistic policies in controlling yam sales.[280,1151] The pharmaceutical companies were already using other methods when a Mexican government yam monopoly raised prices such that pills from the yam were more expensive than those from other sources. The starting materials can be cholesterol, which can be cheaply extracted from wool fat derived from the first step of cleaning wool after shearing, or from fish oil. Various soil microorganisms (fungi and bacteria) were isolated that metabolize and derivatize the cholesterol to materials that can be further chemically derivatized to the requisite hormones.[1151] Sitosterol, stigmasterol, and solasodine from other plants were also being used to manufacture various sterol hormones.[280] The large fruits of *Solanum marginatum,* indigenous to East Africa, contain 5 percent solasodine, and this African species was being cultivated in Ecuador. Radiocarbon dating the steroids in birth control pills allowed scientists to ascertain that a goodly proportion are made by total synthesis from ancient petrochemicals, not recent plant or animal material. All this information about alternatives to wild yams already being used commercially was sequestered in literature that plant scientists were unlikely to find in the days before electronic databases. Still, to this day, papers appear from academia and biotech companies alike, keeping in vitro studies of yam cell and tissue culture technologies alive,[259,279,777] despite three decades of known and used inexpensive alternatives.

Is this all for naught? It depends on how one interprets the Chinese proverb: "If you never change direction—You will end up where you are heading." Those that have learned from the experience have gone on to participate in projects that were not made redundant before they started. Cell cultures are being used to supply human enzymes, antibodies, proteins that cannot be produced in bacteria or yeast because of special glycosylation or incompatibilities, and are too expensive to produce in mammalian cell cultures—a smaller niche than once presumed, but large enough to garner the support of venture capitalists who do have experts that scour the literature and ascertain what is worth doing. Some have been able to elevate levels of products to be commercially feasible by using various tricks of special promoters and cellular compartmentation. Others continue with hopes, even of using diosgenin as a natural "organic" insecticide,[851] despite what we know about how it has been used ethnobotanically in the past as a hormone mimic, and the extensive modern literature on its pharmacological properties.[1178] Will regulatory authorities allow the use of an insecticide that binds to human estrogen receptors, just because it is "natural"? The regulators and environmentalists are down on synthetic pesticides that purportedly affect frog sexuality because they are pseudoestrogens. How will diosgenin pass the frog test? We know how its derivatives affected an earlier type of frog test, the frog test used for detecting pregnancy. Indeed, the derivatized estrogens are not well degraded in municipal sewage facilities (e.g., reference 1154), and their levels in effluents would not be allowed if they emanated from agriculture, but environmentalist-activists have been inactive in writing tracts against the pollution of our water supplies by synthetic estrogens. Is there a double standard?

CHAPTER TWENTY-ONE

Tomato

Bring Back Flavr Savr™—Conceptually

The problems with tomatoes are common to many fruits, and thus the word tomato in this chapter can often be changed to many other soft fruits, at least in the general sense of thinking how one might go about dealing with the reader's favorite perishable fruit, from the timing of ripening after harvest, shipping and handling through to the consumer. Few of the specific solutions being applied at present to tomatoes by the research community may be specific to that crop. It is the concepts that are important at this juncture, because the one "solution" that had been applied to tomato, and had received regulatory approval, is no longer on the market due to quality considerations that are discussed below.

21.1. Tasteless Mushy Tomatoes

Fresh vegetable crops such as tomato (previously *Lycopersicon esculentum* and now reclassified as *Solanum esculentum*) and soft fruits such as peaches, plums, and apricots taste far better, and have considerably higher yields, if they are picked ripe. Conversely, these soft fruits are hardest to pick, ship, and market without damage when picked in their prime. Wastage can become a considerable portion of the crop, in particular, due to local market volatility. Thus, tomatoes are often picked green, shipped thousands of kilometers in refrigerated vehicles, and arrive green in market warehouses. They would slowly ripen although not with the full flavor of field-ripened fruit. As time is money, they are treated with gaseous ethylene in chambers. Ethylene, a natural plant hormone (but produced synthetically for this purpose) hastens a pseudo, incomplete ripening process. Then it is quickly downhill for the tomato, just as with a field-ripened tomato, whether the ethylene is naturally produced in the tomato (termed "climacteric") or applied exogenously. The tomatoes go soft and mushy, whether field ripened or not, because of their short shelf life. This is part of a natural senescence process, but natural is what neither the retailers nor the consumers desire. Indeed, retailers discard large quantities of overripe tomatoes ("spoilage"), because they cannot always gauge sales. Whereas any tomato tastes better if not refrigerated, tomatoes are typically placed in the refrigerator, especially in households where once-a-week shopping is typical, and still many a housewife discards mushy and rotten tomatoes.

Thus, the need is for tomatoes (and other fruit) that can be picked ripe with full flavor (which matters to the consumer) and full size (which matters to the farmer) and can be shipped and stored without costly and fossil-fuel-expensive refrigeration. When the tomatoes thus arrive on market, they should have an extended shelf life, which is better for both the consumer and marketer. If the consumers do not need to refrigerate, the tomatoes will taste better. These problems need to be redefined in scientific terms. Ethylene and many fruits have a climacteric effect; soon after picking a peak occurs in ethylene production by the fruit. The ethylene induces a huge burst in respiration. The final processes of ripening begin, and soon the fruit softens to a stage where it is inedible, even if not attacked by rot fungi, to which overripe fruit are highly susceptible. The pathogens find soft fruit irresistible, because the chemicals that ward off fungal attack also disappear. The early green picking is to get the fruit to the city warehouse before the climacteric arrives. The treatment with ethylene is to sell the fruit as soon as possible, whether it was ready or not, that is, to turn green fruit to red, even if the rest of the flavor has not matured. Such a tomato appears ready to eat but is organoleptically blah. The refrigeration is to slow the softening after the ethylene-induced climacteric. We must remember that people are part of the evolutionary development of tomato dissemination. When seeds are mature, the climacteric naturally starts the fruit senescence mechanism. People act in the evolutionary framework as postsenescence dispersal mechanisms of the seeds, guaranteeing the propagation of the species. The attractive combination of color, texture, taste, and smell draw us to the tomato for the same reason that other mammals are attracted to other ripe fruits. The tomato seeds remain undigested after eating, as witnessed by the luxurious summer thickets of tomatoes naturally growing on the banks of many a municipal sewage-settling pond.

21.1.1. The Challenge

The ideal challenge is to develop tomatoes that ripen without a climacteric, and/or for which the climacteric does not initiate softening. If successful, this would give rise to a tasty, firm, but not hard, aromatic red fruit with an excellent balance of texture, solids, acid, and sweetness that takes weeks to turn into mush, even without refrigeration. The appearance of the enzymes that soften the fruit to mush must be inhibited. A similar extension of shelf life is also important for cut flowers, as discussed in Chapter 22. Some of the solutions are similar, but more complex with tomatoes, where more than physical appearance is desired.

21.2. Technological Solutions

As described above, the technological solutions have mainly been refrigeration and ethylene gassing, with unsatisfactory results. Ethylene does more than facilitate the climacteric and the subsequent ripening. Smaller amounts of ethylene are continually present and hasten the rest of the postripening processes after the climacteric. The greater understanding of the processes has led groups to find chemical solutions. Long ago it was recognized that silver ions inhibit the production of ethylene and delay ripening[118]—the ripening-related mRNAs that are typically turned on by ethylene are not produced.[254] As we spend our silver to buy tomatoes, we prefer not eat this heavy metal, so this solution is academic and indeed is often used in the laboratory environment, especially in tissue cultures, where gaseous ethylene can inhibit development. A commercial replacement for silver soon appeared, gaseous methylcyclopropene (MCP). This compound binds to the ethylene receptors and blocks the effects of naturally produced ethylene.[990] MCP also inhibits ethylene production, and the effect can last well over a week.[505] Treatment of tomato fruit with this compound at various stages delayed the onset of the next stage, including when fruit was firm red ripe,[505] without any major changes in quality, firmness, or aroma.[791] At the molecular level MCP blocked ethylene-inducible levels of the mRNAs encoding phytoene synthase (a key enzyme in carotenoid biosynthesis), expansins (leading to fruit softening), and 1-aminocyclopropane-1-carboxylic acid (ACC) oxidase (leading to the synthesis of more ethylene).[505]

21.3. Breeding Solutions

The breeders have found mutants in two genes that have successfully been used for the producers (but not the consumers). These are the *rin* (ripening inhibitor) and *nor* (nonripening) mutations. Tomatoes homozygous for *rin* or *nor* do not produce ethylene, and are responsive to ethylene only insofar as exogenous ethylene causes the unripe fruit to turn red, without inducing the rest of the ripening processes. There is no softening of homozygous fruit from a woody texture to a firm texture, nor is there development of much flavor due to aroma,[706] but the tomatoes can be stored for quite a while at that induced red color, although a chain saw would slice them easier than a knife. Hybrid tomato varieties heterozygous for these genes are somewhat more acceptable, with a lengthened shelf life, which is not as long as with homozygous fruit. Tomatoes recessive for both genes were of even worse quality,[608] but they were never marketed. The hybrid fruit was at least partially deficient in a whole series of mRNAs relating to fruit ripening further suggesting control by a regulatory gene[590]. The genes for the wild-type *RIN* and *NOR* have been isolated by positional cloning[744,1116] and are regulatory in nature, as expected.

21.4. Biotechnological Solutions

Various solutions are proposed below. They are not mutually exclusive and may well be stacked to achieve the ideal tomato. Readers who prefer peaches, plums, or other soft fruits can easily extrapolate the approaches described to their favorite fruit. This even includes using the *rin* and/or the *nor* genes, but in ways they could not be used in classical breeding.

21.4.1. Preventing Softening of Ripe Fruit

Soon after a tomato vine ripens, processes of softening to mushiness ensue. A series of genes that are related to the ripening process might be modulated to prolong the shelf life of tomatoes, providing a higher-quality, longer-lasting product (Table 31). Concrete progress along these lines has been made.

Two groups independently came up with the same conclusions on how to prevent the softening while retaining the flavor. A key enzyme that turns on soon after vine ripening is pectinase (also called polygalacturonidase), which slowly degrades the pectin glue between cells. By using antisense technology with fruit-specific promoters for one of the plant's own genes, they were able to stop the softening, with fruit remaining ripe for a long time.

One group did this in varieties of tomato to produce tomato paste.[958] The task was not easy as the enzymes are encoded by families of genes, with some redundancies, and some tissue-specific expression systems. Thus, finding the right gene took time. Despite a wide variety of cell wall-degrading enzymes being expressed during softening, transgenically suppressing the synthesis of many of them is without effect on the fruit.[185] The eventually commercialized transgenic tomato paste exhibited a higher quality than normal paste due to higher total solids content and having more pectin, and the resulting pizzas were tastier. Pizza chefs add less carbohydrate thickeners when there is more pectin in the paste. This paste was made in the United States and successfully marketed, legally in British supermarkets, well labeled as transgenic for a considerable period. When the media blitz against transgenics began after the outbreak of mad cow disease, the supermarket chains dropped this material, while explaining to their customers that they wished to give them maximum choice. How the activists were able to convince the public to equate mad cows and transgenics was a public relations coup, even if scientifically untenable. As there are no labeling requirements for such material in the United States, it is not known whether the transgenic tomato paste is marketed there.

A second group put a similar construct into a market tomato variety[610] and began to market the tomatoes under the logo Flavr Savr™. A series of problems ensued. The company did not pick a high-quality tomato variety; the variety they chose did not have much taste to begin with, so there was no flavor to save. They did not realize the regional nature of varieties, that tomato

Table 31. Genes Encoding Ripening that May be Utilizable to Modulate Fruit Ripening to Enhance Market Quality

Gene	Encodes	Form used or might be used	Possible application	Reference
Many	Pectinases	Antisense	Delays softening	744, 1118
LeACO1	Isoform of ACC oxidase	Antisense or RNAi	Prevents ethylene biosynthesis, ethylene applied near consumer	45
LeEXP1	Expansin	Antisense or RNAi softening	Expansin facilitates the slippage between cell walls leading to	44
RIN	Nonripening	RNAi with a promoter that turns on postclimacteric	Could possibly delay the senescence of fruit. Its constitutive natural mutant only reddens, but remains nearly inedible	1116
NOR	Ripening inhibitor	Same as RIN	Same as RIN	744
ACC deaminase	Prevents ethylene formation	Expression of bacterial gene	Would degrade ACC, a precursor of ethylene— best put under a fruit maturation specific promoter	573
SAM hydrolase	Prevents ethylene formation	Expression of a bacterial gene	Reduces ethylene precursor—best put under a fruit maturation specific promoter	573

varieties from one geographic zone do not grow well in another, yet they had transformed only one variety.[697] Instead of going back to the bench and doing it right, or crossing the gene into quality varieties, the project was dropped. This is highly unfortunate; besides potentially providing a high-quality tomato with far less wastage, there should be a large savings in energy costs (no need to spoil tomato flavor by refrigeration, as is done in most "once a week to the supermarket" households).

Thus, those who imagined having tasty heirloom tomatoes that could last a week had their hopes shattered. A lesson to learn: excellent products require excellent horticulture along with genetic engineering. A possible reason for not doing so was the possible realization that it would be best to transform each variety separately, as this is quicker, easier, and more likely to return to the quality bred into each superior variety than traditional backcrossing. Such transformations are now doable with most varieties, but each such RNA interference (RNAi) or antisense transformed variety would have to undergo full regulatory procedures because of the illogical "event-based" regulatory rules (section 4.1).

Other genes can be suppressed and allow shelf life to be extended. The "expansin" proteins act to soften the cell–cell connections, allowing plants to grow. They are also part of the fruit-softening process. Antisensing the expansin gene *LeExp*1 resulted in a more than 95 percent suppression of the protein expression,[165] and the fruit had an extended shelf life of about a week at a temperature of 13°C, and tomato paste made from the pulp was much thicker than from the nontransformed controls.[164]

Not all genes that cause complex carbohydrate depolymerization induce softening, at least not by themselves. Xylanoglucan hemicelluloses are more than 20 percent of the cell wall polymers of tomato, and xylanoglucanases are produced during ripening,[1118] but are not produced in *rin* fruit. When *rin* fruits were transformed with genes encoding xylanoglucanases, these polysaccharides were degraded, but the fruit was no softer,[393] so there was no reason to use RNAi or antisense to suppress this gene.

As noted above, the *rin* and *nor* constitutive mutants led to inedible (but marketed) fruits that can be colored by ethylene treatment. As the genes for the wild-type *RIN* and *NOR* have been isolated and cloned,[744,1116] it may be useful to transform them back into tomatoes in RNAi or antisense form, but under the control of a promoter that turns on *rin* or *nor* only when the fruit is fully ripe, to stop the onset of senescence (softening) but to retain full ripening. One might achieve the same extended shelf life of the homozygous *rin* or *nor* mutants, but have a fruit of higher quality than the heterozygous hybrids. *RIN* is a MADS-box regulatory gene, so it should have such an effect.

The *nr* (never ripe) mutation is due to a mutation in an ethylene receptor gene *LeETR3*, a member of a family of six such genes.[592] These receptors act as negative regulators, suppressing ethylene-responsive genes when there is no ethylene. Klee[592] speculates that overexpression of this gene would bind ethylene during ripening, preventing softening.

One could delay softening by preventing the appearance of the low levels of ethylene that accelerate softening after ripening. Two enzymes encode the biosynthesis of ethylene; ACC synthase and ACC oxidase. It is assumed that the temporal suppression of the expression of either gene after the climacteric with ripening stage-specific promoters could increase shelf life of ripe fruit. These enzymes are encoded by families of genes, and finding the active one is a task. It was recently reported that using a multiplex reverse transcriptase-polymerase chain reaction (RT-PCR) approach with different genotypes of tomato led to the conclusion that *LeACO1* gene is predominant.[45] It has yet to be demonstrated that the RNAi or antisense form of this gene prolongs shelf life of tomatoes.

There is good reason to believe that the anti-softening concept would work in other soft fruits such as peaches and apricots, possibly with different genes or mixes of genes. Very few people are still alive who have tasted the old, real aromatic apricot varieties with short shelf life, long abandoned commercially, which might be resuscitated by such technology. The tasteless, mealy, standard five-pointed apple varieties have been heavily replaced with tasty varieties over the past three decades, much to the consternation of the conventional knowledge of experts who were so sure a premium apple had to have five points on the bottom. The time has come to provide the consumers with tastier soft fruits.

21.4.2. Full Ripening in the Warehouse?

Another approach could be to isolate the genes responsible for flavor development and reengineer them back into tomato under a climacteric-inducible promoter. In this scenario tomatoes could be picked green, and full ripening with flavor could be achieved under the controlled conditions so favored by tomato marketers. As flavor is typically a complex of a large array of aromatics, it may be hard to attain full flavor with such an approach, but the bulk market deserves better than what it is getting at present. Eventually, it might be possible to hit upon the controlling elements of flavor development, and full flavor could be achieved.

21.5. Regulatory Constraints

It will be necessary to transform the successful constructs into regional elite varieties. It is not possible to develop a marketing structure based on a single seasonal and regional variety. It will be exceedingly hard to achieve this by backcrossing; as evidenced by the Flavr Savr debacle. It is not that hard to transform tomatoes. In the current regulatory environment, it will probably not be economically feasible to deal with transformation "event-based" regulation (section 4.1) unless antisense/RNAi transformations are exempted from regulation, as some suggest they should. Shutting down a plant gene can be achieved by mutation, and in one respect antisense/RNAi resembles mutation. It is almost impossible to achieve ripening-stage-specific mutations, but the equivalent can be achieved with antisense/RNAi with ripening-stage-specific promoters. Clearly the governmental regulatory structures must be modified in a case like this when all that will be done is to control the tomato's own genes at times that best fit the consumer and producer.

21.6. Biosafety Considerations

Currently available literature limits the range of wild species interbreeding with tomatoes to small parts of South America. Taxonomists recently preferred to reincorporate *Lycopersicon* back into *Solanum* where they had been a century ago, but that does not mean that they cross with other *Solanum* species outside their center of origin. Tomatoes are self-pollinating, so it is unlikely that there can be consequential pollen flow, that is, any pollen flow to nearby, nontransgenic varieties should be orders of magnitude beneath any current regulatory threshold.

CHAPTER TWENTY-TWO

Orchids

Sustaining Beauty

Orchids are the epitome of beauty and biodiversity, both to be preserved. More than twenty-five thousand species have been described, and crossing wild forms with cultivated forms, or with each other has produced many hybrids. Orchid flowers typically cross-pollinate in the wild, leading to many hybrids as well, a treasure trove for taxonomists who like to name many species ("splitters") and a bane to those who like to have larger groupings ("lumpers"). Insects, often species-specific ones, are usually the pollinators, but birds and bats are involved in pollinating some species.

We only see a minuscule part of this biodiversity in the market place. Some tropical developing countries have established large industries of cultivating and shipping the few orchid species that can withstand the vicissitudes of travel and vase. Other tropical countries dream of establishing such trades with their splendid indigenous orchids, but the species at hand do not have the shelf life that would allow entering the trade. Simple genetic engineering can add to the shelf life. There are a few possible targets in flowers that preserve vase life, and one can extrapolate from orchids to less exotic cut flowers, with the concepts and systems available as well as those yet to be developed.

About seventy genes have already been cloned from seven genera of orchids, as summarized in Table 32.[759] It is surprising how little has been reported so far about the utilization of the isolated genes.

Table 32. Genes (or cDNA) Cloned from Orchid

Putative Function *Gene designations*	Type of gene
Flower induction/transition	
ovg2/DOH1	Homeobox, class I knox
om1, otg7, DOMADS 1,2 and 7	MADS box
ovg 27	Transcriptional repressor
otg 16	Casein kinase
ovg 11, 15, 50, otg40	Somehow involved
Cell division and structure	
ovg 14, otg 4, ovg 30, ovg 29	Cell cycle regulators
otg 2, P-ACT1, ACT 2, profilin	Myosin and actin related
Flower senescence	
Ds-ACS1 and 2, ACS 2 and 3, pOACS 10	ACC synthase
pPEFEA	ACC oxidase
Petr1, Per1	Ethylene receptor/response sensor
Ovule development	
o39	Homeobox protein
o40/CYP78A2	Putative cytochrome P_{450}
o108	Cell cycle regulator
o126	Glycine-rich cell wall proten
o138	Embryo formation
Flower color	
OCH 3, 4, 8	Chalcone synthase
Fht/pCF1	Flavanone hydrolase
ODFR, Dfr	Dihydroflavanol reductases
Disease defense / stress response	
Pal/OPAL1, ovg 43	Phenylalanine ammonia lyase
pBibSy211 and 811, pBBS 1	Bi-benzyl synthase
pahh511	S-Adenosyl homocysteine hydrolase
Primary metabolism	
14 different genes	10 different functions
Other	
cko1	Cytokinin oxidase

Source: condensed from Mudalige and Kuehnle,[759] where primary references and GenBank accession numbers are found.

22.1. Flower Traits in Transgenic Orchids

At least one group is currently bioengineering orchids to produce varieties with unique colors such as deep red and dark blue. The dark blue is probably being transformed with the same or similar genes as are being used in carnations (see section 22.3.3), but other than news releases little information has been published about what has actually been achieved. So far the only report is of little value; orchids were transformed with luciferase, and they glow in the dark when sprayed with luciferin (reported in reference 759).

22.2. Physiologically Extending Orchid Vase Life

Knowledge of the physiology and biochemistry of senescence of the orchid flower can help in understanding what transgenic interventions may assist in extending vase life, if indeed there are not sufficient other interventions that may work. Extending vase life in orchids is akin to extending shelf life in tomatoes (Chapter 21) with many of the same elements involved. Still, the beauty in a tomato is in the flavor of its fruit, and a fruit is the last thing we want in an orchid. Even the commercial propagator is not interested in fruits or seeds; most orchid propagation is performed vegetatively, with parental stocks kept sterile and virus free in tissue culture. Indeed, as far as orchids are concerned, sex is the beginning of senescence, at least the senescence of the flower. Physically keeping pollen off the stigma already extends vase life. When this is performed by removing anthers, the results vary between laboratories and species, as removing anthers can cause injury, which in itself often enhances senescence. Many low-molecular-weight, water-soluble, nonproteinaceous, heat-stable signals have been reported to initiate orchid flower senescence after pollination. These include indole-3-acetic acid (IAA) among the other collected fractions[862], as well as 1-aminocyclopropane-1-carboxylic acid (ACC), the precursor of ethylene.[578] In a case where cut orchids can last two weeks if not pollinated, following pollination a rapid acceleration of the wilting process takes place, which is completed within two days.

The first event detected following pollination was an increase in ethylene sensitivity four hours after pollination. Ethylene production was detected as a climacteric twelve hours after pollination.[861] Treatment of orchids with silver thiosulfate[1097] and methylcyclopropene (MCP),[498] both inhibitors of ethylene action, completely inhibited the pollination-induced increase in ethylene production and considerably extended vase life. MCP application suppresses ACC oxidase activity and ethylene production.[1169] The inhibitors of ethylene biosynthesis, aminooxyacetic acid (AOA) and 2,5-norbornadiene, also delayed premature senescence of pollinated orchid flowers, though AOA was more effective,[579] as was aminoethoxyvinylglycine.[1124] Conversely, treatment of flowers with calcium and its ionophore A23187, which increase ethylene sensitivity and protein phosphorylation, also promoted ethylene production, enhancing senescence.[862] The higher the concentration of ACC in pollen, the greater the rate of senescence.[578] The initiation of ethylene biosynthesis is regulated by the coordinated expression of three distinct ACC synthase genes in orchid flowers. One ACC synthase gene (*Phal*-ACS1) is regulated by ethylene and participates in amplification and inter-organ transmission of the pollination signal. Two additional ACC synthase genes (*Phal*-ACS2 and *Phal*-ACS3) are expressed primarily in the stigma and ovary of pollinated orchid flowers.[168] Treatment with okadaic acid, a specific inhibitor of type 1 or type 2A serine/threonine protein phosphatases, induced senescence correlated with a differential expression pattern of the ACC synthase multigene family. Okadaic acid induced accumulation of *Phal*-ACS1 mRNA in the stigma, labelum, and ovary in contrast to *Phal*-ACS2 and *Phal*-ACS3.[1124] Staurosporine, a protein kinase inhibitor, inhibited the okadaic acid-induced *Phal*-ACS1 expression in the stigma and delayed flower senescence, suggesting that hyperphosphorylation of an unidentified protein(s) is involved in up-regulating the expression of *Phal*-ACS1 gene, resulting in increased ethylene production and accelerated senescence of the orchid flower.[1124] In summary, inhibition of ethylene synthesis or action delays senescence, stimulating either of them induces senescence. Thus, ethylene is key to dealing with senescence and vase life.

22.3. Biotechnological Solutions to Vase Life and Other Constraints to Orchid Production

Orchids are ideal for technologies that will prolong vase life. They are continually propagated from tissue culture into plantlets for the commercial growers, and primary tissue cultures can be transformed. It is obvious that similar technologies can be considered for other vegetatively propagated species cultivated for flowers, even if they are not as beautiful as orchids. Still, the orchid industry may well be happy with the chemical solutions at hand, instead of the natural solutions, affecting the plants own genes, as the chemical cost is inconsequential compared with the value of orchids. Breeding is unlikely to arrive at the same solutions, as orchids are predominantly vegetatively propagated. No grower would take thousands of vegetatively propagated orchids and treat their cut flowers with senescence-inducing ethylene to look for a mutant that has a longer vase life. Still, it is possible to envisage that a transgenic solution can be longer lasting, well after the point of sale, the last point where a chemical can be applied.

Orchids have been transformed by many groups for over a decade, either biolistically[615] and by *Agrobacterium* transformation,[775] but typically with marker genes and not with genes of commercial utility.

22.3.1. Prevention of Fertilization

As fertilization typically induces the loss of beauty in flowers it can be prevented by inducing male sterility. This has been done in other species by using the *barnase* gene under the control of a tepetum-specific promoter,[691] as described in Chapter 4.

22.3.2. Prevention of Senescence

Senescence can be delayed by stimulating the overproduction of senescence-inhibiting hormones such as cytokinins. The cytokinin benzyladenine increases the vase life of many orchid species, often acting synergistically with compounds that inhibit ethylene production.[1169] Thus genes enhancing cytokinin production might enhance vase life, especially if stacked with genes preventing ethylene biosynthesis in the flowers. The first and seemingly key step of de novo biosynthesis of cytokinins is catalyzed by adenosine phosphate-isopentenyltransferase (IPT), which produces isopentenyl adenine nucleotide. In higher plants, *trans*-zeatin, a major cytokinin, is formed by subsequent hydroxylation, which is catalyzed by a cytochrome P_{450} monooxygenase, CYP735A1 or CYP735A2.[933] These genes have all been isolated and cloned. Probably the closest use of such genes was with other monocots, *Lolium* and maize, where plants were transformed with plasmid constructs containing 2 kb of the five prime flanking sequence of *SEE1* (a maize cysteine protease gene showing enhanced expression during senescence) fused to an *Agrobacterium IPT* gene delaying senescence.[648,910]

It may also be possible to increase cytokinins by activating endogenous *ipt* genes. Activation of three different KNOXI proteins using an inducible system resulted in a rapid increase in mRNA levels of isopentenyltransferase 7 (AtIPT7) and in the activation of ARR5, a cytokinin response factor resulting in a rapid and dramatic increase in cytokinin levels.[1170]

The obvious approach is to suppress the ethylene hormone that induces senescence. The best approach for orchids would be to attack the genes encoding ACC synthase and ACC oxidize. Reports have appeared on the internet where groups used antisense constructs of those genes with several species of orchids, demonstrating the effectiveness of prolonging the vase life by this approach, but hard data have not been forthcoming.

22.3.3. Modification of Flower Color

As if orchids were not beautiful enough, there are always those frustrated by a limited range of possible flower colors due to the lack of an anthocyanin biosynthetic gene or the substrate specificity of a key anthocyanin biosynthetic enzyme, dihydroflavonol-4-reductase (DFR). One group wished to change *Cymbidium hybrida* orchid flower cyanidin-type (pink to red) anthocyanins to the pelargonidin-type (orange to brick-red) anthocyanins. They cloned a *Cymbidium DFR* gene and transformed it into a petunia line lacking *DFR*. The *Cymbidium DFR* did not efficiently reduce dihydrokaempferol, which is an essential step for pelargonidin production.[552] Thus, they will have to look elsewhere for the appropriate gene(s).

Various approaches to modifying flower color have been reviewed far more times than actual papers on such modifications.[738,1046] The gene encoding a UDP-glucose:anthocyanin-3'-*O*-glucosyltransferase that has strict substrate specificity for the 3'-hydroxy group of delphinidin-type anthocyanins was cloned from gentian. The specificity was confirmed by expression of the cDNA in transgenic petunia producing the stable blue anthocyanin delphinidin.[364] Carnations were transformed to produce delphinidin, but the pure blue color was not obtained because the native genes encoding the enzymes that glycosylate the native anthocyanins, also modify delphinidin, resulting in a more violet than blue color.[365] Other approaches have also been used to obtain blue flowers.[16] The most bizarre transformation was to obtain flowers that fluoresce green in the dark when excited by ultraviolet light.[724]

22.3.4. Other Traits

Other traits that have received attention in other flower species include floral scent, floral and plant morphology, and disease resistance.[1046] Orchids were transformed with Cymbidium Mosaic Virus coat protein cDNA conferring resistance to the virus, and transformants were then retransformed with sweet pepper ferredoxin-like protein cDNA to also enable expression of *Erwinia carotovora* bacterial disease resistance.[199]

22.4. Biosafety Considerations

Orchid hybridization occurs both naturally and artificially. Many natural hybrids exist. No other family of plants has produced as many hybrids, either in total numbers or in the number of complex crosses. This genetic fluidity of orchids may be associated with their relative recent evolution as a distinct plant group. When two closely related species grow intermingled in an area, the pollen of one may be transferred to the stigma of the other and sometimes a hybrid is formed.

Gene flow between transgenic orchids and wild relatives poses special concerns in centers of diversity and centers of origin. In these places, the probability of gene flow occurring superficially seems relatively high, as commercial orchid greenhouses are usually relatively open structures. There is a broad consensus that the biological and genetic diversity of those regions must be preserved and could be vulnerable to ecological disturbances.

22.4.1. Preventing Transgene Flow from Commercial Production

No wild orchids would be growing in the open orchid commercial production structures. Thus the only risk is to wild orchids at a distance from such structures. Most orchids are exported before anthers open and release pollen. Orchids are insect pollinated, and pollen can move those distances from unharvested flowers. The rare pollen carrying transgenic flower color traits that may escape could give rise to plants whose fitness is neutral in nature, or deleterious. If the movement is rare when compared with natural pollinations, then there should be little change in nature.

Rare pollination of commercial orchids by wild orchids would be inconsequential as the commercial orchids are propagated through tissue culture, and the flowers exported before pollen usually forms. The use of male sterility to enhance vase life (section 22.3.1) will serve to prevent gene flow from commercial production to the wild, even if it does not prove to actually increase vase life. This would be an added fail-safe mechanism for a situation that is hardly critical.

The approaches of preventing senescence through an increase in cytokinins or a decrease in ethylene would probably render a hybrid unfit by preventing seed formation. If the amount of pollen escaping to the wild is small compared with the amount of native pollen in the wild, then a small proportion of unfit individuals should be inconsequential, but this had better be ascertained with a few simple experiments.

Only disease resistance, of the genes so far transformed into orchids, may confer a competitive advantage in nature. As orchids hardly compete with each other, and have trouble competing for niches with other species, the result of such hybrids might be a more beautiful nature, but it is up to the regulatory authorities to decide whether more healthy orchids are desirable in nature.

CHAPTER TWENTY-THREE

Olives

And Other Allergenic, Messy Landscaping Species

The olive (*Olea europea*) has very desirable characters as an ornamental tree. It has a picturesque gnarled trunk topped with willowy shoots bearing green-gray leaves. The long life and fast-growing nature, coupled with the asymmetric beauty and the nonuniformity have caused olives to be heavily chosen for subtropical and desert urban landscaping. Olives have a low water requirement once established and can withstand the lead in gasoline and the hydrocarbons in diesel fuel and other exhaust emissions. The domestication of the wild oleaster to olive is probably near the simplest domestication of any species; it is primarily a function of pruning.[151] An unpruned olive tree in a few short years reverts to being a feral oleaster, albeit there has been some selection for fruit size and oil content among the olives.

One problem with olives took an inordinate time to rediscover; the extreme allergenicity of the pollen (e.g., references 241, 383), which had been known[575] and forgotten. Most physicians dealing in allergies resided in temperate, not subtropical climates. When the extent of allergenicity was realized, some jurisdictions such as Pima County, Arizona, banned the sale and planting of olives.[801] It would have been more conducive to obtaining a solution to the problem had the ban been on pollen-producing olives. A second problem is the mess created by dropped ripe fruit on streets and sidewalks. In some areas (e.g., Australia), birds eat the unpicked ripe fruit, dropping the seeds into natural areas, displacing the native vegetation.[1006] The fertility of the olive, desired by the commercial olive grower, is contraindicated by the landscaper. People with olive allergies have petitioned municipalities to remove olive trees because of their personal problems, and solutions are needed that will satisfy the landscape and the allergic.

Ten separate allergenic olive proteins already have been described by various authors,[913] with some more characterized than others. Other landscaping trees, as well as fruit trees found in urban gardens, have allergenic pollen, and the allergens from one species often cross react with those of other ornamental or weed species (Table 33). Thus, those allergic to olives would get little respite if the trees were chopped down, as suggested, as they will remain allergic to other species because of the cross reactivities.

23.1. Genetic Solutions

Olive trees, like all fruit trees, have a long juvenile phase that has precluded intensive breeding programs, which perforce must be long term. Selection and vegetative propagation of the best types by grafting have been the mode of varietal achievement. Researchers are beginning to apply molecular marker techniques to olives to identify and characterize useful germplasm. Most of the characterized markers deal with oil composition and quality,[478] and alas none of traits listed among them had anything to do with allergenicity.

California researchers looked for a fruitless olive for ornamental plantings in the days before olive allergies were known. Most trees that were claimed to be fruitless came from areas with insufficient winter cold to induce flowering but flowered in normal olive cultivation environments. Part of the problem with olive trees was solved many years ago when researchers brought shoots to California from a 30-year-old fruitless olive mutant from an olive-growing region in Australia, and made it available for landscaping as cv. Swan Hill, named after the town in Australia where it was found.[477] It had to be side-tongue grafted onto rootstocks, obviously, because it produced no seeds. The flowers were all staminate.[477] It was not noted until two decades later that these staminate flowers failed to produce normal pollen.[801] Only 15 percent of the anthers partially dehisce, and shaking does not release a cloud of

Table 33. Landscaping Trees Blow Allergenic Pollen

Common	Latin name	Allergen	Cross reactivities (cr) and comments	Reference
Olive	Olea europea	Ole e 1-10	Ole e 1 cr to ash, lilac and privet	400
			Ole e 1 cr Plantago Pla 11	177
			Ole e 5 Cu/Zn superoxide dismutase (like?)	174
			Ole e 10 colocalizes with callose in pollen tube	91
Lilac	Syringa vulgaris	Syr v 1&3		755
Privet	Ligustrum vulgare			187
Ash	Fraxinus excelsior	Fra e 1	cr Ole e1, Syr v1 (lilac), Vig V1, ash	90, 835
Birch	Betula	Bet v 1-8	cr to alder, hazel, and hornbeam pollen	721
Alder	Alnus	Aln g 1&4	cr to birch, hazel, and hornbeam pollen	755
Hazel	Corylus	Cor a 1,2,8&10	cr to birch, alder, and hornbeam pollen	755
Hornbeam	Carpinus	Car b 1	cr to birch, hazel, and alder pollen	755
White oak	Quercus alba	Que a1		755
Chestnut	Castanea	Cas s 1,5&8		755
Japanese cedar	Cryptomeria	Cry j 1&2		755
Juniper	Juniperus	Jun o 4, v 1, a 2&3		755
Cypress	Cupressus	Cup a 1&3, s 1		755
Pecan				878
Orange	Citrus spp.			525
Date palm		Pho d2	a profilin, cr olive	68

dusty pollen, as the pollen grains remained caked, and the amount in the air is tenfold less than for a common olive variety[801] and remains below background levels of total pollen. The variety Swan Hill is marketed for landscaping in the United States. Such a variety could be used for new ornamental plantings, yet those worried about ornamental plantings of olives in the Mediterranean region have not suggested this as a solution. This is also not a solution for standing trees unless one is willing to top-graft major branches on a cut-back tree.

23.2. Biotechnological Solutions

An interesting use of biotechnology to improve olives has been proposed, and even though it has nothing to do with allergenicity, it is worth recounting. Olive pits are a useless by-product of the production of olive oil, but a use has been proposed to have them produce recombinant proteins. The mesocarp (fruit) of the olive is unlike many oil-bearing tissues insofar as the oil is not stored in oil bodies bounded by oleosin proteins.[478] Oleosin protein oil bodies do occur in the seed. Thus, just as recombinant proteins fused with oleosins are produced in specially cultivated safflowers (Chapter 17), it has been suggested to do the same in olive, as a by-product, so that olive oil production costs are diminished.[478]

23.2.1. Suppress Allergen Production in Pollen?

The recombinant production of olive allergens in bacteria or yeast has allowed the study of problems that could not be previously dealt with because of the paucity of material. Ole e 1 is considered to be the most prevalent olive allergen, as it affects more than 70 percent of those allergic to olive pollen.[913] One could conceive a possibility of antisense or RNAi transformation of the gene(s) encoding the allergens. That may be possible with some of them, but some have claimed that many of the key allergens may have vital functions, although the evidence is not always convincing. For example, claims have been made that allergenic protein Ole e 1[24] and Ole e 10[91] both have some role in pollen tube growth. One can question this statement based on the more than tenfold intervarietal differences in allergenic potency of total pollen extracts[191] and a fourfold difference among varieties in the output of Ole e 1[195], which would suggest that the correlations of expression with pollen tube growth are not cause and effect. This is an important issue to ascertain for two reasons. (1) If an important role exists, then using antisense constructs of these genes in commercial plantations would be contraindicated because such pollen could reduce crop yield. (2) If indeed these proteins are not needed, agricultural olive plantings, which cause allergic reactions among farmers, village dwellers, and inhabitants of nearby urban areas, can in the future be transformed to be less allergenic.

Allergen Ole e 5 is claimed to be a Cu/Zn superoxide dismutase based only on an 80 percent sequence similarity.[174] This is hardly conclusive until an activity assay is performed, and if correct might mean that using antisense constructs to deplete this allergen might delete a truly vital enzyme activity, if this is being expressed in other tissues. It would then remain to be ascertained whether this superoxide dismutase can be replaced by a nonallergenic one.

Allergen Ole e 9 had sequence similarity to and was demonstrated to have actual activity of a β-1,3 glucanase, and it could also be placed in group two of the pathogenesis-related proteins.[913]

It should not matter whether a landscaping olive has poorly viable pollen due to deficient allergen production, as long as the landscaping olives are not near commercial groves. Still, why have them produce any pollen at all? Thus, more drastic concepts have been proposed, as discussed next.

23.2.2. Preventing Pollen and Fruit Formation

Two complimentary biotech solutions are possible, and employing one or both (as a dual safeguard) might result in beautiful cities, without the allergy and mess. The *barnase* (ribonuclease) gene under a tapetal promoter renders flowers of other species male sterile,[691] and a floral ablation gene has rendered poplars to be without flowers.[992] If just a lack of pollen production is used, parent trees in nurseries can be pollinated with pollen from olive trees brought from commercial orchards and the seeds used to produce the landscaping trees. The trees in the urban plantings would then have very few fruits, and would have the desired genetic diversity of seedlings. If both male and female sterility are used, then the trees will have to be vegetatively propagated at an added cost.

23.2.3. Easier Landscaping

It will be interesting to ascertain the effect of non-pollen production on an allergenic species such as lilac (which is closely related to olive). This beautiful spring-flowering, fragrant species has such short-lived flowers. Is that because when they are pollinated, petal abscission occurs? If so, will such a technology extend the flower life, as it might with orchids (Chapter 22) and other flowers? Clearly good reasons exist to try this with flowering ornamentals beyond preventing running noses and asthma.

The transformations described in this section basically only replace using the mutant olive cultivar cv. Swan Hill, which does not produce pollen or fruit. Although this is feasible, this cultivar does not seem to have caught on worldwide. If it would catch on, it would lead to all urban plantings being of a single clone, which is a dangerous lack of diversity. The transgenic approach would allow genetic engineers to use a single construct to engineer a diverse olive material, including nonvarietal hybrid seedlings to ensure that single clones are not released.

While doing any one of the preceding transformations, one could consider adding genes for added auxin production to enhance apical dominance, that is, to reduce the suckering and thus require less pruning. Suckering might be further reduced by antisensing genes for cytokinin biosynthesis.

23.2.4. Should Mature Messy Trees Be Chopped Down?

Use of non-pollen-producing olive cv. Swan Hill, or its potentially more diverse transgenic approach, would not please the allergic because of the extant plantings. Must they be cut down and replaced to please the allergic and displease those who like their esthetics? Perhaps not. There are a few possibilities. The expensive one is to cut back the trees to several major branches and top-graft them with material generated as in the section above. Within a year they would look pretty good, but suckers off the trunk would be of the nontransgenic material and would produce pollen. If the transformed shoots also contain genes to increase apical dominance (enhanced auxin production or antisensed cytokinin production) suckering may be decreased but pruning would probably be required to re-form a beautiful tree.

Several viruses attack olives. Reverse transcriptase-polymerase chain reaction (RT-PCR) probes are available for four of them,[116] and two isolates of the Olive Latent Necrosis Virus (OLV-1) have been completely sequenced.[183,339] None seem to cause serious damage and some are completely symptomless, although nursery stock for fruit plantations must be free of all of them.[116] It is possible that they or others could be attenuated, made to lack coat protein genes so that they could not be transmissible, and could be engineered with the same constructs described above to prevent pollen and fruit formation in mature olive trees used in landscaping. Trees could possibly be inoculated by the sandblasting technology described in Chapter 8, or by the same technologies used to inject nutrients into tree trunks. If this has to be done only once in the lifetime of a tree, or even once every five years, it could deal with the pollen problem, especially if the construct also contained genes that enhance apical dominance and prevent suckering. The treatment would then be very cost-effective because of the reduction in pruning costs. The use of such mature tree transformation would also allow testing of the constructs to be used for the permanent transformations described in the previous section. Otherwise, one would have to wait years to ascertain whether the construct used in the transformation of a given shoot had the desired effect. The transformation of adult trees with a systemic virus will quickly provide that information.

23.2.5. Transformation and Regeneration of Plantlets

Micropropagation of olives is common to preserve pathogen-free grafting material; better techniques are continually being developed to reduce contamination and increase efficiency.[152,1188] Olives can be genetically transformed using *Agrobacterium*,[332,720] although it is not easy. In the latter case, it is claimed that a stress tolerance gene had been successfully engineered into olives,[332] a gene that might be useful as part of a construct for urban plantings, even if it has a yield drag with commercial olive fruit.

23.3 Biosafety Considerations

This exercise with olives is an example of how other trees for landscaping may possibly be modified, in particular, where there are groups worried about urban landscaping species encroaching on wild habitats.[609] It is important that none of the tricks used for diminishing allergenicity of landscaping olives escape to olive plantations, or that techniques used to decrease allergenicity of pollen in plantations affect the pollen production and viability within plantations. There is a direct multiyear correlation between olive pollen production and crop yield.[369] Obviously, if landscape olives are engineered to produce no pollen, the issue of gene outflow does not exist. Likewise, if landscaping olives do not produce fruit, the birds have nothing to carry to the wild, and only landscaping trees themselves can become feral if not pruned, but they would be hard put to disperse. If engineered to produce high auxins and/or antisensed for cytokinin production it is even less likely that unpruned standing trees will become feral oleasters, and the oleaster is a pleasant-enough-looking large bush amenable to landscaping.

If coat protein genes are deleted from the virus vectors carrying the transgene(s) to adult trees, they are not transmissible, but it will remain to be shown if they can recombine with virulent viruses and if the recombinants have a fitness that would render them a threat to commercial plantations.

A similar focus on biosafety issues will have to be made for other tree species under consideration. In many cases the considerations will be simpler. The urban trees have no cultivated or wild relatives. These considerations will be simpler yet for tree species that also have a tendency to spread outside of urban plantings, as a lack of pollen formation would severely limit their sex life, and preclude movement.

Epilogue

Discovery consists of seeing what everybody has seen, and thinking what nobody has thought.   ALBERT SZENT-GYORGYI

The battles rage on. The detractors of biotechnology continue to disseminate disinformation and misinformation, including claims that biotechnology will decrease biodiversity. It is hoped that the reader has seen that, by breaching the genetic glass ceiling, genetic engineering can enhance crop biodiversity as a necessary adjunct to breeding. This should lead to a wider variety of healthier crops, while lowering production costs and decreasing pressures on the land.

It is fascinating that many "life science companies" (the new name for the agrochemical giants that have expanded into the biotech and seed businesses) seem to have an internal marketing conflict; they profit much more from agricultural chemicals than from seed and are loath to lose those lucrative chemical markets. Could there be an unholy alliance in Europe between the ecoterrorists and their passive supporters with the agrochemical marketers? In the last Swiss referendum on agrobiotech not a peep was heard in support of agricultural biotechnology from the Swiss major multinational, a company rarely silent on local affairs.

The long dormant major biotech detractor, Jeremy Rifkin, has recently become active again. He resurrected the belief that any new needed trait could be obtained by marker-assisted breeding.[1020] Rifkin was seconded by the European Union (EU) environment minister.[307] This was to the glee of the few older (in spirit, not necessarily age) breeders in academia who do not use genetic engineering as part of their breeding programs. Rifkin and his new allies have forgotten (or ignored) that if the gene does not exist in the gene pool of a crop, no matter how much or intelligently you cross, you will not find it. Spontaneous generation went out a few centuries ago. The utility of exogenous genes is borne out by the two leading traits on the market, encoded by genes brought in from bacteria: resistance to a cheap, environmentally friendly herbicide and the resistance to insects that replaces so much synthetic insecticide. Marker-assisted breeding can also be of little assistance in obtaining tissue-specific expression or suppression of an endogenous gene, which is quite easy with transgenics. Marker-assisted breeding is an excellent technique to follow recessive genes through a breeding program, but antisense and RNAi mechanisms are simpler yet.

Indeed, marker-assisted breeding can be an enemy of crop biodiversity. It requires large amounts of genomic chromosome mapping. Note how few DNA sequences have been reported for some of the crops discussed in this book (Fig. 18). Who can afford the cost of setting up the genomic databases for such crops? Wise modern breeders use genetic engineering to introduce new traits into a crop, and then classical breeding to distribute these traits into locally adapted elite lines.

Some companies and research groups are finding it expedient to pander to antitransgenic sentiments and claim to have equally good but "cuddlier" technologies that "common people" will readily accept. One company, Cibus Genetics,[224] is promoting chimeraplasty as a "nontransgenic" method to change codons that would require multiple mutations. Their press releases claim that they will introduce herbicide resistance into sorghum, wheat, and rice in this manner. They ignore the fact that these crops all have weedy relatives, and the genes will flow, negating the utility of the trait. One can transgenically contain or mitigate gene flow, but not with chimeraplasty, which only introduces substitutions in existing genes.

Another group has coined the term "cisgenics" for moving genes from relatives to the crops by genetic engineering and request that their technology be free of regulation just as breeding is[956] (except in Canada where all novel trait introduction is regulated). Genes and promoters from relatives are okay, but genes from afar (transgenics) are not. Again, they could not achieve the two major gene products already commercialized from genetic engineering. The promoters of cisgenics request to be free to move toxin genes from *Solanum nigrum* (black nightshade) to *Solanum esculentum* (tomato). This would kill insects, but also mammals. Unfortunately, they are in the Netherlands and not in Canada.

Even some proponents of genetic engineering are afraid the public will not eat the products of genetic engineering. They forget that the British had no problems with transgenic tomato paste until it was removed from supermarket shelves by chains finding it expedient to pander to antitransgene hysteria; the Chinese devour tofu from transgenic soybeans, and the Americans enjoy their transgenic papayas, as did Europeans until the importer was caught.[170]

Unfortunately, a great part of the antibiotech movement is politically motivated, typically coupled with an anti-multinational, anti-globalization, anti-capitalism agenda. They could instead be positive to biotechnology while retaining the same political agenda by calling for more public sector involvement in agricultural biotechnology to provide crop biodiversity and perform the types of research and development in which the multinationals suppose that their stockholders have no economic interest. This political anti-technology movement has public sector researchers in at least one European country barred from developing new products by using genetic engineering. Instead of investing in the productive biotechnology of increasing crop biodiversity, much of public sector funding in biotechnology is for the detection of transgenes in food, but no one has been hurt by transgenes. Little money goes for detection of mycotoxins in food and virtually none for biotech interventions to keep mycotoxins out. Taxpayers should revolt against this misuse of resources. Why is it that those who claim to support biodiversity do not support crop biodiversity? Why do those who claim to support consumer choice do their best to ensure that the consumer cannot choose transgenic products?

Many of the detractors justify their antibiotech tirades by reasoning that farmers will have to buy seed and have turned "farmer saved seed" into a holy mantra. Almost all those well experienced in agriculture know that there is nothing worse for farming than farmer-saved seed. Yields steadily decrease because of the loss of vigor, an increase in disease, and an often massive contamination with weed seeds. Only the very best growers are chosen to grow certified seed, and even they are heavily monitored by the seed companies, who are further regulated by governmental authorities. Why have those who know how bad farmer-saved seed is for agriculture been cowed from countering this politically correct but agronomically incorrect view? Few productive farmers will ever go back to nonhybrid maize seed. If transgenics do as well as hybrids, farmers will buy in; if they have little to offer, farmers will eschew transgenics. Some detractors of biotech want to keep biotech from subsistence farmers, which will perpetuate poverty, whether they mean it or not. It is hoped that the reader discerned from the chapters of this book that biotech has more to offer the rural poor in developing countries than the rich.

What seemed to be science fiction in agricultural biotechnology a few years ago is now reality. It is hoped that much of what seems to be science fiction in this book will soon be a reality in the field, enhancing crop diversity, with the moral and/or physical support of some of the readers of this book. I hope that some readers will take the crops discussed here and further analyze the problems and come up with far more elegant solutions than proposed herein, and will perform the same type exercise with their favorite underutilized crops that have reached a ceiling that is too low.

The human mind is like a parachute—it functions better when it is open.
(ATTRIBUTED TO COLE'S RULES)

References

1. AATF. (2005) African Agricultural Technology Foundation. www.aatf-africa.org.
2. Abate, D., and Gashe, B.A. (1985) Prevalence of *Aspergillus flavus* in Ethiopian cereal grains: A preliminary survey. Ethiopian Medical Journal 23, 143–148.
3. Abayo, G.O., English, T., Eplee, R.E., Kanampiu, F.K., Ransom, J.K. and Gressel. J. (1998) Control of parasitic witchweeds (*Striga* spp.) on corn (*Zea mays*) resis-tant to acetolactate synthase inhibitors. Weed Science 46, 459–466.
4. Abbadi, A., Domergue, F., Bauer, J., Napier, J.A., Welti, R., Zähringer, U., Cirpus, P., and Heinz, E. (2004) Biosynthesis of very-long-chain polyunsaturated fatty acids in transgenic oilseeds: Constraints on their accumulation. Plant Cell 16, 2734–2748.
5. Abbott, A. (2005) Avian flu special: What's in the medicine cabinet. Nature 434, 407–409.
6. Abell, B., Hahn, M., Holbrook, L.A., and Moloney, M.M. (2004) Membrane topology and sequence requirements for oil body targeting of oleosin. Plant Journal 37, 461–470.
7. Adefris, T., and Adugna, W. (2004) Seed filling and oil accumulation in noug (*Guizotia abyssinica*). Ethiopean Journal of Science 27, 25–32.
8. Adhikari, K.N., and Campbell, C.G. (1998) Natural outcrossing in common buckwheat. Euphytica 102, 233–237.
9. Adnew, T., Ketema, S., Tefera, H., and Sridhara, H. (2005) Genetic diversity in tef [*Eragrostis tef* (Zucc.) Trotter] germplasm. Genetic Resources and Crop Evolution 52, 891–902.
10. Aggarwal, R.K., Brar, D.S., and Khush, G.S. (1997) Two new genomes in the *Oryza* complex identified on the basis of molecular divergence analysis using total genomic DNA hybridization. Molecular and General Genetics 254, 1–12.
11. Aguirre, C., Valdes-Rodriguez, S., Mendoza-Hernandez, G., Rojo-Dominguez, A., and Blanco-Labra, A. (2004) A novel 8.7 kDa protease inhibitor from chan seeds (*Hyptis suaveolens* L.) inhibits proteases from the larger grain borer *Prostephanus truncatus* (Coleoptera: Bostrichidae). Comparative Biochemistry and Physiology B-Biochemistry & Molecular Biology 138, 81–89.
12. Agunbiade, S.O., and Longe, O.G. (1999) Essential amino acid composition and biological quality of yambean, *Sphenostylis stenocarpa* (Hochst ex A. Rich) Harms. Nahrung-Food 43, 22–24.
13. Aharoni, A., and Vorst, O. (2001) DNA microarrays for functional plant genomics. Plant Molecular Biology 48, 99–118.
14. Ahmed, N.E., Sugimoto, Y., Babiker, A.G.T., Mohamed, O.E., Ma, Y., Inanaga, S., and Nakajima, H. (2001) Effects of *Fusarium solani* isolates and metabolites on *Striga* germination. Weed Science 49, 354–358.
15. Ahonsi, M.O., Berner, D.K., Emechebe, A.M., Lagoke, S.T., and Sanginga, N. (2003) Potential of ethylene-producing pseudomonads in combination with effective N_2-fixing bradyrhizobial strains as supplements to legume rotation for *Striga hermonthica* control. Biological Control 28, 1–10.
16. Aida, R., Yoshida, K., Kondo, T., Kishimoto, S., and Shibata, M. (2000) Co-pigmentation gives bluer flowers on transgenic torenia plants with the antisense dihydroflavonol-4-reductase gene. Plant Science 160, 49–56.

17. Aigbokhan, E.I., Berner, D.K., and Musselman, L.J. (1998) Reproductive ability of hybrids of *Striga aspera* and *Striga hermonthica*. Phytopathology 88, 563–567.

18. Akbar, M.A. (1989) Resynthesis of *Brassica napus* aiming for improved earliness and carried out by different approaches. Hereditas 111, 239–246.

19. Al-Ahmad, H., and Gressel, J. (2006) Mitigation using a tandem construct containing a selectively unfit gene precludes establishment of *Brassica napus* transgenes in hybrids and backcrosses with weedy *Brassica rapa*. Plant Biotechnology Journal 4, 23–33.

20. Al-Ahmad, H., Galili, S., and Gressel, J. (2005) Poor competitive fitness of transgenically mitigated tobacco in competition with the wild type in a replacement series. Planta 272, 372–385.

21. Al-Ahmad, H., Dwyer, J., Moloney, M.M., and Gressel, J. (2006) Mitigation of establishment of *Brassica napus* transgenes in volunteers using a tandem construct containing a selectively unfit gene. Plant Biotechnology Journal 4, 7–21.

22. Al-Ahmad, H.I., Galili, S., and Gressel, J. (2004) Tandem constructs mitigate risks of transgene flow from crops: tobacco as a model. Molecular Ecology 13, 687–710.

23. Al-Khatib, K., Baumgartner, J.R., Peterson, D.E., and Currie, R.S.F. (1998) Imazethapyr resistance in common sunflower (*Helianthus annuus*). Weed Science 46, 403–407.

24. Alche, J.D., M'Rani-Alaoui, M., Castro, A.J., and Rodriguez-Garcia, M.I. (2004) Ole e 1, the major allergen from olive (*Olea europaea* L.) pollen, increases its expression and is released to the culture medium during in vitro germination. Plant and Cell Physiology 45, 1149–1157.

25. Aldryhim, Y.N., and Adam, E.E. (1999) Efficacy of gamma irradiation against *Sitophilus granarius* L. Journal of Stored Products Research 35, 225–232.

26. Aleman, A., Sastre, J., Quirce, S., de las Heras, M., Carnes, J., Fernandez-Caldas, E., Pastor, C., Blazquez, A.B., Vivanco, F., and Cuesta-Herranz, J. (2004) Allergy to kiwi: a double-blind, placebo-controlled food challenge study in patients from a birch-free area. Journal of Allergy and Clinical Immunology 113, 543–550.

27. Alemaw, G., and Wold, A.T. (1995) An agronomic and seed-quality evaluation of noug (*Guizotia abyssinica* Cass) germplasm in Ethiopia. Plant Breeding 114, 375–376.

28. Alfonso-Rubi, J., Ortego, F., Castanera, P., Carbonero, P., and Diaz, I. (2003) Transgenic expression of trypsin inhibitor CMe from barley in indica and japonica rice, confers resistance to the rice weevil *Sitophilus oryzae*. Transgenic Research 12, 23–31.

29. Alizadeh, H., Teymouri, F., Gilbert, T.I., and Dale, B.E. (2005) Pretreatment of switchgrass by ammonia fiber explosion (AFEX). Applied Biochemistry and Biotechnology 121, 1133–1141.

30. Alkire, M.L. (1996) M.Sc. Thesis. An investigation into the bird repellancy of sorghum grain variety Arkansas 3048, Purdue University, Lafayette, IN.

31. Allah, E.M.F. (1998) Occurrence and toxigenicity of *Fusarium moniliforme* from freshly harvested maize ears with special reference to fumonisin production in Egypt. Mycopathologia 140, 99–103.

32. Altpeter, F., Diaz, I., McAuslane, H., Gaddour, K., Carbonero, P., and Vasil, I.K. (1999) Increased insect resistance in transgenic wheat stably expressing trypsin inhibitor CMe. Molecular Breeding 5, 53–63.

33. Aminigo, E.R., and Metzger, L.E. (2005) Pretreatment of African yam bean (*Sphenostylis stenocarpa*): Effect of soaking and blanching on the quality of African yam bean seed. Plant Foods for Human Nutrition 60, 165–171.

34. Amirhusin, B., Shade, R.E., Koiwa, H., Hasegawa, P.M., Bressan, R.A., Murdock, L..L, Zhu-Salzman,. K. (2004) Soyacystatin N inhibits proteolysis of wheat alpha-amylase inhibitor and potentiates toxicity against cowpea weevil. Journal of Economic Entomology 97, 2095–2100.

35. Ammann, K. (2004) The role of science in the application of the precautionary approach. In Molecular Farming, Plant-made Pharmaceuticals and Technical Proteins (eds. Fischer, R., and Schillberg, S.), 291–302. Wiley-VCH, Weinheim.

36. Ammann, K. (2004) Biosafety in agriculture: Is it justified to compare directly with natural habitats? In Frontiers in Ecology, Ecological Society of America, www.frontiersinecology.org.

37. Ammann, K., Jacot, Y., and Rufener Al Mazyad, P. (2000) Ecological risk assessment of vertical gene flow. In Safety of Genetically Engineered Crops (ed. Custers, R.). Flanders Interuniversity Institute for Biotechnology, Zwijinarde, Belgium.

38. Ammann, K., Jacot, Y., and Rufener Al Mazyad, P. (2005) The ecology and detection of plant ferality in the historic records. In Crop Ferality and Volunteerism (ed. Gressel, J.), 31–43. CRC Press, Boca Raton, FL.

39. Ammann, K. (2007) Reconciling traditional knowledge with modern agriculture: A guide for building bridges. In Intellectual Property Management in Health and Agricultural Innovation: a Handbook of Best Practices (eds. A. Krattiger et al.), 1539–1559. MIHR and PIPRA, Oxford, United Kingdom, and Davis CA.

40. Amsellem, Z., Cohen, B.A., and Gressel, J. (2002) Engineering hypervirulence in an inundative mycoherbicidal fungus for efficient weed control. Nature Biotechnology 20, 1035–1039.

41. Amsellem, Z., Kleifeld, Y., Kerenyi, Z., Hornok, L., Goldwasser, Y., and Gressel, J. (2001) Isolation, identification, and activity of mycoherbicidal pathogens from juvenile broomrape plants. Biological Control 21, 274–284.

42. Amsellem, Z. Barghouthi, S., Cohen, B., Goldwasser, Y., Gressel, J., Hornok, L., Kerenyi, Z., Kleifeld, Y., Kroschel, J., Sauerborn, J., Muller-Stover, D., Thomas, H., Vurro, M., and Zonno, M-C. (2001) Recent advances in the biocontrol of *Orobanche* broomrape species. BioControl 46, 211–228.

43. Anderson, J.A., and Kolmer, J.A. (2005) Rust control in glyphosate tolerant wheat following application of the herbicide glyphosate. Plant Disease 89, 1136–1142.

44. Anjanasree, K.N., and Bansal, K.C. (2003) Isolation and characterization of ripening-related expansin cDNA from tomato. Journal of Plant Biochemistry and Biotechnology 12, 31–35.

45. Anjanasree, K.N., Verma, P.K., and Bansal, K.C. (2005) Differential expression of tomato ACC oxidase gene family in relation to fruit ripening. Current Science 89, 1394–1399.

46. Anonymous. (1994) The biology of *Brassica napus* L (canola/rapessed). Directive Dir. 94-09, Plant Products Division, Agriculture and Agri-Food Canada, Nepean Ontario (www.cfiaacia. agr.ca/English/food/pbo/dir9409.html).

47. Anonymous. (1994) Assessment criteria for determining environmental safety of plants with novel traits. Directive Dir. 94–08, Plant Products Division, Agriculture and Agri-Food Canada, Nepean, Ontario (www.cfia-acia.agr.ca/English/food/pbo/dir9408.html).

48. Anonymous. (1995) Determination of environmental safety of Monsanto Canada Inc's Roundup herbicide tolerant *Brassica napus* Canola line GT73. Decision Document 95-02. Plant Products Division, Agriculture and Agri-Food Canada, Nepean Ontario (.www.cfia-acia.agr. ca/English/food/pbo/dd9502.html).

49. Anonymous. (1995) Determination of environmental safety of Pioneer Hi-Bred International Inc's imidazolinone-tolerant Canola line. Decision Document DD95-03. Plant Products Division, Agriculture and Agri-Food Canada, Nepean Ontario (www.cfia-acia.agr.ca/English/food/pbo/ dd9503e.html).

50. Anonymous. (1996) Tef. In Lost Crops of Africa. Vol. I: Grains, 215–236. National Academy Press, Washington, DC.

51. Anonymous. (1996) Determination of environmental safety of AgrEvo Canada Inc's glufosinate-ammonium herbicide tolerant *Brassica napus* Canola line, Decision Document 96-11, 7 pp, Plant Products Division, Agriculture and Agri-Food Canada, Nepean Ontario (www.cfiaacia.agr.ca/English.food/pbo/dd9503e.html).

52. Anonymous. (1997) Consensus document on the biology of *Brassica napus* L (oilseed rape) series on the harmonization of regulatory oversight in biotechnology No. 7. Environmental Directorate, Organization for Economic Co-operation and Development, Paris.

53. Anonymous. (2000) Consensus document on the biology of *Glycine max* (L.) Merr. (soybean). Series on Harmonization of Regulatory Oversight in Biotechnology No. 15OECD, Paris.

54. Anonymous. (2002) Consensus Document on Compositional Considerations for New Varieties of Potatoes: Key Food and Feed Nutrients, Anti-Nutrients and Toxicants ENV/JM/MONO (2002)5OECD, Paris.

55. Anterola, A.M., and Lewis, N.G. (2002) Trends in lignin modification: a comprehensive analysis of the effects of genetic manipulations/mutations on lignification and vascular integrity. Phytochemistry 61, 221–294.

56. APHIS (Animal and Plant Health Inspection Service). (2006) Field tests of oilseed rape. www.isb.vt.edu/cfdocs/fieldtests3.cfm.

57. APHIS (Animal and Plant Health Inspection Service). (2006) *www.aphis.usda.gov/brs/ph_permits.html.*

58. Arazi, T., Huang, P.L., Zhang, L., Shiboleth, Y.M., Gal-On, A., and Lee-Huang, S. (2002) Production of antiviral and antitumor proteins MAP30 and GAP31 in cucurbits using the plant virus vector ZYMV-A. Biochemical and Biophysical Research Communications 292, 441–448.

59. Arber, W. (2002) Roots, strategies and prospects of functional genomics. Current Science 83, 826–828.

60. Arber, W. (2004) Biological evolution: Lessons to be learned from microbial population biology and genetics. Research in Microbiology 155, 297–300.

61. Aresa-Biodetection-ApS. (2005) www.aresa.dk.

62. Arriola, P.E., and Ellstrand, N.C. (1996) Crop-to-weed gene flow in the genus *Sorghum* (Poaceae): Spontaneous interspecific hybridization between johnsongrass, *Sorghum halepense,* and crop sorghum, *S. bicolor.* American Journal of Botany 83, 1153–1160.

63. Arriola, P.E., and Ellstrand, N.C. (1997) Fitness of interspecific hybrids in the genus *Sorghum:* Persistence of crop genes in wild populations. Ecological Applications 7, 512–518.

64. Ashri, A., and Knowles, P.F. (1960) Cytogenetics of safflower (*Carthamus* L.) species and their hybrids. Agronomy Journal 52, 11–17.

65. Ashri, A., and Rudich, J. (1965) Unequal reciprocal natural hybridization rates between two *Carthamus* L. Crop Science 5, 190–191.

66. Askew, S.D., Shaw, D.R., and Street, J.E. (1998) Red rice (*Oryza sativa*) control and seedhead reduction with glyphosate. Weed Technology 12, 504–506.

67. Assefa, K., Tefera, H., and Merker, A. (2002) Variation and inter-relationships of quantitative traits in tef (*Eragrostis tef* (Zucc.) Trotter) germplasm from western and southern Ethiopia. Hereditas 136, 116–125.

68. Asturias, J.A., Ibarrola, I., Fernandez, J., Arilla, M.C., Gonzalez-Rioja, R., and Martinez, A. (2005) Pho d 2, a major allergen from date palm pollen, is a profilin: cloning, sequencing, and immunoglobulin E cross-reactivity with other profilins. Clinical and Experimental Allergy 35, 374–381.

69. Asuzu, I.U. (1986) Pharmacological evaluation of the folklore use of *Spheno-stylis stenocarpa.* Journal of Ethnopharmacology 16, 263–267.

70. Asuzu, I.U,. and Undie, A. (1986) Some observations on the toxic effects of the seed extract of *Sphenostylis stenocarpa* (Hochst ex A Rich) Harms on intestinal muscle. Qualitas Plantarum-Plant Foods for Human Nutrition 36, 3–9.

71. Aswidinnoor, H., Nelson, R.J., and Gustafson, J.P. (1995) Genome-specific repetitive DNA probes detect introgression of *Oryza minuta* genome into cultivated rice, *Oryza sativa.* Asia Pacific Journal of Molecular Biology 3, 215–223.

72. Attasart, P., Charoensilp, G., Kertbundit, S., Panyim, S., and Juricek, M. (2002) Nucleotide sequence of a Thai isolate of papaya ringspot virus type W. Acta Virologica 46, 241–246.

73. Attieh, J.M., Hanson, A.D., and Saini, H.S. (1995) Purification and characterization of a novel methyltransferase responsible for biosynthesis of halomethanes and methanethiol in *Brassica oleracea.* Journal of Biological Chemistry 270, 9250–9257.

74. Attieh, J.M., Djiana, R., Koonjul, P., Etienne, C., Scarace, S.A., and Saini, H.S. (2002) Cloning and functional expression of two plant thiolmethyltransferases; a new class of enzymes involved in the biosynthesis of sulfur volatiles. Plant Molecular Biology 50, 511–521.

75. Aviv, D., Amsellem, Z., and Gressel, J. (2002) Transformation of carrots with mutant aceto-lactate synthase for *Orobanche* (broomrape) control. Pest Management Science 58, 1187–1193.

76. Avni, A., and Edelman, M. (1991) Direct selection for paternal inheritance of chloroplasts in sexual progeny of *Nicotiana*. Molecular & General Genetics 225, 273–277.

77. Ayele, M., and Nguyen, H.T. (2000) Evaluation of amplified fragment length polymorphism markers in tef, *Eragrostis tef* (Zucc.) Trotter, and related species. Plant Breeding 119, 403–409.

78. Ayele, M., Blum, A., and Nguyen, H. (2001) Diversity for osmotic adjustment and root depth in tef [*Eragrostis tef* (Zucc) Trotter]. Euphytica 121, 237–249.

79. Azam-Ali, S.N., and Massawe, F.J. (2003) Re-evaluating underutilized crops and the Green Revolution failure. Ponte 59, 53–71.

80. Azpiroz, R., Wu, Y., LoCascio, J.C., and Feldmann, K.A. (1998) An *Arabidopsis* brassinosteroid-dependent mutant is blocked in cell elongation. Plant Cell 10, 219–230.

81. Babiker, A.G.T., Butler, L.G., Ejeta, G., and Woodson, W.R. (1994) Enhancement of ethylene biosynthesis and germination by cytokinins and 1 aminocyclopropane-1-carboxylic acid in *Striga asiatica* seeds. Physiologia Plantarum 89, 21–26.

82. Baertlein, D.A., Lindow, S.E., Panopoulos, N.J., Lee, S.P., Mindrinos, M.N., and Chen, T.H.H. (1992) Expression of a bacterial ice nucleation gene in plants. Plant Physiology 100, 1730–1736.

83. Bai, G.H., Ayele, M., Tefera, H., and Nguyen, H.T. (2000) Genetic diversity in tef [*Eragrostis tef* (Zucc) Trotter] and its relatives as revealed by Random Amplified Polymorphic DNAs. Euphytica 112, 15–22.

84. Baird, W.V., and Riopel, J.L. (1985) Surface characteristics of root and haustorial hairs of parasitic Scrophulariaceae. Botanical Gazette 146, 63–69.

85. Baker, H.G. (1974) The evolution of weeds. Annual Reviews of Ecological Systems 5, 1–24.

86. Baker, H.G. (1991) The continuing evolution of weeds. Economic Botany 45, 445–449.

87. Baltazar, A.M., and Janiya, J.D. (1999) Weedy rice in the Philippines. In Wild and Weedy Rices (eds. Baki, B.B., Chin, D.V., and Mortimer, M.), 74–78. International Rice Research Institute, Manilla, Phillipines.

88. Bandman, O., Shah, P., Guegler, K.J., and Corley, N.C. (2003) Human aflatoxin B1 aldehyde reductase. U.S. Patent Application 2003/0013853.

89. Baniya, B.K., Riley, K.W., Dongol, D.M.S., and Shershand, K.K. (1992) Characterization and evaluation of Nepalese buckwheat (*Fagopyrum* spp.) landraces. In Proceedings of the Fifth International Symposium on Buckwheat (eds. Lin, R., Zhou, M.-D., Tao, Y., Li, J., and Zhang, Z.), 64–74. Agricultural Publishing House, Taiyuan, China.

90. Barderas, R., Purohit, A., Papanikolaou, I., Rodriguez, R., Pauli, G., and Vill-alba, M. (2005) Cloning, expression, and clinical significance of the major allergen from ash pollen, Fra e 1. Journal of Allergy and Clinical Immunology 115, 351–357.

91. Barral, P., Suarez, C., Batanero, E., Alfonso, C., Alche, J.de D., Rodriguez-Garcia, M.I., Villalba, M., Rivas, G., and Rodriguez, R. (2005) An olive pollen protein with allergenic activity, Ole e 10, defines a novel family of carbohydrate-binding modules and is potentially implicated in pollen germination. Biochemical Journal 390, 77–84.

92. Barrett, S.S.H. (1983) Crop mimicry in weeds. Economic Botany 37, 255–282.

93. Barriere, Y., Guillet, C., and Goffner, D. (2003) Genetic variation and breeding strategies for improved cell wall digestibility in annual forage crops. Animal Research 52, 193–228.

94. Barriere, Y., Ralph, J., Mechin, V., Guillaumie, S., Grabber, J.H., Argillier, O., Chabbert, B., and Lapierre, C. (2004) Genetic and molecular basis of grass cell wall biosynthesis and degradability. II. Lessons from brown-midrib mutants. Comptes Rendus Biologies 327, 847–860.

95. Baruteau, J., Sadani, G., Jourdan, C., Morelle, K., Broué-Chabbert, A., and Rancé F. (2005) Buckwheat allergy: case report and review of the literature. Revue Francaise D Allergologie et d Immunologie Clinique 45, 422–425.

96. Bashir, K., Husnain, T., Fatima, T., Riaz, N., Makhdoom, R., and Riazzudin, S. (2005) Novel indica basmati line (B-370) expressing two unrelated genes of *Bacillus thuringiensis* is highly resistant to two lepidopteran insects in the field. Crop Protection 24, 870–879.

97. Bateson, M.F., Lines, R.E., Revill, P., Chaleeprom, W., Ha, C.V., Gibbs, A.J., and Dale, J.L. (2002) On the evolution and molecular epidemiology of the potyvirus papaya ringspot virus. Journal of General Virology 83, 2575–2585.

98. Bau, H.J., Cheng, Y.H., Yu, T.A., Yang, J.S., Liou, P.C., Hsiao, C.H., Lin, C.Y., and Yeh, S.D. (2004) Field evaluation of transgenic papaya lines carrying the coat protein gene of Papaya ringspot virus in Taiwan. Plant Disease 88, 594–599.

99. Baumgertel, A., Grimm, R., Eisenbeiss, W., and Kreis, W. (2003) Purification and characterization of a flavonol 3-O-beta-heterodisaccharidase from the dried herb of *Fagopyrum esculentum* Moench. Phytochemistry 64, 411–418.

100. Beckie, H.J., Warwick, S.I., Nair, H., and Seguin-Swartz, G. (2003) Gene flow in commercial fields of herbicide-resistant canola (*Brassica napus*). Ecological Applications 13, 1276–1294.

101. Beckie, H.J., Séguin-Swartz, G., Nair, H., Warwick, S.I., and Johnson, E. (2004) Multiple herbicide-resistant canola can be controlled by alternative herbicides. Weed Science 52, 152–157.

102. Beebe, S., Toro, C.O., González, A.V., Chacón, M.I., and Debouck, D.G. (1997) Wild-weed-crop complexes of common bean (*Phaseolus vulgaris* L., Fabaceae) in the Andes of Peru and Colombia, and their implications for conservation and breeding. Genetic Resources and Crop Evolution 44, 73–91.

103. Beesley, J.S.S., and Lee, P.G. (1979) The assessment of bird resistance in sorghum cultivars in Botswana. PANS 25, 391–393.

104. Beetham, P.R., Kipp, P.B., Sawycky, X.L., Arntzen, C.J., and May, G.D. (1999) A tool for functional plant genomics: Chimeric RNA/DNA oligonucleotides cause in vivo gene-specific mutations. Proceedings of the National Academy of Sciences USA 96, 8774–8778.

105. Bekele, E., Klock, G., and Zimmermann, U. (1995) Somatic embryogenesis and plant regeneration from leaf and root explants and from seeds of *Eragrostis tef* (Gramineae). Hereditas 123, 183–189.

106. Belay, G., Tefera, H., Tadesse, B., Metaferia, G., Jarra, D., and Tadesse, T. (2006) Participatory variety selection in the Ethiopian cereal tef (*Eragrostis tef*). Experimental Agriculture 42, 91–101.

107. Belton, P.S., and Taylor, J.R.N. (eds.) (2003) AFRIPRO Workshop on the proteins of sorghums and millets: Enhancing the nutritional and functional properties for Africa, Online www.afripro.org.uk/, Pretoria, South Africa.

108. Bennett, J.W., and Klich, M. (2003) Mycotoxins. Clinical Microbiology Reviews 16, 497–516.

109. Bennett, L.E., Burkhead, J.L., Hale, K.L., Terry, N., Pilon, M., and Pilon-Smits, E.A. (2003) Analysis of transgenic Indian mustard plants for phytoremediation of metal-contaminated mine tailings. Journal of Environmental Quality 32, 432–440.

110. Bennett, R.N., and Wallsgrove, R.M. (1994) Secondary metabolites in plant defence mechanisms. New Phytologist 127, 617–633.

111. Bennetzen, J., Sorrells, M., and Tefera, H. (2004) Allelic diversity for plant height in *Eragrostis tef.* www.cerealsgenomics.org/proposalparagraphs/allelic.htm.

112. Berg, P., Baltimore, D., Brenner, S., Roblin, R.O., and Singer, M.F. (1975) Summary statement of Asilomar conference on recombinant DNA molecules. Proceedings of the National Academy of Sciences USA 72, 1981–1984.

113. Bergelson, J., and Purrington, C.B. (1996) Surveying patterns in the cost of resistance in plants. American Naturalist 148, 536–558.

114. Berner, D.K., and Williams, O.A. (1998) Germination stimulation of *Striga gesnerioides* seeds by hosts and nonhosts. Plant Disease 82, 1242–1247.

115. Berner, D.K., Schaad, N.W., and Volksch, B. (1999) Use of ethylene-producing bacteria for stimulation of *Striga* spp. seed germination. Biological Control 15, 274–282.

116. Bertolini, E., Olmos, A., Lopez, M.M., and Cambra, M. (2003) Multiplex nested reverse transcription-polymerase chain reaction in a single tube for sensitive and simultaneous detection of four RNA viruses and *Pseudomonas savastanoi* pv. *savastanoi* in olive trees. Phytopathology 93, 286–292.

117. Bervillé, A., Muller, M.H., Poinso, B., and Serieys, H. (2005) Ferality: Risks of gene flow between sunflower and other *Helianthus* species. In Crop Ferality and Volunteerism (ed. Gressel, J.), 209–230. CRC Press, Boca Raton, FL.

118. Beyer, E. (1976) Silver ion: a potent antiethylene agent in cucumber and tomato. HortScience 11, 195–196.

119. Bezuidenhout, S.C., Gelderblom, W.C.A., Gorstallman, C.P., Horak, R.M., Marasas, W.F.O., Spiteller, G., and Vleggaar, R. (1988) Structure elucidation of the fumonisins, mycotoxins from *Fusarium moniliforme*. Journal of the Chemical Society-Chemical Communications, 743–745.

120. Bhagya, S., and Sastry, M.C.S. (2003) Chemical, functional and nutritional properties of wet dehulled niger (*Guizotia abyssinica* Cass.) seed flour. Lebensmittel-Wissenschaft und-Technologie (Food Science and Technology) 36, 703–708.

121. Bhatnagar, D., Cleveland, T., Linz, J. and Payne, G. (1995) Molecular biology to eliminate aflatoxins. INFORM 6, 262–271.

122. Biebel, R., Rametzhofer, E., Klapal, H., Polheim, D. and Viernstein, H. (2003) Action of pyrethrum-based formulations against grain weevils. International Journal of Pharmaceutics 256, 175–171.

123. Bilyeu, K.D., Cole, J.L., Laskey, J.G., Riekhof, W.R., Esparza, T.J., Kramer, M.D., and Morris, R.O. (2001) Molecular and biochemical characterization of a cytokinin oxidase from maize. Plant Physiology 125, 378–386.

124. Binding, H., Jain, S.M., Finger, J., Mordhorst, G., Nehls, R., and Gressel, J. (1982) Somatic hybridization of an atrazine-resistant biotype of *Solanum nigrum* with *S. tuberosum:* I. Clonal variation in morphology and in atrazine sensitivity. Theoretical and Applied Genetics 63, 273–277.

125. Blackwell, B.A., Gilliam, J.T., Savard, M.E., Miller, J.D., and Duvick, J.P. (1999) Oxidative deamination of hydrolyzed fumonisin B-1 (AP(1)) by cultures of *Exophiala spinifera*. Natural Toxins 7, 31–38.

126. Blevins, L.G., and Cauley, T.H. (2005) Fine particulate formation during switchgrass/coal cofiring. Journal of Engineering for Gas Turbines and Power-Transactions of the ASME 127, 457–463.

127. Bock, R. (2001) Transgenic plastids in basic research and plant biotechnology. Journal of Molecular Biology 312, 425–438.

128. Boeke, S.J., Baumgart, I.R., van Loon, J.J.A., van Huis, A., Dicke, M., and Kossou, D.K. (2004) Toxicity and repellence of African plants traditionally used for the protection of stored cowpea against *Callosobruchus maculatus*. Journal of Stored Products Research 40, 423–438.

129. Bohn, M., Groh, S., Khairallah, M.M., Hoisington, D.A., Utz, H.F., and Melchinger A.E. (2001) Reevaluation of the prospects of marker-assisted selection for improving insect resistance against *Diatraea* spp. in tropical maize by cross validation and independent validation. Theoretical and Applied Genetics 103, 1059–1067.

130. Bommert, P., Sataoh-Nagasawa, N., Jackson, D., and Hirano, H.-Y. (2005) Genetics and evolution of inflorescence and flower development in grasses. Plant and Cell Physiology 46, 69–78.

131. Bommert, P., Lunde, C., Nardmann, J., Vollbrecht, E., Running, M., Jackson, D., Hake, S., and Werr, W. (2005) *thick tassel dwarf1* encodes a putative maize ortholog of the *Arabidopsis CLAVATA1* leucine-rich repeat receptor-like kinase. Development 132, 1235–1245.

132. Bond, J.M., Mogg, R.J., Squire, G.R., and Johnstone, C. (2004) Microsatellite amplification in *Brassica napus* cultivars: Cultivar variability and relationship to a long-term feral population. Euphytica 139, 173–178.

133. Borevitz, J.O., Xia, Y.J., Blount, J., Dixon, R.A., and Lamb, C. (2000) Activation tagging identifies a conserved MYB regulator of phenylpropanoid biosynthesis. Plant Cell 12, 2383–2393.

134. Bouda, H., Tapondjou, L.A., Fontem, D.A., and Gumedzoe, M.Y.D. (2001) Effect of essential oils from leaves of *Ageretum conyzoides, Lantana camara,* and *Chromolaena odorata* on the mortality of *Sitophilus zeamais* (Coleoptera, Curculionidae). Journal of Stored Products Research 37, 103–109.

135. Boudry, P., Mirchen, M., Saumitou-Laprade, P., Vernet, H., and Van Dijk, H. (1993) The origin and evolution of weed beets: consequences for the breeding and release of herbicide-resistant transgenic sugar beets. Theoretical and Applied Genetics 87, 471–478.

136. Boulter, D. (1993) Insect pest control by copying nature using genetically engineered crops. Phytochemistry 34, 1453–1466.

137. Bouquin, T., Meier, C., Foster, R., Nielsen, M.E., and Mundy, J. (2001) Control of specific gene expression by gibberellin and brassinosteroid. Plant Physiology 127, 450–458.

138. Bouvier-Nave, P., Benveniste, P., Oelkers, P., Sturley, S.L., and Schaller, H. (2000) Expression in yeast and tobacco of plant cDNAs encoding acyl CoA: diacylglycerol acyltransferase. European Journal of Biochemistry 267, 85–96.

139. Bouwmeester, H.J., Matusova, R., Zhongkui, S., and Beale, M.H. (2003) Secondary metabolite signalling in host-parasite plant interactions. Current Opinion in Plant Biology 6, 358–364.

140. Bowman, D.T., May, O.L., and Creech, J.B. (2003) Genetic uniformity of the US upland cotton crop since the introduction of transgenic cotton. Crop Science 43, 515–518.

141. Boxall, R.A. (2001) Post-harvest losses to insects-a world overview. International Biodeterioration & Biodegradation 48, 137–152.

142. Boxall, R.A. (2002) Damage and loss caused by the larger grain borer *Prostephanus truncatus*. Integrated Pest Management Reviews 7, 105–121.

143. Bradford, K.J., Van Deynze, A., Gutterson, N., Parrott, W., and Strauss, S.H. (2005) Regulating transgenic crops sensibly: lessons from plant breeding, biotechnology and genomics. Nature Biotechnology 23, 439–444.

144. Bradshaw, A.D. (1982) Evolution of heavy metal resistance-an analogy for herbicide resistance? In Herbicide Resistance in Plants (eds. LeBaron, H.M., and Gressel, J.), 293–307. Wiley, New York.

145. Brahmbhatt, S., Brahmbhatt, R.M., and Boyages, S.C. (2000) Thyroid ultrasound is the best prevalence indicator for assessment of iodine deficiency disorders: a study in rural/tribal schoolchildren from Gujarat (western India). European Journal of Endocrinology 143, 37–46.

146. Brahmbhatt, S.R., Fearnley, R., Brahmbhatt, R.M., Eastman, C.J., and Boyages, S.C. (2001) Study of biochemical prevalence indicators for the assessment of iodine deficiency disorders in adults at field conditions in Gujarat (India). Asia Pacific Journal of Clinical Nutrition 10, 51–57.

147. Branch, W.D. (2002) Variability among advanced gamma-irradiation induced large-seeded mutant breeding lines in the 'Georgia Browne' peanut cultivar. Plant Breeding 121, 275–277.

148. Brar, D.S., and Khush, G.S. (1997) Alien introgression in rice. Plant Molecular Biology 35, 35–47.

149. Breitler, J.C., Vassal, J.M., Catala, M.D., Meynard, D., Marfa, V., Melé, E., Royer, M., Murillo, I., San Segundo, B., Guiderdoni, E., and Messeguer, J. (2004) Bt rice harbouring *cry* genes controlled by a constitutive or wound-inducible promoter: protection and transgene expression under Mediterranean field conditions. Plant Biotechnology Journal 2, 417–430.

150. Bres-Patry, C., Lorieux, M., Clement, G., Bangratz, M., and Ghesquiere, A. (2001) Heredity and genetic mapping of domestication-related traits in a temperate japonica weedy rice. Theoretical and Applied Genetics 102, 118–126.

151. Breton, C., Médail, F., Pinatel, C., and Bervillé, A. (2005) Olives. In Crop Ferality and Volunteerism (ed. Gressel, J.), 233–234. CRC Press, Boca Raton, FL.

152. Brhadda, N., Abousalim, A., and Loudiyi, D.E.W. (2003) Effet du milieu de culture sur le microbouturage de l'olivier (*Olea europeae* L.) cv Picholine Marocaine. Biotechnologie Agronomie Societe et Environment 7, 177–182.

153. Briddon, R.W., Phillips, S., Brunt, A., and Hull, R. (1999) Analysis of the sequence of *Dioscorea alata* Bacilliform Virus; Comparison to other members of the Badnavirus group. Virus Genes 18, 277–283.

154. Broothaerts, W., Mitchell, H.J., Weir, B., Kaines, S., Smith, L.M., Yang, W., Mayer, J.E., Roa-Rodriguez, C., and Jefferson, R.A. (2005) Gene transfer to plants by diverse species of bacteria. Nature 433, 632–633.

155. Broun, P., and Somerville, C. (1997) Accumulation of ricinoleic, lesquerolic, and densipolic acids in seeds of transgenic *Arabidopsis* plants that express a fatty acyl hydroxylase cDNA from castor bean. Plant Physiology 113, 933–942.

156. Broun, P., Gettner, S., and Somerville, C. (1999) Genetic engineering of plant lipids. Annual Review of Nutrition 19, 197–216.

157. Brown, E., and Jacobson, M.F. (2005) Cruel Oil-(available online at http://cspinet.org/palmoilreport/PalmOilReport.pdf). Center for Science in the Public Interest (CSPI), Washington, DC.

158. Brown, J., Brown, A.P., Davis, J.B. and Erickson, D. (1997) Intergeneric hybridization between *Sinapis alba* and *Brassica napus*. Euphytica 93, 163–168.

159. Brown, R.A., Rosenberg, N.J., Hays, C.J., Easterling, W.E., and Mearns, L.O. (2000) Potential production and environmental effects of switchgrass and traditional crops under current and greenhouse-altered climate in the central United States: a simulation study. Agriculture Ecosystems and Environment 78, 31–47.

160. Brown, R.L., Brown-Jenco, C.S., Bhatnagar, D., and Payne, G.A. (2003) Construction and preliminary evaluation of an *Aspergillus flavus* reporter gene construct as a potential tool for screening aflatoxin resistance. Journal of Food Protection 66, 1927–1931.

161. Bruce, D.M., Hobson, R.N., Morgan, C.L., and Child, R.D. (2001) Threshability of shatter-resistant seed pods in oilseed rape. Journal of Agricultural Engineering Research 80, 343–350.

162. Bruce, D.M., Farrent, J.W., Morgan, C.L., and Child, R.D. (2002) Determining the oilseed rape pod strength needed to reduce seed loss due to pod shatter. Biosystems Engineering 81, 179–184.

163. Brucker, G., Mittmann, F., Hartmann, E., and Lamparter, T. (2005) Targeted site-directed mutagenesis of a heme oxygenase locus by gene replacement in the moss *Ceratodon purpureus*. Planta 220, 864–874.

164. Brummell, D.A., Howie, W.J., Ma, C., and Dunsmuir, P. (2002) Postharvest fruit quality of transgenic tomatoes suppressed in expression of a ripening-related expansin. Postharvest Biology and Technology 25, 209–220.

165. Brummell, D.A., Harpster, M.H., Civello, P.M., Palys, J.M., Bennett, A.B, and Dunsmuir, P. (1999) Modification of expansin protein abundance in tomato fruit alters softening and cell wall polymer metabolism during ripening. Plant Cell 11, 2203–2216.

166. Buckland, P.C. (1981) The early dispersal of insect pests of stored grain products as indicated by archaeological records. Journal of Stored Products Research 17, 1–12.

167. Budak, H., Pedraza, F., Cregan, P.B., Baenziger, P.S., and Dweikat, I. (2003) Development and utilization of SSRs to estimate the degree of genetic relationships in a collection of pearl millet germplasm. Crop Science 43, 2284–2290.

168. Bui, A.Q., and O'Neill, S.D. (1998) Three 1-aminocyclopropane-1-carboxylate synthase genes regulated by primary and secondary pollination signals in orchid flowers. Plant Physiology 116, 419–428.

169. Burnet, M.W., Kanampiu, F., and Gressel, J. (2004) A slow release agrochemicals dispenser and method of use. PCT patent application WO 2004/004453 A2.

170. Busch, U., Pecoraro, S., Posthoff, K., and Estendorfer-Rinner, S. (2004) First time detection of a genetically modified papaya in Europe-Official complaint of a non authorized genetically modified organism within the EU. Deutsche Lebensmittel-Rundschau 100, 377–380.

171. Busk, P.K., and Moller, B.L. (2002) Dhurrin synthesis in sorghum is regulated at the transcriptional level and induced by nitrogen fertilization in older plants. Plant Physiology 129, 1222–1231.

172. Butenko, R.G., Vorobev, A.S., Nosov, A.M., and Knyazkov, I.E. (1992) Synthesis, accumulation, and location of steroid glycosides in cells of different strains of *Dioscorea deltoidea* Wall. Soviet Plant Physiology 39, 763–768.

173. Butler, D. (2005) Drugs could head off a flu pandemic-but only if we respond fast enough. Nature 436, 614–615.

174. Butteroni, C., Afferni, C., Barletta, B., Iacovacci, P., Corinti, S., Brunetto, B., Tinghino, R., Ariano, R., Panzani, R.C., Pini, C., and Di Felice, G. (2005) Cloning and expression of the *Olea europaea* allergen Ole e 5, the pollen Cu/Zn superoxide dismutase. International Archives of Allergy and Immunology 137, 9–17.

175. Cabrera Ponce, J.L., Vegas Garcia, A., and Herrera Estrella, L. (1995) Herbicide-resistant transgenic papaya plants produced by an efficient particle bombardment transformation method. Plant Cell Reports 15, 1–7.

176. Cahoon, E.B., Carlson, T.J., Ripp, K.G., Schweiger, B.J., Cook, G.A, Hall, S.E., and Kinney, A.J. (1999) Biosynthetic origin of conjugated double bonds: Production of fatty acid components of high-value drying oils in transgenic soybean embryos. Proceedings of the National Academy of Sciences of the United States of America, 96, 12935–12940.

177. Calabozo, B., Diaz-Perales, A., Salcedo, G., Barber, D., and Polo, F. (2003) Cloning and expression of biologically active *Plantago lanceolata* pollen allergen Pla l 1 in the yeast *Pichia pastoris*. Biochemical Journal 372, 889–896.

178. CAMBIA. (2005) www.cambia.org.

179. Campbell, C.G. (1997) Grass pea. *Lathyrus sativus L.*International Plant Genetic Resources Institute, Rome, Italy.

180. Campbell, C.G. (1997) Buckwheat. *Fagopyrum esculentum* Moench. Promoting the conservation and use of underutilized and neglected crops. 19. International Plant Genetic Resources Institute, Rome, Italy.

181. Canizares, M.C., Nicholson, L., and Lomonossoff, G.P. (2005) Use of viral vectors for vaccine production in plants. Immunology and Cell Biology 83, 263–270.

182. Cao, Q., Lu, B.R., Xia, H., Rong, J., Sala, F., Spada, A., and Grassi, F. (2006) Genetic diversity and origin of weedy rice (*Oryza sativa* f. *spontanea*) populations found in north-eastern China revealed by simple sequence repeat (SSR) markers. Annals of Botany 98, 1241–1252.

183. Cardoso, J.M.S., Felix, M.R., Clara, M.I.E., and Oliveira, S. (2005) The complete genome sequence of a new necrovirus isolated from *Olea europaea* L. Archives of Virology 150, 815–823.

184. Cardwell, K.F., Kling, J.G., Mazia-Dixon, B., and Bosque-Perez, N.A. (2000) Interaction between *Fusarium verticilloides, Aspergillus flavus*, and insect infestation in four maize genotypes in lowland Africa. Phytopathology 90, 276–284.

185. Carey, A.T., Smith, D.L., Harrison, E., Bird, C.R., Gross, K.C., Seymour, G.B., and Tucker, G.A. (2001) Down-regulation of a ripening-related beta-galactosidase gene (*TBG1*) in transgenic tomato fruits. Journal of Experimental Botany 52, 663–668.

186. Carey, V.F.I., Hoagland, R.E., and Talbert, R.E. (1997) Resistance mechanisms in propanil-resistant barnyardgrass. II. In vivo metabolism of the propanil molecule. Pesticide Science 49, 333–338.

187. Carinanos, P., Alcazar, P., Galan, C., and Dominguez, E. (2002) Privet pollen (*Ligustrum* sp.) as potential cause of pollinosis in the city of Cordoba, south-west Spain. Allergy 57, 92–97.

188. Carlson, P.S., Fahey, J.W., and Flynn, J.L. (1992) Modified plant containing a bacterial inoculant. U.S. Patent 5,157,207.

189. Carlson, R. (2003) The pace and proliferation of biological technologies. Biosecurity and Bioterrorism: Biodefense Strategy, Practice, and Science 1, 203–214.

190. Carmel-Goren, L., Liu, Y.S., Lifschitz, E., and Zamir, D. (2003) The *SELF-PRUNING* gene family in tomato. Plant Molecular Biology 52, 1215–1222.

191. Carnes-Sanchez, J., Iraola, V.M., Sastre, J., Florido, F., Boluda, L., and Fernandez-Caldas, E. (2002) Allergenicity and immunochemical characterization of six varieties of *Olea europaea*. Allergy 57, 313–318.

192. Carrol, L. (1960) Alice's Adventures in Wonderland and through the Looking-Glass. New American Library, New York.

193. Carrozzi, N.B., Karamer, V.C., Warren, G.W., and Koziel, M.G. (1993) *Bacillus thuringiensis* strains against Coleopteran insects. U.S. Patent 5,294,100.

194. Castle, L.A., Siehl, D.L., Gorton, R., Patten, P.A., Chen, Y.H., Bertain, S., Cho, H.J., Duck, N., Wong, J., Liu, D., and Lassner, M.W. (2004) Discovery and directed evolution of a glyphosate tolerance gene. Science 304, 1151–1154.

195. Castro, A.J., Alche, J.D., Cuevas, J., Romero, P.J., Alche, V., and Rodriguez-Garcia, M.I. (2003) Pollen from different olive tree cultivars contains varying amounts of the major allergen ole e 1. International Archives of Allergy and Immunology 131, 164–173.

196. Cazzonelli, C.I., McCallum, E.J., Lee, R., and Botella, J.R. (2005) Characterization of a strong, constitutive mung bean (*Vigna radiata* L.) promoter with a complex mode of regulation in planta. Transgenic Research 14, 941–967.

197. CBD. (1992) United Nations Convention on Biological Diversity. www.biodiv.org/doc/publications/guide.asp.

198. Champ, B.R., and Dyte, C.E. (1976) Report of the FAO global survey of pesticide susceptibility of stored grain pests. In Plant Production and Protection Series No. 5. Food and Agriculture Organization of the United Nations (FAO), Rome, Italy.

199. Chan, Y.L., Lin, K.H., Sanjaya, Liao, L.J., Chen, W.H., and Chan, M.T. (2005) Gene stacking in *Phalaenopsis* orchid enhances dual tolerance to pathogen attack. Transgenic Research 14, 279–288.

200. Chandler, J., Corbesier, L., Spielmann, P., Dettendorfer, J., Stahl, D., Apel, K., and Melzer, S. (2005) Modulating flowering time and prevention of pod shatter in oilseed rape. Molecular Breeding 15, 87–94.

201. Chang, T.T. (1995) Rice. In Evolution of Crop Plants (eds. Smartt, J., and Simmonds, N.W.), 147–155. Longman Group, Harlow, United Kingdom.

202. Charlesworth, D. (1988) Evidence for pollen competition in plants and its relationship to progeny fitness: A comment. American Naturalist 132, 298–302.

203. Chatterjee, R., Duvick, J., and English, J. (2005) AP1 amine oxidase variants. World Patent WO2005003289 .

204. Chauvaux, N., Child, R., John, K., Ulvskov, P., Borkhardt, B., Prinsen, E., and Van Onckelen, H.A. (1997) The role of auxin in cell separation in the dehiscence zone of oilseed rape pods. Journal of Experimental Botany 48, 1423–1429.

205. Cheke, R.A., Rosenberg, L.J., and Kieser, M.E. (eds.) (2000) Workshop on research priorities for migrant pests of agriculture in southern Africa. Natural Resources Institute, Chatham, United Kingdom.

206. Chelule, P.K., Gqaleni, N., Dutton, M.F., and Chuturgoon, A.A. (2001) Exposure of rural and urban populations in KwaZulu Natal, South Africa to fumonisin B_1 in maize. Environmental Health Perspectives 109, 253–256.

207. Chemo, R., Maoz, I., and Yarden, G. (2003) Multi-barrel plant inoculation gun. U.S. Patent 6,644,341.

208. Chen, G., Ye, C.M., Huang, J.C., Yu, M., and Li, B.J. (2001) Cloning of the papaya ringspot virus (PRSV) replicase gene and generation of PRSV-resistant papayas through the introduction of the PRSV replicase gene. Plant Cell Reports 20, 272–277.

209. Chen, Q., Jahier, J., and Cauderon, Y. (1993) The B-chromosome system of inner Mongolian *Agropyron* Gaertn.3. Cytogenetical evidence for B-A pairing at metaphase-I. Hereditas 119, 53–58.

210. Chen, Q.F., Hsam, S.L.K., and Zeller, F.J. (2004) A study of cytology, isozyme, and interspecific hybridization on the big-achene group of buckwheat species (*Fagopyrum*, Polygonaceae). Crop Science 44, 1511–1518.

211. Chen, S.B., Liu, X., Peng, H. Y., Gong, W. K., Wang, R., Wang, F. and Zhu, Z. (2004) Cre/lox-mediated marker gene excision in elite indica rice plants transformed with genes conferring resistance to lepidopteran insects. Acta Botanica Sinica 46, 1416–1423.

212. Chen, W.S., Lennox, S.J., Palmer, K.E., and Thomson, J.A. (1998) Transformation of *Digitaria sanguinalis:* A model system for testing maize streak virus resis-tance in Poaceae. Euphytica 104, 25–31.

213. Chen, Z.-Y., Brown, R.L., Lax, A.R., Cleveland, T.E., and Russin, J.S. (1999) Inhibition of plant-pathogenic fungi by a corn trypsin inhibitor overexpressed in *Escherichia coli*. Applied and Environmental Microbiology 65, 1320–1324.

214. Chiang, Y.J., Stushnoff, C., and McSay, A.E. (2005) Overexpression of mannitol-1-phosphate dehydrogenase increases mannitol accumulation and adds protection against chilling injury in petunia. Journal of the American Society for Horticultural Science 130, 605–610.

215. Child, R.D., Summers, J.E., Babij, J., Farrent, J.W., and Bruce, D.M. (2003) Increased resistance to pod shatter is associated with changes in the vascular structure in pods of a resynthesized Brassica napus line. Journal of Experimental Botany 54, 1919–1930.

216. Cho, Y.-C., Cung, T.-Y., and Suh, H.-S. (1995) Genetic characteristics of Korean weedy rice (*Oryza sativa* L.) by RFLP analysis. Euphytica 86, 103–110.

217. Choe, S.B.P., Dilkes, B.P., Fujioka, S., Takatsuto, S., Sakurai, A., and Feldmann, K.A. (1998) The DWF4 gene of *Arabidopsis* encodes a cytochrome P450 that mediates multiple 22 α-hydroxylation steps in brassinosteroid biosynthesis. Plant Cell 10, 231–243.

218. Choi, I.-R., Stander, D.C., Morris, T.J., and French, R. (2000) A plant virus vector for systemic expression of foreign genes in cereals. Plant Journal 23, 547–555.

219. Chourey, P.S., Taliercio, E.W., Carlson, S.J., and Ruan, Y.-L. (1998) Genetic evidence that the two isozymes of sucrose synthase present in developing maize endosperm are critical, one for cell wall integrity and the other for starch biosynthesis. Molecular & General Genetics 259, 88–96.

220. Christou, P., and Klee, H.J. (eds.) (2004) Handbook of Plant Biotechnology. Wiley, New York.

221. Chuck, G., Muszynski, M., Kellogg, E., Hake, S., and Schmidt, R.J. (2002) The control of spikelet meristem identity by the branched silkless1 gene in maize. Science 298, 1238–1241.

222. Chung, Y.C., Bakalinsky, A., and Penner, M.H. (2005) Enzymatic saccharification and fermentation of xylose-optimized dilute acid-treated lignocellulosics. Applied Biochemistry and Biotechnology 121, 947–961.

223. Chèvre, A.M., Eber, F., Darmency, H., Fleury, A., Picault, H., Letanneur, J.C., and Renard, M. (2000) Assessment of interspecific hybridization between transgenic oilseed rape and wild radish under normal agronomic conditions. Theoretical Applied Genetics 100, 1233–1239.

224. Cibus. (2007) www.cibusllc.com.

225. Ciotola, M., DiTommaso, A., and Watson, A.K. (2000) Chlamydospore production, inoculation methods and pathogenicity of *Fusarium oxysporum* M12-4A, a biocontrol for *Striga hermonthica*. Biocontrol Science and Technology 10, 129–145.

226. Clemetsen, M., and van Laar, J. (2000) The contribution of organic agriculture to landscape quality in the Sogn og Fjordane region of western Norway. Agriculture, Ecosystems & Environment 77, 125–141.

227. Cobb, D., Feber, R., Hopkins, A., Stockdale, L., O'Riordan, T., Clements, R.O., Firbank, L., Goulding, K., Jarvis, S., and MacDonald, D. (1999) Integrating the environmental and economic consequences of converting to organic agriculture: evidence from a case study. Land Use Policy 16, 207–221.

228. Coca, M., Penas, G., Gomez, J., Campo, S., Bortolotti, C., Messeguer, J., and Segundo, B.S. (2006) Enhanced resistance to the rice blast fungus *Magnaporthe grisea* conferred by expression of a cecropin A gene in transgenic rice. Planta 223, 392–406.

229. Cohen, A., Amsellem, Z., Maor, R., Sharon, A., and Gressel, G. (2002) Transgenically-enhanced expression of IAA confers hypervirulence plant pathogens. Phytopathology 92, 590–596.

230. Cohen, M.B., Jackson, M.T., Lu, B.R., Morin, S.R., Mortimer, A.M., Pham, J. L., and Wade, l.J. (1999) Predicting the environmental impact of transgene outcrossing in wild and weedy rices in Asia. In Gene Flow in Agriculture: Relevance for Transgenic Crops, 151–157. British Crop Protection Council, Farnham, United Kingdom.

231. Colbach, N., Molinari, N., Meynard, J.M., and Messean, A. (2005) Spatial aspects of gene flow between rapeseed varieties and volunteers. Agronomy for Sustainable Development 25, 355–368.

232. Colton, R. (1980) Harvesting and Windrowing Rapeseed. New South Wales Department of Agriculture, Sydney, Australia.

233. Conner, A.J., and Dale, P.J. (1996) Reconsideration of pollen dispersal data from field trials of transgenic potatoes. Theoretical and Applied Genetics 92, 505–508.

234. Cooper, L.D., Doss, R.P., Price, R., Peterson, K., and Oliver, J.E. (2005) Application of Bruchin B to pea pods results in the up-regulation of *CYP93C18,* a putative isoflavone synthase gene, and an increase in the level of pisatin, an isoflavone phytoalexin. Journal of Experimental Botany 56, 1229–1237.

235. Crameri, A., Bermudez, E., Raillard, S., and Stemmer, W.P.C. (1998) DNA shuffling of a family of genes from diverse species accelerates directed evolution. Nature 3, 284–290.

236. Crameri, A., Dawes, G., Rodriguez, E., Silver, S., and Stemmer, W.P. (1997) Molecular evolution of an arsenate detoxification pathway by DNA shuffling. Nature Biotechnology 15, 436–438.

237. Crouch, M.L. (1998) How the terminator terminates: An explanation for the non-scientist of a remarkable patent for killing second generation seeds of crop plants. www.edmonds-institute.org/crouch.html. The Edmonds Institute, Edmond WA.

238. Cunliffe, K. (2005) The ryegrass complex. In Crop Ferality and Volunteerism (ed. Gressel, J.), 236–242. CRC Press, Boca Raton, FL.

239. Curtis, I.S., Nam, H.G., Yun, J.Y., and Seo, K.-H. (2002) Expression of an antisense GIGANTEA (GI) gene fragment in transgenic radish causes delayed bolting and flowering. Transgenic Research 11, 249–256.

240. Curtis, I.S., Hanada, A., Yamaguchi, S., and Kamiya, Y. (2005) Modification of plant architecture through the expression of GA 2-oxidase under the control of an estrogen inducible promoter in *Arabidopsis thaliana* L. Planta 222, 957–967.

241. Cárdaba, B., Cortegano, I., Florido, F., Arrieta, I., Aceituno, E., del Pozo, V., Gallardo, S., Rojo, M., Palomino, P., and Lahoz, C. (2000) Genetic restrictions in olive pollen allergy. Journal of Allergy and Clinical Immunology 105, 292–298.

242. D'Mello, J.P.F., Macdonald, A.M.C., Postel, D., Dijiksman, W.T.P., Dujardin, A., and Placinta, C.M. (1998) Pesticide use and mycotoxin production in *Fusarium* and *Aspergillus* phytopathogens. European Journal of Plant Pathology 104, 741–751.

243. Dai, W.M., Zhang, K.-Q., Duan, B -W., Zheng, K.-L., Zhuang, J.-Y., and Cai, R. (2005) Genetic dissection of silicon content in different organs of rice. Crop Science 45, 1345–1352.

244. Dajue, L., and Mündel, H.-H. (1996) Safflower. *Carthamus tinctorius L,* International Plant Genetic Resources Institute, Rome, Italy.

245. Damania, A.B., Valkoun, J., Willcox, G., and Qualset, C.O. (eds.) (1998) The Origins of Agriculture and Crop Domestication. ICARDA (International Center for Agricultural Research in the Dry Areas), Aleppo, Syrian Arab Republic.

246. Daniell, H. (2002) Molecular strategies for gene containment in transgenic crops. Nature Biotechnology 20, 581–586.

247. Daniell, H., Datta, R., Varma, S., Gray, S., and Lee, S.B. (1998) Containment of herbicide resistance through genetic engineering of the chloroplast genome. Nature Biotechnology 16, 345–348.

248. Daniell, H., Chebolu, S., Kumar, S., Singleton, M., and Falconer, R. (2005) Chloroplast-derived vaccine antigens and other therapeutic proteins. Vaccine 23, 1779–1783.

249. Dark, P., and Gent, H. (2001) Pests and diseases of prehistoric crops: A yield 'honeymoon' for early grain crops in Europe? Oxford Journal of Archaeology 20, 59–78.

250. Darmency, H. (1994) Genetics of herbicide resistance in weeds and crops. In Herbicide Resistance in Plants: Biology and Biochemistry (eds. Powles, S.B., and Holtum, J.A.M.), 263–298. Lewis, Boca Raton, FL.

251. Darmency, H. (2005) Oats. In Crop Ferality and Volunteerism (ed. Gressel, J.), 231–233. CRC Press, Boca Raton, FL.

252. Darmency, H. (2005) Incestuous relations of foxtail millet (*Setaria italica*) with its parents and cousins. In Crop Ferality and Volunteerism (ed. Gressel, J.), 81–96. CRC Press, Boca Raton, FL.

253. Datta, S.K. (1999) Transgenic cereals: *Oryza sativa* (rice). In Molecular Improvement of Cereal Crops (ed. Vasil, I.K.), 149–187. Kluwer Academic Publishers, United Kingdom.

254. Davies, K.M., Hobson, G.E., and Grierson, D. (1990) Differential effect of silver ions on the accumulation of ripening-related mRNAs in tomato fruit. Journal of Plant Physiology 135, 708–713.

255. Davis, M.A. (2003) Biotic globalization: Does competition from introduced species threaten biodiversity? BioScience 53, 481–489.

256. Dawit, W., and Andnew, Y. (2005) The study of fungicides application and sowing date, resistance, and maturity of *Eragrostis tef* for the management of teff rust [*Uromyces eragrostidis*]. Canadian Journal of Plant Pathology 27, 521–527.

257. Day, J.S. (2000) Anatomy of capsule dehiscence in sesame varieties. Journal of Agricultural Science 134, 45–53.

258. Dayan, F.E., Kagan, I.A., and Rimando, A.M. (2003) Elucidation of the biosynthetic pathway of the allelochemical sorgoleone using retrobiosynthetic NMR analysis. Journal of Biological Chemistry 278, 28607–28611.

259. De, D., and De, B. (2005) Elicitation of diosgenin production in *Dioscorea floribunda* by ethylene-generating agent. Fitoterapia 76, 153–156.

260. de la Campa, R., Hooker, D.C., Miller, J.D., Schaafsma, A.W., and Hammond, B.G. (2005) Modeling effects of environment, insect damage, and Bt genotypes on fumonisin accumulation in maize in Argentina and the Philippines. Mycopathologia 159, 539–552.

261. de Sousa-Majer, M.J., Turner, N.C., Hardie, D.C., Morton, R.L., Lamont, B., and Higgins, T.J. (2004) Response to water deficit and high temperature of transgenic peas (*Pisum sativum* L.) containing a seed-specific alpha-amylase inhibitor and the subsequent effects on pea weevil (*Bruchus pisorum* L.) survival. Journal of Experimental Botany 55, 497–505.

262. de Wet, J.M.J. (1978) Systematics and evolution of sorghum sect. Sorghum (Gramineae). American Journal of Botany 65, 477–484.

263. De Wet, J.M.J. (1995) Finger millet. In Evolution of Crop Plants (eds. Smartt, J., and Simmonds, N.W.), 137–140. Longman, Harlow, United Kingdom.

264. Deal, B., Farello, C., Lancaster, M., Kompare, T., and Hannon, B. (2000) A dynamic model of the spatial spread of an infectious disease: the case of fox rabies in Illinois. Environmental Modeling and Assessment 5, 46–62.

265. Dehesh, K., Jones, A., Knutzon, D.S., and Voelker, T.A. (1996) Production of high levels of 8:0 and 10:0 fatty acids in transgenic canola by overexpression of *Ch FatB2*, a thioesterase cDNA from *Cuphea hookeriana*. Plant Journal 9, 167–172.

266. Desjardins, A.E., and Plattner, R.D. (2000) Fumonisin B-1-nonproducing strains of *Fusarium verticillioides* cause maize (*Zea mays*) ear infection and ear rot. Journal of Agricultural and Food Chemistry 48, 5773–5780.

267. Devitt, L.C., Sawbridge, T., Holton, T.A., Mitchelson, K., and Dietzgen, R.G. (2006) Discovery of genes associated with fruit ripening in *Carica papaya* using expressed sequence tags. Plant Science 170, 356–363.

268. Devlin, P.F., Patel, S.R., and Whitelam, G.C. (1998) Phytochrome E influences internode elongation and flowering time in *Arabidopsis*. Plant Cell 10, 1479–1487.

269. De Francischi, M.L.P., Salgado, J.M., and Leitão, R.F.F. (1994) Chemical, nutritional and technological characteristics of buckwheat and non-prolamine buckwheat flours in comparison with wheat flour. Plant Foods for Human Nutrition 46, 323–329.

270. Dhakal, M.R., and Pandey, A.K. (2001) Storage potential of niger (*Guizotia abyssinica* Cass.) seeds under ambient conditions. Seed Science and Technology 29, 205–213.

271. Dhliwayo, T., and Pixley, K.V. (2003) Divergent selection for resistance to maize weevil in six maize populations. Crop Science 43, 2043–2049.

272. Dhliwayo, T., Pixley, K.V., and Kazembe, V. (2005) Combining ability for resistance to maize weevil among 14 southern African maize inbred lines. Crop Science 45, 662–667.

273. Diamond, J. (1996) Guns, Germs and Steel: The Fates of Human Societies. W. W. Norton, New York.

274. Dias, S.C., Franco, O.L., Magalhaes, C.P., de Oliveira-Neto, O.B., Laumann, R.A., Figueira, E.L.Z., Melo, F.R., and Grossi-de-Sa M.F. (2005) Molecular cloning and expression of an alpha-amylase inhibitor from rye with potential for controlling insect pests. Protein Journal 24, 113–123.

275. Diawara, M.M., Trumble, J.T., and Quiros, C.F. (1993) Linear furanocoumarins of 3 celery breeding lines-implications for integrated pest-management. Journal of Agricultural and Food Chemistry 41, 819–824.

276. Diaz, J., Schmiediche, and Austin, D.F. (1996) Polygon of crossability between eleven species of *Ipomoea:* Section Batatas (Convolvulaceae). Euphytica 88, 189–200.

277. Diepenbrock, W., and Leon, J. (1988) Quantative effects of volunteer plants on glucosinolate content in double-low rapeseed (*Brassica napus* L)—A theoretical approach. Agronomie 8, 373–377.

278. Dinneny, J.R., and Yanofsky, M.F. (2005) Drawing lines and borders: how the dehiscent fruit of *Arabidopsis* is patterned. Bioessays 27, 42–49.

279. Dixit-Sharma, S., Ahuja-Ghosh, S., Mandal, B.B., and Srivastava, P.S. (2005) Metabolic stability of plants regenerated from cryopreserved shoot tips of *Dioscorea deltoidea*-an endangered medicinal plant. Scientia Horticulturae 105, 513–517.

280. Djerassi, C. (1976) The manufacture of steroidal contraceptives: technical vs. political aspects. Proceedings of the Royal Society London B 195, 175–186.

281. Djerassi, C. (1979) The chemical history of the pill. In The Politics of the Pill (ed. Djerassi, C.), 226–255. Norton, New York.

282. Doebley, J. (2004) The genetics of maize evolution. Annual Review of Genetics 38, 37–59.

283. Doebley, J., and Stec, A. (1991) Genetic analysis of the morphological differences between maize and teosinte. Genetics 129, 285–295.

284. Doebley, J., and Stec, A. (1993) Inheritance of the morphological differences between maize and teosinte: comparison of results for two F_2 populations. Genetics 134, 559–570.

285. Doggett, H. (1988) Sorghum. Longman, Harlow, United Kingdom.

286. Doko, M.B., Rapior, S., Visconti, A., and Schjoth, J.E. (1995) Incidence and levels of fumonisin contamination in maize genotypes grown in Europe and Africa. Journal of Agricultural and Food Chemistry 43, 429–434.

287. Domergue, F., Abbadi, A., and Heinz, E. (2005) Relief for fish stocks: oceanic fatty acids in transgenic oilseeds. Trends in Plant Science 10, 112–116.

288. Domoney, C., Welham, T., Ellis, N., Mozzanega, P., and Turner, L. (2002) Three classes of proteinase inhibitor gene have distinct but overlapping patterns of expression in *Pisum sativum* plants. Plant Molecular Biology 48, 319–329.

289. Donald, P.L. (1998) Changes in the abundance of invertebrates and plants on British farmland. British Wildlife. 9, 279–289.

290. Dong, C.M., Beetham, P., Vincent, K., and Sharp, P. (2006) Oligonucleotide-directed gene repair in wheat using a transient plasmid gene repair assay system. Plant Cell Reports 25, 457–465.

291. Donson, J., Fang, Y., Espiritu-Santo, G., Xing, W., Salazar, A., Miyamoto, S., Armendarez, V., and Volkmuth, W. (2002) Comprehensive gene expression analysis by transcript profiling. Plant Molecular Biology 48, 75–97.

292. Douglas, S.J., and Riggs, C.D. (2005) Pedicel development in *Arabidopsis thaliana:* Contribution of vascular positioning and the role of the *BREVIPEDICELLUS* and *ERECTA* genes. Developmental Biology 284, 451–463.

293. Duke, S.O. (ed.) (1996) Herbicide Resistant Crops: Agricultural, Environmental, Economic, Regulatory, and Technical Aspects. CRC Press, Boca Raton, FL.

294. Dulloo, M.E., Guarino, L., and Ford-Lloyd, B.V. (1997) A bibliography and review of genetic diversity studies of African germplasm using protein and DNA markers. Genetic Resources and Crop Evolution 44, 447–470.

295. Dunaevsky, Y.E., Elpidina, E.N., Vinokurov, K.S., and Belozersky, M.A. (2005) Protease inhibitors: Use to increase plant resistance to pathogens and insects. Molecular Biology 39, 608–613.

296. Durieu, P., and Ochatt, S.J. (2000) Efficient intergeneric fusion of pea (*Pisum sativum* L.) and grass pea (*Lathyrus sativus* L.) protoplasts. Journal of Experimental Botany 51, 1237–1242.

297. Dushenkov, S., Skarzhinskaya, M., Glimelius, K., Gleba, D., and Raskin, I. (2002) Bioengineering of a phytoremediation plant by means of somatic hybridization. International Journal of Phytoremediation 4, 117–126.

298. Duvick, J. (2001) Prospects for reducing fumonisin contamination of maize through genetic modification. Environmental Health Perspectives 109, 337–342.

299. Duvick, J., Rood, T., Maddox, J., and Gilliam, J. (1998) Detoxification of mycotoxins in planta as a strategy for improving grain quality and disease resistance. In Molecular Genetics of Host Specific Toxins in Plant Diseases (eds. Kohmoto, K., and Yoder, O.C.), 369–381. Kluwer, Dordrecht, The Netherlands.

300. Duvick, J., Maddox, J., Gilliam, J.T., Crasta, O., and Folkerts, O. (2001) Amino polyolamine oxidase polynucleotides and related polypeptides. U.S. Patent 6,211,435.

301. Duvick, J., Gilliam, J. T., Maddox, J., Rao, A.G., Crasta, O., and Folkerts, O. (2005) Amino polyol amine oxidase polynucleotides and related polypeptides and methods of use thereof. U.S. Patent 6,943,279.

302. Dwivedi, U.N., Campbell, W.H., Yu, J., Datla, R.S.S., Bugos, R.C., Chiang, V.L., and Podila, G.K. (1994) Modification of lignin biosynthesis in transgenic *Nicotiana* through expression of an antisense *O*-methyltransferase gene from *Populus*. Plant Molecular Biology 26, 61–71.

303. Dyer, J.M., and Mullen, R.T. (2005) Development and potential of genetically engineered oilseeds. Seed Science Research 15, 255–267.

304. Dörr, I. (1997) How striga parasitizes its host: A TEM and SEM study. Annals of Botany 79, 463–472.

305. Eccleston, V.S., and Ohlrogge, J.B. (1998) Expression of lauroyl-acyl carrier protein thioesterase in *Brassica napus* seeds induces pathways for both fatty acid oxidation and biosynthesis and implies a set point for triacylglycerol accumulation. Plant Cell 10, 613–621.

306. Edem, D.O. (2002) Palm oil: biochemical, physiological, nutritional, hematological, and toxicological aspects: a review. Plant Foods and Human Nutrition 57, 319–341.

307. Editorial. (2006) Parallel universes? Nature Biotechnology 24, 1178.

308. Edwards, A.L., Jenkins, R.L., Davenport, L.J., and Duke, J.A. (2002) Presence of diosgenin in *Dioscorea batatas* (Dioscoreaceae). Economic Botany 56, 204–206.

309. Edwards-Jones, G., and Howells, O. (2001) The origin and hazard of inputs to crop protection in organic farming systems: are they sustainable? Agricultural Systems 67, 31–47.

310. Eijlander, R., and Stiekema, W.J. (1994) Biological containment of potato (*Solanum tuberosum*): outcrossing to the related wild species black nightshade (*Solanum nigrum*) and bittersweet (*Solanum dulcamara*). Sexual Plant Reproduction 7, 29–40.

311. Ejeta, G. (2005) Integrating biotechnology, breeding and agronomy in the control of the parasitic weed *Striga* spp. in sorghum. In In the Wake of the Double Helix: From the Green Revolution to the Gene Revolution (eds. Tuberosa, R., Phillips, R.L., and Gale, M.), 239–251. Avenue Media, Bologna, Italy.

312. Ejeta, G., and Grenier, C. (2005) Sorghum and its weedy hybrids. In Crop Ferality and Volunteerism (ed. Gressel, J.), 125–135. CRC Press, Boca Raton, FL.

313. Ejeta, G., Hassen, M.M., and Mertz, E.T. (1987) Invitro digestibility and amino-acid-composition of pearl-millet (*Pennisetum typhoides*) and other cereals. Proceedings of the National Academy of Sciences USA 84, 6016–6019.

314. Ejeta, G., Mohammed, A., Rich, P., Melake-Berhan, A., Housley, T.L., and Hess, D.E. (2000) Selection for specific mechanisms of resistance to *Striga* in sorghum. In Breeding for *Striga* Resistance in Cereals (eds. Haussmann, B.I.G., Koyama, M.L., Grivet, L., Rattunde, H.F., and Hess, D.E.), 29–37. Margraf, Weikersheim, Germany.

315. Eleftherohorinos, I.G., Dhima, K.V., and Vasilakoglou, I.B. (2002) Interference of red rice in rice grown in Greece. Weed Science 50, 167–172.

316. Elhag, E.A. (2000) Deterrent effects of some botanical products on oviposition of the cowpea bruchid *Callosobruchus maculatus* (F.) (Coleoptera: Bruchidae). International Journal of Pest Management 46, 109–113.

317. Ellis, R.T., Stockhoff, B.A., Stamp, L., Schnepf, H.E., Schwab, G.E., Knuth, M., Russell, J., Cardineau, G.A., and Narva, K.E. (2002) Novel *Bacillus thuringiensis* binary insecticidal crystal proteins active on western corn rootworm, *Diabrotica virgifera virgifera* LeConte. Applied and Environmental Microbiology 68, 1137–1145.

318. Ellstrand, N.C. (2003) Dangerous Liaisons—When Cultivated Plants Mate with Their Wild Relatives. Johns Hopkins University Press, Baltimore, MD.

319. Elnour, A., Hambraeus, L., Eltom, M., Dramaix, M., and Bourdoux, P. (2000) Endemic goiter with iodine sufficiency: a possible role for the consumption of pearl millet in the etiology of endemic goiter. American Journal of Clinical Nutrition 71, 59–66.

320. Ene-Obong, H.N., and Obizoba, I.C. (1996) Effect of domestic processing on the cooking time, nutrients, antinutrients and *in vivo* protein digestibility of the African yambean (*Sphenostylis stenocarpa*). Plant Foods for Human Nutrition 49, 43–52.

321. Epstein, E. (1994) The anomaly of silicon in plant biology. Proceedings of the National Academy of Sciences USA 91, 11–17.

322. Epstein, E. (1999) Silicon. Annual Review of Plant Physiology and Plant Mo-lecular Biology 50, 641–664.

323. Estorninos, L.E., Gealy, D.R., and Talbert, R.E. (2002) Growth response of rice (*Oryza sativa*) and red rice (*O. sativa*) in a replacement series study. Weed Technology 16, 401–406.

324. Esumi, T., Tao, R., and Yonemori, K. (2005) Isolation of *LEAFY* and *TERMINAL FLOWER* 1 homologues from six fruit tree species in the subfamily Maloideae of the Rosaceae. Sexual Plant Reproduction 17, 277–287.

325. Facciotti, M.T., Bertain, P.B., and Yuan, L. (1999) Improved stearate phenotype in transgenic canola expressing a modified acyl-acyl carrier protein thioesterase. Nature Biotechnology 17, 593–597.

326. Fahey, J.W., and Anders, J. (1995) Delivery of beneficial clavibacter microorganisms to seeds and plants. U.S. Patent 5,415,672.

327. Fahey, J.W., Dimock, M.B., Tomasino, S.F., Taylor, J.M., and Carlson, P.S. (1991) Genetically engineered endophytes as biocontrol agents: a case study from industry. In Microbial Ecology of Leaves (eds. Andrews, J.H., and Hirano, S.S.), 401–411. Springer-Verlag, New York.

328. Fakhoury, A.M., and Woloshuk, C.P. (2001) Inhibition of growth of *Aspergillus flavus* and fungal alpha-amylases by a lectin-like protein from *Lablab purpureus*. Mo-lecular Plant-Microbe Interactions 14, 955–961.

329. FAOStat. (2005) FAO statistical database. http://faostat.fao.org.

330. Farquhar, T., Wood, J.Z., and van Beem, J. (2000) The kinematics of wheat struck by a wind gust. Journal of Applied Mechanics 67, 496–502.

331. Farquhar, T., Zhou, J., and Wood, W.H. (2002) Competing effects of buckling and anchorage strength on optimal wheat stalk geometry. Journal of Biomechanical Engineering-Transactions of the ASME 124, 441–449.

332. Farzaneh, M., Jazi, F.R., and Motamed, N. (2005) Application of dotblotting for detecting the expression of *p5cs* gene in transgenic olive plantlets. FEBS Journal 272, 547–547.

333. Fauci, A.S. (2005) Race against time. Nature 435, 423–424.

334. Fauteux, F., Remus-Borel, W., Menzies, J.G., and Belanger, R.R. (2005) Silicon and plant disease resistance against pathogenic fungi. FEMS Microbiology Letters 249, 1–6.

335. Favret, E.A. (1962) Contributions of radio-genetics to plant breeding. International Journal of Applied Radiation and Isotopes 13, 445–453.

336. Fedoroff, N., and Brown, N.L. (2004) Mendel in the Kitchen: A Scientist's View of Genetically Modified Food. Joseph Henry Press, Washington, DC.

337. Feldman, M. (2001) The origin of cultivated wheat. In The World Wheat Book (eds. Bonjean, A., and Angus, W.), 3–56. Lavousier Tech & Doc, Paris.

338. Feldman, M., Liu, B., Segal, G., Abbo, S., Levy, A.A., and Vega, J.M. (1997) Rapid elimination of low-copy DNA sequences in polyploid wheat: a possible mechanism for differentiation of homoeologous chromosomes. Genetics 147, 1381–1387.

339. Felix, M.R., Cardoso, J.M.S., Oliveira, S., and Clara, M.I.E. (2005) Viral properties, primary structure and phylogenetic analysis of the coat protein of an Olive latent virus 1 isolate from *Olea europaea* L. Virus Research 108, 195–198.

340. Feng, P.C.C., Baley, G.J., Clinton, W.P., Bunkers, G.J., Alibhai, M.F., Paulitz, T.C., and Kidwell, K.K. (2005) Glyphosate inhibits rust diseases in glyphosate-resis-tant wheat and soybean. Proceedings of the National Academy of Sciences USA 102, 17290–17295.

341. Fermin, G., Inglessis, V., Garboza, C., Rangel, S., Dagert, M., and Gonsalves, D. (2004) Engineered resistance against papaya ringspot virus in Venezuelan transgenic papayas. Plant Disease 88, 516–522.

342. Ferrandiz, C., Liljegren, S.J., and Yanofsky, M.F. (2000) Negative regulation of *SHATTERPROOF* genes by *FRUITFULL* during *Arabidopsis* fruit development. Science 289, 436–438.

343. Ferreira, S.A., Pitz, K.Y., Manshardt, R., Zee, F., Fitch, M., and Gonsalves, D. (2002) Virus coat protein transgenic papaya provides practical control of papaya ringspot virus in Hawaii. Plant Disease 86, 101–105.

344. Ferry, N., Jouanin, L., Ceci, L.R., Mulligan, E.A., Emami, K., Gatehouse, J.A., and Gatehouse, A.M.R. (2005) Impact of oilseed rape expressing the insecticidal serine protease inhibitor, mustard trypsin inhibitor-2 on the beneficial predator *Ptero-stichus madidus*. Molecular Ecology 14, 337–349.

345. Figueira, E.L.Z., Hirooka, E.Y., Mendiola-Olaya, E. and Blanco-Labra, A. (2003) Characterization of a hydrophobic amylase inhibitor from corn (*Zea mays*) seeds with activity against amylase from *Fusarium verticillioides*. Phytopathology 93, 917–922.

346. Firbank, L.G. (2005) Striking a new balance between agricultural production and biodiversity. Annals of Applied Biology 146, 163–175.

347. Fischer, A.J., and Ramirez, A. (1993) Red rice (*Oryza sativa*)—Competition studies for management decisions. International Journal of Pest Management 39, 133–138.

348. Fischer, A.J., Comfort, M.A., Bayer, D.E., and Hill, J.E. (2000) Herbicide-resistant *Echinochloa oryzoides* and *E. phyllopogon* in California *Oryza sativa* fields. Weed Science 48, 225–230.

349. Fischer, A.J., Bayer, D.E., Carriere, M.D., Ateh, C., and Yim, K.-O. (2001) Mechanisms of resistance to bispyribac-sodium in an *Echinochloa phyllopogon* accession. Pesticide Biochemistry and Physiology 68, 156–165.

350. Fitch, M.M.M., Manshardt, R.M., Gonsalves, D., and Slightom, J.L. (1993) Transgenic papaya plants from *Agrobacterium*-mediated transformation of somatic embryos. Plant Cell Reports 12, 245–249.

351. Fitzgerald, H.A., Canlas, P.E., Chern, M.S., and Ronald, P.C. (2005) Alteration of TGA factor activity in rice results in enhanced tolerance to *Xanthomonas oryzae* pv. *oryzae*. Plant Journal 43, 335–347.

352. Flaherty, J.E., and Woloshuk, C.P. (2004) Regulation of fumonisin biosynthesis in *Fusarium verticillioides* by a zinc binuclear cluster-type gene, *ZFR1*. Applied and Environmental Microbiology 70, 2653–2659.

353. Flannery, M.-L., Meade, C., and Mullins, E. (2005) Employing a composite gene-flow index to numerically quantify a crop's potential for gene flow: an Irish perspective. Environmental Biosafety Research 4, 29–43.

354. Flensburg, J., Tangen, A., Prieto, M., Hellman, U., and Wadensten, H. (2005) Chemically-assisted fragmentation combined with multi-dimensional liquid chromatography and matrix-assisted laser desorption/ionization post source decay, matrix-assisted laser desorption/ionization tandem time-of-flight or matrix-assisted laser desorption/ionization tandem mass

spectrometry for improved sequencing of tryptic peptides. European Journal of Mass Spectrometry 11, 169–179.

355. Fliessbach, A., and Mader, P. (2000) Microbial biomass and size-density fractions differ between soils of organic and conventional agricultural systems. Soil Biology and Biochemistry 32, 757–768.

356. Forner, M.D.M.C. (1995) Chemical and cultural-practices for red rice control in rice fields in Ebro Delta (Spain). Crop Protection 14, 405–408.

357. Fox, J.L. (2003) Puzzling industry response to ProdiGene fiasco. Nature Biotechnology 21, 3–4.

358. Franco, O.L., Rigden, D.J., Melo, F.R., Bloch, C. Jr., Silva, C.P., and Grossi de Sa, M.F. (2000) Activity of wheat alpha-amylase inhibitors towards bruchid alpha-amylases and structural explanation of observed specificities. European Journal of Biochemistry 267, 2166–2173.

359. Franco, O.L., dos Santos, R.C., Batista, J.A., Mendes, A.C., de Araujo, M.A., Monnerat, R.G., Grossi-de-Sa, M.F., and de Freitas, S.M. (2003) Effects of black-eyed pea trypsin/chymotrypsin inhibitor on proteolytic activity and on development of *Anthonomus grandis*. Phytochemistry 63, 343–349.

360. Franco, O.L., Melo, F.R., Mendes, P.A., Paes, N.S., Yokoyama, M., Coutinho, M.V., Bloch, C. Jr., and Grossi-de-Sa, M.F. (2005) Characterization of two *Acanthoscelides obtectus* alpha-amylases and their inactivation by wheat inhibitors. Journal of Agricultural and Food Chemistry 53, 1585–1590.

361. Frello, S., Hansen, K.R., Jensen, J., and J rgensen, R.B. (1995) Inheritance of rapeseed (*Brassica napus*)-specific RAPD markers and a transgene in the cross *B. juncea* x (*B. juncea x B. napus)*. Theoretical and Applied Genetics 91, 236–241.

362. Frugoli, J., and Harris, J. (2001) *Medicago truncatula* on the move! Plant Cell 13, 458–463.

363. Fu, Z.W., Zhang, C.X., and Qian, Y.X. (2001) The determination of butachlor resistant *Echinochloa crus-galli* (L.) in paddy rice and study of the resistant mechanism. Proceedings 18th Asian-Pacific Weed Science Society Conference—Beijing, 488–493.

364. Fukuchi-Mizutani, M., Okuhara, H., Fukui, Y., Nakao, M., Katsumoto, Y., Yonekura-Sakakibara, K., Kusumi, T., Hase, T., and Tanaka, Y. (2003) Biochemical and molecular characterization of a novel UDP-glucose: anthocyanin 3 '-O-glucosyltransferase, a key enzyme for blue anthocyanin biosynthesis, from gentian. Plant Physiology 132, 1652–1663.

365. Fukui, Y., Tanaka, Y., Kusumi, T., Iwashita, T., and Nomoto, K. (2003) A rationale for the shift in colour towards blue in transgenic carnation flowers expressing the flavonoid 3',5'-hydroxylase gene. Phytochemistry 63, 15–23.

366. Funatsuki, H., Maruyama-Funatsuki, W., Fujino, K., and Agatsuma, M. (2000) Ripening habit of buckwheat. Crop Science 40, 1103–1108.

367. Funk, T., Wenzel, G., and Schwarz, G. (2006) Outcrossing frequencies and distribution of transgenic oilseed rape (*Brassica napus* L.) in the nearest neighbourhood. European Journal of Agronomy 24, 26–34.

367a. Gadir, W.S.A., and Adam, S.E.I. (2000) Effects of pearl millet (*Pennisetum typhoides*) and fermented and processed fermented millet on Nubian goats. Veterinary and Human Toxicology 42, 133–136.

368. Gaitan, E., Cooksey, R.C., Legan, J., and Lindsay, R.H. (1995) Antithyroid effects in-vivo and in-vitro of vitexin-a C-glucosylflavone in millet. Journal of Clinical Endocrinology and Metabolism 80, 1144–1147.

369. Galan, C., Vazquez, L., Garcia-Mozo, H., and Dominguez, E. (2004) Forecasting olive (*Olea europaea*) crop yield based on pollen emission. Field Crops Research 86, 43–51.

370. Galun, E. (2005) RNA silencing. World Scientific Publishing, Singapore.

371. Galun, E., and Breiman, A. (1997) Transgenic Plants. Imperial College Press, London.

372. Galvano, F., Galofaro, V., and Galvano, G. (1996) Occurrence and stability of aflatoxin M(1) in milk and milk products: A worldwide review. Journal of Food Protection 59, 1079–1090.

373. Gan, J., Yates, S.R., Ohr, H.D., and Sims, J.J. (1998) Production of methyl bromide by terrestrial higher plants. Geophysical Research Letters 25, 3595–3598.

374. Garcia-Lara, S., Bergvinson, D.J., Burt, A.J., Ramputh, A.I., Diaz-Pontones, D.M., and Arnason, J.T. (2004) The role of pericarp cell wall components in maize weevil resistance. Crop Science 44, 1546–1552.

375. Garcia-Torres, L., and Lopez-Granados, F. (1991) Control of broomrape (*Orobanche crenata* Forsk) in broad bean (*Vicia faba* L) with imidazolinones and other herbicides. Weed Research 31, 227–235.

376. Gealy, D.R. (2005) Gene movement between rice (*Oryza sativa*) and weedy rice (*Oryza sativa*): a U.S. temperate rice perspective. In Crop Ferality and Volunteerism (ed. Gressel, J.), 323–354. CRC Press, Boca Raton, FL.

377. Gealy, D.R., Tai, T.H., and Sneller, C.H. (2002) Identification of red rice, rice, and hybrid populations using microsatellite markers. Weed Science 59, 333–339.

378. Gebhardt, Y., Witte, S., Forkmann, G., Lukacin, R., Matern, U., and Martens, S. (2005) Molecular evolution of flavonoid dioxygenases in the family Apiaceae. Phytochemistry 66, 1273–1284.

379. Gelderblom, W.C.A., Kriek, N.P.J., Marasas, W.F.O., and Thiel, P.G. (1991) Toxicity and carcinogenicity of the *Fusarium moniliforme* metabolite, fumonisin-B1, in rats. Carcinogenesis 12, 1247–1251.

380. Gelderblom, W.C.A., Marasas, W.F.O., Lebepe-Mazur, S., Swanevelder, S., Vessey, C.J., and Hall, P.D. (2002) Interaction of fumonisin B-1 and aflatoxin B-1 in a short-term carcinogenesis model in rat liver. Toxicology 171, 161–173.

381. Gelderblom, W.C.A., Jaskiewicz, K., Marasas, W.F.O., Thiel, P.G., Horak, R.M., Vleggaar, R., and N. P. J. Kriek. (1988) Fumonisins-novel mycotoxins with cancer-promoting activity produced by *Fusarium moniliforme*. Applied and Environmental Microbiology 54, 1806–1811.

382. Gelderblom, W.C.A., Snyman, S.D., Abel, S., Lebepe-Mazur, S., Smuts, C.M., Van der Westhuizen, L., Marasas, W.F.O., Victor, T.C., Knasmüller, S., and Huber, W. (1996) Hepatotoxicity and carcinogenicity of the fumonisins in rats. A review regarding mechanistic implications for establishing risk in humans. Advances in Experimental Medicine and Biology 392, 279–296.

383. Geller-Bernstein, C., Arad, G., Keynan, N., Lahoz, C., Cardaba, B., and Waisel, Y. (1996) Hypersensitivity to pollen of *Olea europaea* in Israel. Allergy 51, 356–359.

384. Gepts, P. (2001) Origins of plant agriculture and major crop plants. in Our fragile world: Challenges and opportunities for sustainable development (ed. Tolba, M.K.), 627–637. EOLSS Publishers, Oxford.

385. Gepts, P., and Papa, R. (2003) Possible effects of (trans)gene flow from crops on the genetic diversity from landraces and wild relatives. Environmental Biosafety Research 2, 89–103.

386. Gertz, J.M.J., Vencill, W.K., and Hill, N.S. (1999) Tolerance of transgenic soybean (*Glycine max*) to heat stress. Brighton Conference-Weeds, 835–840.

387. Getahun, H., Lambein, F., and Vanhoorne, M. (2002) Neurolathyrism in Ethiopia: assessment and comparison of knowledge and attitude of health workers and rural inhabitants. Society for Science and Medicine 54, 1513–1524.

388. Giavalisco, P., Nordhoff, E., Kreitler, T., Kloppel, K.D., Lehrach, H., Klose, J., and Gobom, J. (2005) Proteome analysis of Arabidopsis thaliana by two-dimensional gel electrophoresis and matrix-assisted laser desorption/ionisation-time of flight mass spectrometry. Proteomics 5, 1902–1913.

389. Gibb, M.C., Carberry, J.T., and Carter, R.G. (1974) Hydrocyanic acid poisoning of cattle associated with sudan grass. New Zealand Veterinary Journal 22, 127.

390. Giga, D.P., and Mazarura, U.W. (1991) Levels of resistance to the maize weevil, *Sitophilus zeamais* (Motsch) in exotic, local open-pollinated and hybrid maize germplasm. Insect Science and its Application 12, 159–169.

391. Giga, D.P., Mutemerewa, S., Moyo, G., and Neeley, D. (1991) Assessment and control of losses caused by insect pests in small farmers stores in Zimbabwe. Crop Protection 10, 287–292.

392. Gilbertson, L. (2003) Cre-lox recombination: Cre-ative tools for plant biotechnology. Trends in Biotechnology 21, 550–555.

393. Giovannoni, J., DellaPenna, D., Bennett, A.B., and Fischer, R.L. (1989) Expression of chimeric polygalacturonidase gene in transgenic *rin* (ripening inhibitor) tomato fruit results in polyurinide degradation but not fruit softening. Plant Cell 1, 53–63.

394. Gleeson, D., Lelu-Walter, M.A. and Parkinson, M. (2005) Overproduction of proline in transgenic hybrid larch (*Larix* x *leptoeuropaea* (Dengler)) cultures renders them tolerant to cold, salt and frost. Molecular Breeding 15, 21–29.

395. Goklany, I. (2002) The ins and out of organic farming. Science 298, 1889–1890.

396. Goldman, J.J., Hanna, W.W., Fleming, G., and Ozias-Akins, P. (2003) Fertile transgenic pearl millet [*Pennisetum glaucum* (L.) R. Br.] plants recovered through microprojectile bombardment and phosphinothricin selection of apical meristem-, inflorescence-, and immature embryo-derived embryogenic tissues. Plant Cell Reports 21, 999–1009.

397. Gomes, A.D.G., Dias, S.C., Bloch, C., Melo, F.R., Furtado, J.R., Monnerat, R.G., Grossi-de-Sa, M.F., and Franco, O.L. (2005) Toxicity to cotton boll weevil *Anthonomus grandis* of a trypsin inhibitor from chickpea seeds. Comparative Biochemistry and Physiology B-Biochemistry & Molecular Biology 140, 313–319.

398. Gomez, P., Ortuna, A., and Del Rio, J.A. (2004) Ultrastructural changes and diosgenin content in cell suspensions of *Trigonella foenum-graecum* L. by ethylene treatment. Plant Growth Regulation 44, 93–99.

399. Gong, Y.Y., Cardwell, K., Hounsa, A., Egal, S., Turner, P.C., Hall, A.J., and Wild, C.P. (2002) Dietary aflatoxin exposure and impaired growth in young children from Benin and Togo: cross sectional study. British Medical Journal 325, 20–21.

400. Gonzalez, E.M., Villalba, M., Quiralte, J., Batanero, E., Roncal, F., Albar, J.P., and Rodriguez, R. (2006) Analysis of IgE and IgG B-cell immunodominant regions of Ole e 1, the main allergen from olive pollen. Molecular Immunology 43, 570–578.

401. Gould, F. (1991) The evolutionary potential of crop pests. American Scientist 79, 496–507.

402. Grabber, J.H., Panciera, M.T., and Hatfield, R.D. (2002) Chemical composition and enzymatic degradability of xylem and nonxylem walls isolated from alfalfa internodes. Journal of Agricultural and Food Chemistry 50, 2595–2600.

403. Grau, A., Ortega Dueñas, R., Nieto Cabrera, C., and Hermann, M. (eds.) (2003) Mashua (*Tropaeolum tuberosum* Ruíz & Pav.), International Plant Genetic Resources Institute, Rome.

404. Greaves, J.A., Rufener, G.K., Chang, M.T., and Koehler, P.H. (1993) Development of resistance to pursuit herbicide in corn-the IT gene. In Proceedings of the 48th Annual Corn and Sorghum Industry Research Conference (ed. Wilkinson, D.), 104–118. American Seed Trade Association, Washington, DC.

405. Green, A.G., and Marshall, D.R. (1984) Isolation of induced mutants in linseed (*Linum usitatissimum*) having reduced linolenic acid content. Euphytica 33, 321–328.

406. Grenier, A.M., Mbaiguinam, M., and Delobel, B. (1997) Genetical analysis of the ability of the rice weevil *Sitophilus oryzae* (Coleoptera, Curculionidae) to breed on split peas. Heredity 79, 15–23.

407. Gressel, J. (1992) The needs for new herbicide-resistant crops. in Achievements and developments in combating pesticide resistance (eds. Denholm, I., Devonshire, A.L., and Hollomon, D.W.), 283–294. Elsevier, London.

408. Gressel, J. (1992) Indiscriminate use of selectable markers-sowing wild oats? Trends in Biotechnology 10, 382.

409. Gressel, J. (1997) Can herbicide resistant oilseed rapes from commodity shipments potentially introgress with local *Brassica* weeds, endangering agriculture in importing countries? Pest Resistance Management Winter, 2–5.

410. Gressel, J. (1999) Fulfilling needs for transgenic herbicide-resistant rice while preventing and mitigating introgression into weedy rice. In 17th Asian Pacific Weed Science Society Conference 21–38 Asian Pacific Weed Science Society, Bangkok, Thailand.

411. Gressel, J. (1999) Tandem constructs: preventing the rise of superweeds. Trends in Biotechnology 17, 361–366.

412. Gressel, J. (2002) Molecular Biology of Weed Control. Taylor & Francis, London.
413. Gressel, J. (2004) Two choices for agchem/agbiotech industries: the only way to go is down. Outlooks on Pest Management October, 209–210.
414. Gressel, J. (2004) Transgenic mycoherbicides; needs and safety considerations. In Handbook of Fungal Biotechnology (ed. Arora, D.K.), 549–564. Dekker, New-York.
415. Gressel, J. (2005) The challenges of ferality. In Crop Ferality and Volunteerism (ed. Gressel, J.), 1–7. CRC Press, Boca Raton, FL.
416. Gressel, J. (2005) Problems in qualifying and quantifying assumptions in plant protection models: Resultant simulations can be mistaken by a factor of million. Crop Protection 24, 1007–1015.
417. Gressel, J. (ed.) (2005) Crop Ferality and Volunteerism. CRC Press, Boca Raton, FL.
418. Gressel, J., and Kleifeld, Y. (1994) Can wild species become problem weeds because of herbicide resistance? *Brachypodium distachyon:* a case study. Crop Protection 13, 563–566.
419. Gressel, J., and Baltazar, A. (1995) Herbicide resistance in rice: status, causes, and prevention. Iin Weed Management in Rice (eds. Auld, B.A., and Kim, K.U.), 195–238. FAO, Rome.
420. Gressel, J., and Joel, D.M. (1997) Use of glyphosate salts in seed dressing herbicidal compositions. US Patent 6,096,686.
421. Gressel, J., and Levy, A. (2000) Giving *Striga hermonthica* the DT's. In Breeding for *Striga* Resistance in Cereals (eds. Haussmann, B.I.G., Hess, D.E., Koyama, M.L., Grivet, L., Rattunde, H.F.W., and Geiger, H.H.), 207–224. Margraf Verlag, Weikersheim.
422. Gressel, J., and Rotteveel, T. (2000) Genetic and ecological risks from biotechnologically-derived herbicide resistant crops: decision trees for risk assessment. Plant Breeding Reviews 18, 251–303.
423. Gressel, J., and Zilberstein, A. (2003) Let them eat (GM) straw. Trends in Biotechnology 21, 525–530.
424. Gressel, J., and Al-Ahmad, H. (2005) Molecular containment and mitigation of genes within crops: prevention of gene establishment in volunteer offspring and feral strains. In Crop Ferality and Volunteerism (ed. Gressel, J.), 371–388. CRC Press, Boca Raton, FL.
425. Gressel, J., Cohen, N., and Binding, H. (1984) Somatic hybridization of an atrazine-resistant biotype of *Solanum nigrum* with *Solanum tuberosum*. II: Segregation of plastomes. Theoretical and Applied Genetics 67, 131–134.
426. Gressel, J., Segel, L.E., and Ransom, J.K. (1996) Managing the delay of evolution of herbicide resistance in parasitic weed. International Journal of Pest Management 42, 113–129.
427. Gressel, J., Zhang, Z., and Galun, E. (1997) DNA molecules conferring dalapon resistance to plants and plants transformed thereby. Israel Patent Application 122,270.
428. Gressel, J., Regev, Y., Malkin, S., and Kleifeld, Y. (1983) Characterization of a s-triazine resistant biotype of *Brachypodium distachyon*. Weed Science 31, 450–456.
429. Gressel, J., Gianessi, L., Darby, C.P., and Seth, A. (1994) Herbicide-resistant weeds: A threat to wheat production in India. Discussion Paper PS-94-3, National Center for Food and Agricultural Policy, Washington, DC.
430. Gressel, J., Vered, Y., Bar-Lev, S., Milstein, O., and Flowers, H.M. (1983) Partial suppression of cellulase action by artificial lignification of cellulose. Plant Science Letters 32, 349–353.
431. Gressel, J., Hanafi, A., Head, G., Marasas, W., Obilana, B., Ochanda, J., Souissi, T., and Tzotzos, G. (2004) Major heretofore intractable biotic constraints to African food security that may be amenable to novel biotechnological solutions. Crop Protection 23, 661–689.
432. Griffith, M., and Yaish, M.W.F. (2004) Antifreeze proteins in overwintering plants: A tale of two activities. Trends In Plant Science 9, 399–405.
433. Griffiths, S., Sharp, R., Foote, T.N., Bertin, I., Wanous, M., Reader, S., Colas, I., and Moore, G. (2006) Molecular characterization of *Ph*1 as a major chromosome pairing locus in polyploid wheat. Nature 439, 749–752.
434. Grigliatti, T.A., Pfeifer, T.A., and Meister, G.A. (2001) TAC-TICS: transposon-based insect control systems. in Enhancing biocontrol agents and handling risks (eds. Vurro, M. et al.), 201–216. IOS Press, Amsterdam.

435. Groopman, J.D., Scholl, P., and Wang, J.-S. (1996) Epidemiology of human aflatoxin exposures and their relation to liver cancer. in Genetics and cancer susceptibility: implications for risk assesment (eds. Walker, C., Groopman, J., Slaga, T., and Klein-Szanto, A.), 211–222. Wiley-Liss, New York.

436. Grossi de Sa, M.F., Mirkov, T.E., Ishimoto, M., Colucci, G., Bateman, K.S., and Chrispeels, M.J. (1997) Molecular characterization of a bean alpha-amylase inhibitor that inhibits the alpha-amylase of the Mexican bean weevil *Zabrotes subfasciatus*. Planta 203, 295–303.

437. Gu, Q., Ferrandiz, C., Yanofsky, M.F., and Martienssen, R. (1998) The *FRUITFULL* MADS-box gene mediates cell differentiation during *Arabidopsis* fruit development. Development 125, 1509–1517.

438. Gu, X.-Y. Kianian, S.F., and Foley, M.E. (2004) Multiple loci and epistases control genetic variation for seed dormancy in weedy rice (*Oryza sativa*). Genetics 166, 1503–1516.

439. Gueye, M.T., and Delobel, A. (1999) Relative susceptibility of stored pearl millet products and fonio to insect infestation. Journal of Stored Products Research 35, 277–283.

440. Gulden, R.H., Shirtliffe, S.J., and Thomas, A.G. (2003) Harvest losses of canola (*Brassica napus*) cause large seedbank inputs. Weed Science 51, 83–86.

441. Guo, W., Hou, Y.L., Wang, S.G., and Zhu, Y.G. (2005) Effect of silicate on the growth and arsenate uptake by rice (*Oryza sativa* L.) seedlings in solution culture. Plant and Soil 272, 173–181.

442. Gurney, A.L., Press, M.C., and Scholes, J.D. (2002) Can wild relatives of sorghum provide new sources of resistance or tolerance against *Striga* species? Weed Research 42, 317–324.

443. Gurney, A.L., Slate, J., Press, M.C., and Scholes, J.D. (2006) A novel form of resistance in rice to the angiosperm parasite *Striga hermonthica*. New Phytologist 169, 199–208.

444. Gurung, A.B. (2003) Insects-a mistake in God's creation? Yharu farmers' perception and knowledge of insects: A case study of the Gobardiha Village Development Committee, Dang-Deukhuri, Nepal. Agricultural and Human Values 20, 337–370.

445. Guthman, J. (1998) Regulating meaning, appropriating nature: The codification of California organic agriculture. Antipode 30, 135–154.

446. Haapalainen, M.L., Kobets, N., Piruzian, E., and Metzler, M.C. (1998) Integrative vector for stable transformation and expression of a [beta]-1,3-glucanase gene in *Clavibacter xyli* subsp. *cynodontis*. FEMS Microbiology Letters 162, 1.

447. Haas, H., and Streibig, J.C. (1982) Changes in weed distribution patterns as a result of herbicide use and other agronomic factors. In Herbicide Resistance in Plants (eds. LeBaron, H.M., and Gressel, J.), 57–80. Wiley, New York.

448. Hails, R.S., and Morley, K. (2005) Genes invading new populations: a risk assessment perspective. Trends in Ecology and Evolution 20, 245–252.

449. Haimanot, R.T., Kidane, Y., Wuhib, E., Kassina, A., Endeshaw, Y., Alemu, T., and Spencer, P.S. (1993) The epidemiology of lathyrism in north and central Ethiopia. Ethiopean Medical Journal 31, 15–24.

450. Halbrecq, B., Romedenne, P., and Ledent, J.F. (2005) Evolution of flowering, ripening and seed set in buckwheat (*Fagopyrum esculentum* Moench): Quantitative analysis. European Journal of Agronomy 23, 209–224.

451. Hall, L.M., Habibur Rahman, M., Gulden, R.H., and Thomas, A.G. (2005) Volunteer oilseed rape: will herbicide-resistance traits assist ferality? In Crop Ferality and Volunteerism (ed. Gressel, J.), 59–79. CRC Press, Boca Raton, FL.

452. Hall, L.M., Topinka, K., Huffman, J., Davis, L., and Good, A. (2000) Pollen flow between herbicide-resistant *Brassica napus* is the cause of multiple resistant *B. napus* volunteers. Weed Science 48, 688–694.

453. Halpin, C. (2004) Investigating and manipulating lignin biosynthesis in the postgenomic era. Advances in Botanical Research 41, 63–106.

454. Halpin, C., Holt, K., Chojecki, J., Oliver, D., Chabbert, B., Monties, B., Edwards, K., Barakate, A., and Foxon, G.A. (1998) Brown-midrib maize (*bm1*)—a mutation affecting the cinnamyl alcohol dehydrogenase gene. Plant Journal 14, 545–553.

455. Hamamouch, N., Westwood, J.H., Banner, I., Cramer, C.L., Gepstein, S., and Aly, R. (2005) A peptide from insects protects transgenic tobacco from a parasitic weed. Transgenic Research 14, 227–236.

456. Hammer, K. (2003) A paradigm shift in the discipline of plant genetic resources. Genetic Resources and Crop Evolution 50, 3–10.

457. Hammer, K., Heller, J., and Engels, J. (2001) Monographs on underutilized and neglected crops. Genetic Resources and Crop Evolution 48, 3–5.

458. Hammerschmidt, R. (2005) Silicon and plant defense: The evidence continues to mount. Physiological and Molecular Plant Pathology 66, 117–118.

459. Han, Y., Jiang, J.F., Liu, H.L., Ma, Q.B., Xu, W.Z., Xu, Y.Y., Xu, Z.H., and Chong, K. (2005) Overexpression of *OsSIN*, encoding a novel small protein, causes short internodes in *Oryza sativa*. Plant Science 169, 487–495.

460. Hannink, N., Rosser, S.J., French, C.E., Basran, A., Murray, J.A.H., Nicklin, S., and Bruce, N.C. (2001) Phytodetoxification of TNT by transgenic plants expressing a bacterial nitroreductase. Nature Biotechnology 19, 1168–1172.

461. Hanzawa, Y., Money, T., and Bradley, D. (2005) A single amino acid converts a repressor to an activator of flowering. Proceedings of the National Academy of Sciences USA 102, 7748–7753.

462. Haq, S.K., Atif, S.M., and Khan, R.H. (2004) Protein proteinase inhibitor genes in combat against insects, pests, and pathogens: natural and engineered phytoprotection. Archives of Biochemistry and Biophysics 431, 145–159.

463. Haque, A., Hossain, M., Wouters, G., and Lambein, F. (1996) Epidemiological study of lathyrism in northwestern districts of Bangladesh. Neuroepidemiology 15, 83–91.

464. Haque, A., Hossain, M., Lambein, F., and Bell, E.A. (1997) Evidence of osteo-lathyrism among patients suffering from neurolathyrism in Bangladesh. Natural Toxins 5, 43–46.

465. Harberd, N.P., and Freeling, M. (1989) Genetics of dominant gibberellin-insensitive dwarfism in maize. Genetics 121, 827–838.

466. Hardie, D.C., and Clement, S.L. (2001) Development of bioassays to evaluate wild pea germplasm for resistance to pea weevil (Coleoptera: Bruchidae). Crop Protection 20, 517–522.

467. Harker, K.N., and O'Sullivan, P.A. (1991) Effect of imazamethabenz on green foxtail, Tartary buckwheat and wild oat at different growth-stages. Canadian Journal of Plant Science 71, 821–829.

468. Harlan, J.R. (1971) Agricultural origins-centers and noncenters. Science 174, 468–474.

469. Harlan, J.R. (1991) Disseminating agriculture. Science 31, 4.

470. Harlan, J.R. (1992) Origin and processes of domestication. In Grass Evolution and Domestication (ed. Chapman, G.P.), 159–175. Cambridge University Press, Cambridge.

471. Harlan, J.R. (1995) Barley. In Evolution of Crop Plants (eds. Smartt, J., and Simmonds, N.W.), 140–147. Longman, Harlow, United Kingdom.

472. Harlan, J.R., and Macneish, R.S. (1994) The origins of agriculture and settled life. Journal of Interdisciplinary History 24, 517–518.

473. Harper, D.B., and Hamilton, J.T.G. (2003) The global cycles of naturally-occurring monohalomethanes. In Natural Production of Organohalogen Compounds (ed. Gribble, G.), 17–41. Springer, Berlin.

474. Harper, R.J., Almirall, J.R., and Furton, K.G. (2005) Identification of dominant odor chemicals emanating from explosives for use in developing optimal training aid combinations and mimics for canine detection. Talanta 67, 313–327.

475. Harries, H.C. (1995) Coconut. In Evolution of Crop Plants (eds. Smartt, J., and Simmonds, N.W.), 389–398. Longman, Harlow, United Kingdom.

476. Harrison, L.R., Colvin, B.M., Green, J.T., Newman, L.E., and Cole, J.R. (1990) Pulmonary edema and hydrothorax in swine produced by fumonisin B1, a toxic metabolite of *Fusarium moniliforme*. Journal of Veterinary Diagnosis and Investigation 2, 217–221.

477. Hartmann, H.T. (1967) Swan Hill-a new ornamental fruitless olive for California. California Agriculture 21, 4.

478. Hatzopoulos, P., Banilas, G., Gianoulia, K., Gazis, F., Nikoloudakis, N., Milioni, D. and Haralampidis, K. (2002) Breeding, molecular markers and molecular biology of the olive tree. European Journal of Lipid Science and Technology 104, 574–586.

479. Hauck, C., Muller, S., and Schilknecht, H. (1992) A germination stimulant for parasitic flowering plants from *Sorghum bicolor*, a genuine host plant. Journal of Plant Physiology 139, 474–478.

480. Hawkins, D.J., and Kridl, J.C. (1998) Characterization of acyl-ACP thioesterases from mangosteen (*Garcinia mangostana*) seeds and high levels of stearate production in transgenic canola. Plant Journal 13, 743–752.

481. Haygood, R., Ives, A.R., and Andow, D.A. (2003) Consequences of recurrent gene flow from crops to wild relatives. Proceedings of the Royal Society of London Series B 270, 1879–1886.

482. Haygood, R., Ives, A.R., and Andow, D.A. (2004) Population genetics of transgene containment. Ecology Letters 7, 213–220.

483. Heap, I.M. (2006) International survey of herbicide-resistant weeds. www.weedscience.org.

484. Heath-Pagliuso, S., and Rappaport, L. (1990) Somaclonal variant UC-T3-the expression of *Fusarium*-wilt resistance in progeny arrays of celery, *Apium graveolens* L. Theoretical and Applied Genetics 80, 390–394.

485. Hedden, P., and Kamiya, Y. (1997) Gibberellin biosynthesis: Enzymes, genes and their regulation. Annual Review of Plant Physiology and Plant Molecular Biology 48, 431–460.

486. Hegedus, A., Erdei, S., Janda, T., Tóth, E., Horváth, V.G., and Dudits, D. (2004) Transgenic tobacco plants overproducing alfalfa aldose/aldehyde reductase show higher tolerance to low temperature and cadmium stress. Plant Science 166, 1329–1333.

487. Helbeck, H. (1958) Grautall mandens sidste Maltid (Danish with English summary, cited in Holm 1977). Kuml. for Tysk arkaelogisk Silskab (Aarhus), 83–116.

488. Hell, K., Cardwell, K.F., Setamou, M., and Schulthess, F. (2000) Influence of insect infestation on aflatoxin contamination of stored maize in four agroecological regions in Benin. African Entomology 8, 169–177.

489. Hell, K., Cardwell, K.F., Setamou, M., and Poehling, H.M. (2000) The influence of storage practices on aflatoxin contamination in maize in four agroecological zones of Benin, west Africa. Journal of Stored Products Research 36, 365–382.

490. Heller, J., Begemann, F., and Mushonga, J. (eds.) (1997) Bambara groundnut *Vigna subterranea* (L.) Verdc. International Plant Genetic Resources Institute, Rome.

491. Helliwell, C.A., Sheldon, C.C., Olive, M.R., Walker, A.R., Zeevaart, J.A.D., Peacock, W.J., and Dennis, E.S. (1998) Cloning of the *Arabidopsis ent*-kaurene oxidase gene *GA3*. Proceedings of the National Academy of Science USA 95, 9019–9024.

492. Hema, M.V., and Prasad, D.T. (2004) Comparison of the coat protein of a south Indian strain of PRSV with other strains from different geographical locations. Journal of Plant Pathology 86, 35–42.

493. Hendrickse, R.G. (1999) Of sick turkeys, kwashiorkor, malaria, perinatal mortality, heroin addicts and food poisoning: research on the influence of aflatoxins on child health in the tropics. Annals of Tropical Paediatrics 19, 229–235.

494. Henikoff, S., Till, B.J., and Comai, L. (2004) TILLING. Traditional mutagenesis meets functional genomics. Plant Physiology 135, 630–636.

495. Herling, C., Svendsen, U.G., and Schou, C. (1995) Identification of important allergenic proteins in extracts of the granary weevil (*Sitophilus granarius*). Allergy 50, 441–446.

496. Herman, E.M., Helm, R.M., Jung, R., and Kinney, A.J. (2003) Genetic modification removes an immunodominant allergen from soybean. Plant Physiology 132, 36–43.

497. Herms, D.A., and Mattson, W.J. (1992) The dilemma of plants to grow or to defend. Quarterly Reviews of Biology 67, 283–335.

498. Heyes, J.A., and Johnston, J.W. (1998) 1-Methylcyclopropene extends *Cymbidium* orchid vaselife and prevents damaged pollinia from accelerating senescence. New Zealand Journal of Crop and Horticultural Science 26, 319–324.

499. Hidaka, T., Nugaliyadde, L., and Samanmalie, L.G.I. (2000) Integrated rice pest management in Sri Lanka. Agrochemicals Japan 77, 21–29.

500. Hildebrand, M., Higgins, D.R., Busser, K., and Volcani, B.E. (1993) Silicon-responsive cDNA clones isolated from the marine diatom *Cylindrotheca fusiformis*. Gene 132, 213–218.

501. Hildebrand, M., Volcani, B.E., Gassmann, W., and Schoroeder, J.I. (1997) A gene family of silicon transporters. Nature 385, 688–689.

502. Hilu, K.W., M'Ribu, K., Liang, H., and Mandelbaum, C. (1997) Fonio millets: Ethnobotany, genetic diversity and evolution. South African Journal of Botany 63, 185–190.

503. Hinneburg, I., and Neubert, R.H.H. (2005) Influence of extraction parameters on the phytochemical characteristics of extracts from buckwheat (*Fagopyrum esculentum*) Herb. Journal of Agricultural Food Chemistry 53, 3–7.

504. Hiremath, S.P., Badami, S., Hunasagatta, S.K., and Patil, S.B. (2000) Antifertility and hormonal properties of flavones of *Striga orobanchioides*. European Journal of Pharmacology 391, 193–197.

505. Hoeberichts, F.A., Van der Plas, L.H.W., and Woltering, E.J. (2002) Ethylene perception is required for the expression of tomato ripening-related genes and associated physiological changes even at advanced stages of ripening. Postharvest Biology and Technology 26, 125–133.

506. Hole, D.G., Perkins, A.J., Wilson, J.D., Alexander, I.H., Grice, F., and Evans, A.D. (2005) Does organic farming benefit biodiversity? Biological Conservation 122, 113–130.

507. Holm, L., Pancho, J.V., Herberger, J.P., and Plucknett, D.L. (1979) A Geographical Atlas of World Weeds. Wiley, New York.

508. Holm, L., Doll, J., Holm, J., Pancho, E., and Herberger, J. (eds.) (1997) Worlds Weeds: Natural Histories and Distribution. Wiley, New York.

509. Holm, L.G., Plucknett, J.D., Pancho, L.V., and Herberger, J.P. (1977) The World's Worst Weeds, Distribution and Biology. University Press of Hawaii, Honolulu.

510. Kramer, K.J., Morgan, T.D., Throne, J.E., Dowell, F.E., Bailey, M., and Howard, J.A. (2000) Transgenic avidin maize is resistant to storage insect pests. Nature Biotechnology 18, 670–674.

511. Hu, W.-J., Kawaoka, A., Tsai, C.J., Lung, J., Osakabe, K., Ebinuma, H., and Chiang, V.L. (1998) Compartmentalized expression of two structurally and functionally distinct 4-coumarate: CoA ligase genes in aspen (*Populus tremuloides*). Proceedings of the National Academy of Sciences USA 95, 5407–5412.

512. Hu, W.-J., Harding, S.A., Lung, J., Popko, J.L., Ralph, J., Stokke, D.D., Tsai, C.J., and Chiang, V.L. (1999) Repression of lignin biosynthesis promotes cellulose accumulation and growth in transgenic trees. Nature Biotechnology 17, 808–812.

513. Hu, Y.L., Jia, W., Wang, J., Zhang, Y., Yang, L., and Lin, Z. (2005) Transgenic tall fescue containing the *Agrobacterium tumefaciens ipt* gene shows enhanced cold tolerance. Plant Cell Reports 23, 705–709.

514. Huang, B.-Q., and Gressel, J. (1997) Barnyardgrass (*Echinochloa crus-galli*) resistance to both butachlor and thiobencarb in China. Resistant Pest Management 9, 5–7.

515. Huang, G.Z., Allen, R., Davis, E.L., Baum, T.J., and Hussey, R.S. (2006) Engineering broad rootknot resistance in transgenic plants by RNAi silencing of a conserved and essential root-knot nematode parasitism gene. Proceedings of the National Academy of Sciences USA 103, 14302–14306.

516. Hundera, F., Nelson, L.A., Baenziger, P.S., Bechere, E., and Tefera, H. (2000) Association of lodging and some morpho-agronomic traits in tef [*Eragrostis tef* (Zucc.) Trotter]. Tropical Agriculture 77, 169–173.

517. Husken, A., Baumert, A., Milkowski, C., Becker, H.C., Strack, D., and Mollers, C. (2005) Resveratrol glucoside (Piceid) synthesis in seeds of transgenic oilseed rape (*Brassica napus* L.). Theoretical and Applied Genetics 111, 1553–1562.

518. Hyde, M.B., Bakler, A.A., Ross, A.C., and Lopez-Cesar, O. (1973) Airtight grain storage. Food and Agricultural Organization of the United Nations, Rome.

519. Ibeh, I.N., Uraih, N., and Ogonor, J.I. (1991) Dietary exposure to aflatoxin in Benin-City, Nigeria-a possible public-health concern. International Journal of Food Microbiology 14, 171–174.

520. Ikegami, F., Yamamoto, A., Kuo, Y.H., and Lambein, F. (1999) Enzymatic formation of 2,3-diaminopropionic acid, the direct precursor of the neurotoxin beta-ODAP, in *Lathyrus sativus*. Biological Pharmacology Bulletin 22, 770–771.

521. Ikegami, F., Itagaki, S., Ishikawa, T., Ongena, G., Kuo, Y.H., Lambein, F., and Murakoshi, I. (1991) Biosynthesis of beta-(isoxazolin-5-on-2-yl)alanine, the precursor of the neurotoxic amino acid beta-N-oxalyl-L-alpha,beta-diaminopropionic acid. Chemical Pharmacology Bulletin (Tokyo) 39, 3376–3377.

522. Ingram, A.L., and Doyle, J.J. (2003) The origin and evolution of *Eragrostis tef* (Poaceae) and related polyploids: Evidence from nuclear waxy and plastid *rps*16. American Journal of Botany 90, 116–122.

523. Iogen. (2006) Cellulose ethanol brochure. www.iogen.ca/cellulose_ethanol/what_is_ethanol/cellulose_ethanol.pdf.

524. IPGRI. (2003) Fonio: West Africa's treasure. In Annual Report International Plant Genetic Resources Institute (IPGRI), Rome.

525. Iraneta, S.G., Seoane, M.A., Laucella, S.A., Apicella, C., Alonso, A., and Duschak, V.G. (2005) Antigenicity and immunocrossreactivity of orange tree pollen and orange fruit allergenic extracts. International Archives of Allergy and Immunology 137, 265–272.

526. Irie, K., Hosoyama, H., Takeuchi, T., Iwabuchi, K., Watanabe, H., Abe, M., Abe, K., and Arai, S. (1996) Transgenic rice established to express corn cystatin exhibits strong inhibitory activity against insect gut proteinases. Plant Molecular Biology 30, 149–157.

527. Isaac, P.G., Roberts, J.A., and Coupe, S.A. (1999) Control of plant abscission and pod dehiscence. U.S. Patent 5,907,081.

528. Ishida, B.K. (1988) Improved diosgenin production in *Dioscorea deltoidea* cell-cultures by immobilization in polyurethane foam. Plant Cell Reports 7, 270–273.

529. Islam, A.K.M.R., and Powles, S.B. (1991) Attempts to transfer paraquat resistance from barley grass (*Hordeum glaucum* Steud.) to barley and wheat. Weed Research 31, 395–399.

530. Ito, Y., Katsura, K., Maruyama, K., Taji, T., Kobayashi, M., Seki, M., Shinozaki, K., and Yamaguchi-Shinozaki, K. (2006) Functional analysis of rice DREB1/CBF-type transcription factors involved in cold-responsive gene expression in transgenic rice. Plant and Cell Physiology 47, 141–153.

531. Jach, G., Gornhardt, B., Mundy, J., Logemann, J., Pinsdorf, E., Leah, R., Schell, J., and Maas, C. (1995) Enhanced quantitative resistance against fungal disease by combinatorial expression of different barley antifungal proteins in transgenic tobacco. Plant Journal 8, 97–109.

532. Jackson, M.W., Stinchcombe, J.R., Korves, T.M., and Schmitt, J. (2004) Costs and benefits of cold tolerance in transgenic *Arabidopsis thaliana*. Molecular Ecology 13, 3609–3615.

533. Jahan, K., and Ahmad, K. (1993) Studies on neurolathyrism. Environmental Research 60, 259–266.

534. Jain, R.K., Sharma, J., Sivakumar, A.S., Sharma, P.K., Byadgi, A.S., Verma, A.K., and Varma, A. (2004) Variability in the coat protein gene of papaya ringspot virus isolates from multiple locations in India. Archives of Virology 149, 2435–2442.

535. Jamalian, J., and Ghorbani, M. (2005) Extraction of favism-inducing agents from whole seeds of faba bean (*Vicia faba* L var major). Journal of the Science of Food and Agriculture 85, 1055–1060.

536. Jampates, R., and Dvorak, J. (1986) Location of the *Ph*1 locus in the metaphase chromosome map and the linkage map of the 5Bq arm of wheat. Canadian Journal of Genetics and Cytology 28, 511–519.

537. Jang, C.S., Kamps, T.L, Skinner, D.N., Schulze, S.R., Vencill, W.K., and Paterson, A.H. (2006) Functional classification, genomic organization, putatively cis-acting regulatory elements, and relationship to quantitative trait loci, of sorghum genes with rhizome-enriched expression. Plant Physiology 142, 1148–1159.

538. Janick, J., and Whipkey, J., eds. (2002) Trends in New Crops and New Uses. ASHS Press, Alexandria, VA.

539. Jarvis, D.I., and Hodgins, T. (1999) Wild relatives and crop cultivars: detecting natural introgression and farmer selection of new genetic combinations in agro-ecosystems. Molecular Ecology 8 S, 159–173.

540. Jaworski, J., and Cahoon, E.B. (2003) Industrial oils from transgenic plants. Current Opinion in Plant Biology 6, 178–184.

541. Jeebhay, M., Baatjies, R., and Lopata, A. (2005) Work-related allergy associated with sensitisation to storage pests and mites among grain-mill workers. Current Allergy 18, 72–76.

542. Jefferson, P.G., McCaughey, W.P., May, K., Woosaree, J., and McFarlane, L. (2004) Potential utilization of native prairie grasses from western Canada as ethanol feedstock. Canadian Journal of Plant Science 84, 1067–1075.

543. Jennings, D.L. (1995) Cassava. In Evolution of Crop Plants (eds. Smartt, J., and Simmonds, N.W.). Longman, Harlow, United Kingdom.

544. Jensen, C.S., Salchert, K., Gao, C., Andersen, C.H., Didion, T., and Nielsen, K.K. (2004) Floral inhibition in red fescue (*Festuca rubra* L.) through expression of a heterologous flowering repressor from *Lolium*. Molecular Breeding 13, 37–48.

545. Jenszewski, E., Ronfort, J., and Chevre, A.M. (2003) Crop-to-wild gene flow, introgression, and possible fitness effects of transgenes. Environmental Biosafety Research 2, 9–24.

546. Jepson, I. (1997) Inducible herbicide resistance. EU Patent Application WO 97/06269 (EP0843730).

547. Jepson, I., Martinez, A., and Sweetman, J.P. (1998) Chemical-inducible gene expression systems for plants-a review. Pesticide Science 54, 360–367.

548. Jiang, G.H., Xu, C.G., Tu, J.M., Li, Y.X., He, H.Q., and Zhang, Q.F. (2004) Pyramiding of insect-and disease-resistance genes into an elite indica, cytoplasm male sterile restorer line of rice, 'Minghui 63'. Plant Breeding 123, 112–116.

549. Jideani, I.A. (1999) Traditional and possible technological uses of *Digitaria exilis* (acha) and *Digitaria iburua* (iburu): A review. Plant Foods for Human Nutrition 54, 363–374.

550. Joel, D.M., Kleifeld, Y., Losner-Goshen, D., Herzlinger, G., and Gressel, J. (1995) Transgenic crops against parasites. Nature 374, 220–221.

551. Johnson, D., Mortimer, M., Orr, A., and Riches, C. (2003) Weeds, rice, poor people in South Asia. Natural Resources Institute, Chatham, United Kingdom.

552. Johnson, E.T., Yi, H., Shin, B, Oh, BJ, Cheong, H., and Choi ,G. (1999) *Cymbidium hybrida* dihydroflavonol 4-reductase does not efficiently reduce dihydrokaempferol to produce orange pelargonidin-type anthocyanins. Plant Journal 19, 81–85.

553. Johnston, A.M., Tanaka, D.L., Miller, P.R., Brandt, S.A,. Nielsen, D.C., Lafond, G.P., and Riveland, N.R. (2002) Oilseed crops for semiarid cropping systems in the northern Great Plains. Agronomy Journal 94, 231–240.

554. Joshi, B.D., and Paroda, R.S. (1991) Buckwheat in India. National Bureau of Plant Genetic Resources, New Delhi.

555. Jung, H.J.G., and Vogel, K.P. (1992) Lignification of switchgrass (*Panicum virgatum*) and big bluestem (*Andropogon gerardii*) plant-parts during maturation and its effect on fiber degradability. Journal of the Science of Food and Agriculture 59, 169–176.

556. Jørgensen, R.B., and Andersen, B. (1994) Spontaneous hybridization between oilseed rape (*Brassica napus*) and weedy *B. campestris* (Brassicaceae): a risk of growing genetically modified oilseed rape. American Journal of Botany 81, 1620–1626.

557. Kadkol, G.P., Macmillan, R.H., Burrow, R.P., and Halloran, G.M. (1984) Evaluation of *Brassica* genotypes for resistance to shatter. 1. Development of a laboratory test. Euphytica 33, 63–73.

558. Kahn, R.A., Bak, S., Svendsen, I., Halkier, B.A., and Moller, B.L. (1997) Isolation and reconstitution of cytochrome P450 ox and in vitro reconstitution of the entire biosynthetic pathway of the cyanogenic glucoside dhurrin from sorghum. Plant Physiology 115, 1661–1670.

559. Kanampiu, F.K., Ransom, J.K., and Gressel, J. (2001) Imazapyr seed dressings for *Striga* control on acetolactate synthase target-site resistant maize. Crop Protection 20, 885–895.

560. Kanampiu, F.K., Ransom, J.K., Friesen, D., and Gressel, J. (2002) Imazapyr and pyrithiobac movement in soil and from maize seed coats controls *Striga* while allowing legume intercropping. Crop Protection 21, 611–619.

561. Kanampiu, F.K., Kabambe, V., Massawe, C., Jasi, L., Friesen, D., Ransom, J.K., and Gressel, J. (2003) Multi-site, multi-season field tests demonstrate that herbicide seed-coating herbicide-resistance maize controls *Striga* spp. and increases yields in several African countries. Crop Protection 22, 679–706.

562. Kandarakov, O., Titel, C., Volkova, L., Nosov, A., and Ehwald, R. (2000) Additional phosphate stabilises uninterrupted growth of a *Dioscorea deltoidea* cell culture. Plant Science 157, 209–216.

563. Kandel, H.J., Porter, P.M., Johnson, B.L., Henson, R.A., Hanson, B.K., Weisberg, S., and LeGare, D.G. (2004) Plant population influences niger seed yield in the northern great plains. Crop Science 44, 190–197.

564. Kaplinsky, N.J., and Freeling, M. (2003) Combinatorial control of meristem identity in maize inflorescences. Development 130, 1149–1158.

565. Karlovsky, P. (1999) Biological detoxification of fungal toxins and its use in plant breeding, feed and food production. Natural Toxins 7, 1–23.

566. Kaushal, P., and Ravi. (1998) Crossability of wild species of *Oryza* with *O. sativa* cvs PR 106 and Pusa Basmati 1 for transfer of bacterial leaf blight resistance through interspecific hybridization. Journal of Agricultural Science 130, 423.

567. Kawaoka, A., and Ebinuma, H. (2001) Transcriptional control of lignin biosynthesis by tobacco LIM protein. Phytochemistry 57, 1149–1157.

568. Kebebew, A., Gaj, M.D., and Maluszynski, M. (1998) Somatic embryogenesis and plant regeneration in callus culture of tef, *Eragrostis tef* (Zucc.) Trotter. Plant Cell Reports 18, 154–158.

569. Kebebew, F., and McNeilly, T. (1995) Variation in response of accessions of minor millets, *Pennisetum americanum* (L) Leeke (pearl-millet) and *Eleusine coracana* (L) Gaertn (finger-millet), and *Eragrostis tef* (Zucc) Trotter (tef) to salinity in early seedling growth. Plant and Soil 175, 311–321.

570. Kedera, C.J., Plattner, R.D., and Desjardins, A.E. (1999) Incidence of *Fusarium* spp. and levels of fumonisin B-1 in maize in western Kenya. Applied and Environmental Microbiology 65, 41–44.

571. Kellerman, T.S., Marasas, W.F.O., Thiel, P.G., Gelderblom, W.C.A., Cawood, M., and Coetzer, J.A.W. (1990) Leukoencephalomalacia in two horses induced by oral dosing of fumonisin-B1. Onderstepoort Journal of Veterinary Research 57, 269–275.

572. Kelley, M.J., Glaser, E.M., Herndon, J.E. II, Becker, F., Bhagat, R., Zhang, Y.J., Santella, R.M., Carmella, S.G., Hecht, S.S., Gallot, L., Schilder, L., Crowell, J.A., Perloff, M., Folz, R.J., and Bergan, R.C. (2005) Safety and efficacy of weekly oral oltipraz in chronic smokers. Cancer Epidemiology Biomarkers and Prevention 14, 892–899.

573. Kepczynski, J., and Kepczynska, E. (2005) Manipulation of ethylene biosynthesis. Acta Physiologiae Plantarum 27, 213–220.

574. Kermicle, J. (1997) Cross compatibility within the genus *Zea*. in Gene flow among maize landraces, improved maize varieties, and teosinte: Implications for transgenic maize (eds. Serratos, J.A., Willcox, M.C., and Castillo-González, F.), 40–43. CIMMYT, Mexico.

575. Kessler, A. (1958) Sensitivity to olive pollen (*Olea europea*) as the cause for allergic disease. Dapim Refuiim 17, 3.

576. Ketema, S. (1993) Tef (*Eragrostis tef*): breeding, genetic resources, agronomy, utilization and role in Ethiopian agriculture. Institute of Agricultural Research, Addis Ababa, Ethiopia.

577. Ketema, S. (1997) Tef. *Eragrostis tef* (Zucc.) Trotter. International Plant Genetic Resources Institute, Rome.

578. Ketsa, S., and Luangsuwalai, K. (1996) The relationship between 1-aminocyclopropane-1-carboxylic acid content in pollinia, ethylene production and senescence of pollinated *Dendrobium* orchid flowers. Postharvest Biology and Technology 8, 57–64.

579. Ketsa, S., and Rugkong, A. (2000) Ethylene production, senescence and ethylene sensitivity of *Dendrobium* 'Pompadour' flowers following pollination. Journal of Horticultural Science and Biotechnology 75, 149–153.

580. Khan, M.S., and Maliga, P. (1999) Fluorescent antibiotic resistance marker for tracking plastid transformation in higher plants. Nature Biotechnology 17, 910–915.

581. Khan, Z.R., Hassanali, A., Overholt, W., Khamis, T.M., Hooper, A.M., Pickett, J.A., Wadhams, L.J., and Woodcock, C.M. (2002) Control of witchweed *Striga hermonthica* by intercropping with *Desmodium* spp., and the mechanism defined as allelopathic. Journal of Chemical Ecology 28, 1871–1885.

582. Khush, G.S. (1997) Origin, dispersal, cultivation and variation of rice. Plant Molecular Biology 35, 25–34.

583. Kiang, A.-S., Connolly, V., McConnell, D.J., and Kavavagh, T.A. (1994) Paternal inheritance of mitochondria and chloroplasts in *Festuca pratensis-Lolium perenne* intergeneric hybrids. Theoretical and Applied Genetics 87, 681–688.

584. Kim, S.L., Kim, S.K., and Park, C.H. (2004) Introduction and nutritional evaluation of buckwheat sprouts as a new vegetable. Food Research International 37, 319–327.

585. Kim, Y.O., Kim, J.S., and Kang, H. (2005) Cold-inducible zinc finger-containing glycine-rich RNA-binding protein contributes to the enhancement of freezing tolerance in *Arabidopsis thaliana*. Plant Journal 42, 890–900.

586. King, N., Helm, R., Stanley, J.S., Vieths, S., Luttkopf, D., Hatahet, L., Sampson, H., Pons, L., Burks, W., and Bannon, G.A. (2005) Allergenic characteristics of a modified peanut allergen. Molecular Nutrition & Food Research 49, 963–971.

587. Kinney, A.J. (1996) Development of genetically engineered soybean oils for food applications. Journal of Food Lipids 3, 272–292.

588. Kinney, A.J., Knowlton, S., Cahoon, E.B., and Hitz, W.D. (1998) Reengineering oilseed crops to produce industrially useful fatty acids. in Advances in plant lipid research (eds. Sanchez, J., Cerda-Olmedo, E., and Martinez-Force, E.), 623–628. University of Seville Press, Seville, Spain.

589. Kislev, M.E. (1989) Origins of the cultivation of *Lathyrus sativus* and *Lathyrus cicera* (Fabaceae). Economic Botany 43, 262–270.

590. Kitagawa, M., Ito, H., Shiina, T., Nakamura, N., Inakuma, T., Kasumi, T., Ishiguro, Y., Yabe, K., and Ito. Y.(2005) Characterization of tomato fruit ripening and analysis of gene expression in F_1 hybrids of the ripening inhibitor (*rin*) mutant. Physiologia Plantarum 123, 331–338.

591. Klaus, D., Ohlrogge, J.B., Neuhaus, H.E., and Dormann, P. (2004) Increased fatty acid production in potato by engineering of acetyl-CoA carboxylase. Planta 219, 389–396.

592. Klee, H.J. (2001) Control of ethylene-mediated processes in tomato at the level of receptors. Journal of Experimental Botany 53, 2057–2063.

593. Kline, A.D. (1991) We have not yet identified the heart of the moral issues in agricultural biotechnology. Journal of Agricultural and Environmental Ethics 4, 217–222.

594. Kling, J. (1996) Agricultural ecology—Could transgenic supercrops one day breed superweeds? Science 274, 180–181.

595. Kluh, I., Horn, M., Hyblova, J., Hubert, J., Doleckova-Maresova, L., Voburka, Z., Kudlikova, I., Kocourek, F., and Mares, M. (2005) Inhibitory specificity and insecticidal selectivity of alpha-amylase inhibitor from *Phaseolus vulgaris*. Phytochemistry 66, 31–39.

596. Kmiec, E.B., Johnson, C., and May, G.D. (2001) Chloroplast lysates support directed mutagenesis via modified DNA and chimeric RNA/DNA oligonucleotides. Plant Journal 27, 267–274.

597. Knutzon, D.S., Thompson, G.A., Radke, S.E., Johnson, W.B., Knauf, V.C., and Kridl, J.C. (1992) Modification of brassica seed oil by antisense expression of a stearoyl-acyl carrier protein desaturase gene. Proceedings of the National Academy of Sciences USA 89, 2624–2628.

598. Kochevenko, A., and Willmitzer, L. (2003) Chimeric RNA/DNA oligonucleotide-based site-specific modification of the tobacco acetolactate syntase gene. Plant Physiology 132, 174–184.

599. Kohli, A., Twyman, R.M., Abranches, R., Wegel, E., Stoger, E., and Christou, P. (2003) Transgene integration, organization and interaction in plants. Plant Mo-lecular Biology 52, 247–258.

600. Koinange, E.M.K., Singh, S.P., and Gepts, P. (1996) Genetic control of the domestication syndrome in common bean. Crop Science 36, 1037–1045.

601. Kojima, M., Arai, Y., Iwase, N., Shirotori, K., Shioiri, H., and Nozue, M. (2000) Development of a simple and efficient method for transformation of buckwheat plants (*Fagopyrum esculentum*) using *Agrobacterium tumefaciens*. Bioscience Biotechnology and Biochemistry 64, 845–847.

602. Konde, M., Ingenbleek, Y., Daffe, M., Sylla, B., Barry, O., and Diallo, S. (1994) Goitrous endemic in Guinea. Lancet 344, 1675–1678.

603. Konig, A., Cockburn, A., Crevel, R.W.R., Debruyne, E., Grafstroem, R., Hammerling, U., Kimber, I., Knudsen, I., Kuiper, H.A., Peijnenburg, A.A., Penninks, A.H., Poulsen, M., Schauzu, M., and Wal, J.M. (2004) Assessment of the safety of foods derived from genetically modified (GM) crops. Food and Chemical Toxicology 42, 1047–1088.

604. Konishi, S., Lin, S.Y., Fukuta, Y., Izawa, T., Sasaki, T., and Yano, M. (2005) Molecular cloning of a major QTL, *QSH-1*, controlling seed shattering habit in rice. Plant and Cell Physiology 46, S198S198.

605. Konishi, S., Izawa, T., Lin, S.Y., Ebana, K., Fukuta, Y., Sasaki, T., and Yano, M. (2006) An SNP caused loss of seed shattering during rice domestication. Science 312, 1392–1396.

606. Konishi, T., Yasui, Y., and Ohnishi, O. (2005) Original birthplace of cultivated common buckwheat inferred from genetic relationships among cultivated populations and natural populations of wild common buckwheat revealed by AFLP analysis. Genes and Genetic Systems 80, 113–119.

607. Koo, M., Bendahmane, M., Lettieri, G.A., Paoletti, A.D., Lane, T.E., Fitchen, J.H., Buchmeier, M.J., and Beachy, R.N. (1999) Protective immunity against murine hepatitis virus (MHV) induced by intranasal or subcutaneous administration of hybrids of tobacco mosaic virus that carries an MHV epitope. Proceedings of the National Academy of Science USA 96, 7774–7779.

608. Kopeliovitch, E., Mizrahi, Y., Rabinowitch, H.D., and Kedar, N. (1982) Effect of the fruit-ripening mutant-genes *rin* and *nor* on the flavor of tomato fruit. Journal of the American Society for Horticultural Science 107, 361–364.

609. Kowarik, I. (2005) Urban ornamentals escaped from cultivation. In Crop Ferality and Volunteerism (ed. Gressel, J.), 97–121. CRC Press, Boca Raton, FL.

610. Kramer, M., Sanders, R., Bolkan, H., Waters, C., Sheehy, R.E., and Hiatt, R.W. (1992) Postharvest evaluation of transgenic tomatoes with reduced levels of polygalacturonidase: processing, firmness, and disease resistance. Postharvest Biology and Technology 1, 241–255.

611. Kramer, U., Smith, R.D., Wenzel, W.W., Raskin, I., and Sal, t.D.E. (1997) The role of metal transport and tolerance in nickel hyperaccumulation by *Thlaspi goesingense* Halacsy. Plant Physiology 115, 1641–1650.

612. Kresovich, S., Barbazuk, B., Bedell, J., et al. (2005) Toward sequencing the sorghum genome. A U.S. National Science Foundation-sponsored workshop report. Plant Physiology 138, 1898–1902.

613. Kristensen, C., Morant, M., Olsen, C.E., Ekstr m, C.T., Galbraith, D.W., M ller, B.L., and Bak, S. (2005) Metabolic engineering of dhurrin in transgenic *Arabidopsis* plants with marginal inadvertent effects on the metabolome and transcriptome. Proceedings of the National Academy of Sciences USA 102, 1779–1784.

614. Kucerova, Z., and Stejskal, V. (1994) Susceptibility of wheat cultivars to postharvest losses caused by *Sitophilus granarius* (L.) (Coleoptera, Curculionidae). Zeitschrift fur Pflanzenkrankheiten und Pflanzenschutz 101, 641–648.

615. Kuehnle, A.R., and Sugii, N. (1992) Transformation of *Dendrobium* orchid using particle bombardment of protocorms. Plant Cell Reports 11, 484–488.

616. Kuhn, E. (2001) From library screening to microarray technology: Strategies to determine gene expression profiles and to identify differentially regulated genes in plants. Annals of Botany 87, 139–155.

617. Kuiken, T., Fouchier, R., Rimmelzwaan, G., Osterhaus, A., and Roeder, P. (2006) Feline friend or potential foe? Nature 440, 741–742.

618. Kuo, Y.H., Khan, J.K., and Lambein, F. (1994) Biosynthesis of the neurotoxin beta-ODAP in developing pods of *Lathyrus sativus*. Phytochemistry 35, 911–913.

619. Kusaba, S., Kano-Murakami, Y., Matsuoka, M., Tamaoki, M., Sakamoto, T., Yamaguchi, I., and Fukumoto, M. (1998) Alteration of hormone levels in transgenic tobacco plants overexpressing the rice homeobox gene *OSH1*. Plant Physiology 116, 471–476.

620. Kuta, D.D., Kwon-Ndung, E., Dachi, S., Ukwungwu, M., and Imolehin, E.D. (2003) Potential role of biotechnology tools for genetic improvement of "lost crops of Africa": The case of fonio (*Digitaria exilis* and *Digitaria iburua*). African Journal of Biotechnology 2, 580–585.

621. Kuta, D.D., Kwon-Ndung, E., Dachi, S., Bakare, O., and Ogunkanmi, L.A. (2005) Optimization of protocols for DNA extraction and RAPD analysis in West African fonio (*Digitaria exilis* and *Digitaria iburua*) germplasm characterization. African Journal of Biotechnology 4, 1368–1371.

622. Kuvshinov, V., Anissimov, A., and Yahya, B.M. (2004) Barnase gene inserted in the intron of GUS-a model for controlling transgene flow in host plants. Plant Science 167, 173–182.

623. Kuvshinov, V., Koivu, K., Kanerva, A., and Pehu, E. (2001) Molecular control of transgene escape from genetically modified plants. Plant Science 160, 517–522.

624. Kwon, S.L., Smith, R.J., and Talbert, R.E. (1991) Interference durations of red rice (*Oryza sativa*) in rice (*Oryza sativa*). Weed Science 39, 363–368.

625. Kwon, S.L., Smith, R.J., and Talbert, R.E. (1992) Comparative growth and development of red rice (*Oryza sativa*) and rice (*O. sativa*). Weed Science 40, 57–62.

626. Kwon, Y.S., Kim, K.M., Eun, M.Y., and Sohn, J.K. (2001) Quantitative trait loci mapping associated with plant regeneration ability from seed derived calli in rice (*Oryza sativa* L.). Molecules and Cells 11, 64–67.

627. Kühnel, B., Alcantara, J., Boothe, J., van Rooijen, G., and Moloney, M.M. (2003) Precise and efficient cleavage of recombinant fusion proteins using mammalian aspartic proteases. Protein Engineering 16, 777–783.

628. Ladizinsky, G. (1998) Plant Evolution under Domestication. Kluwer Academic Publishers, New York.

629. Lam, T.B.T., Iiyama, K., and Stone, B.A. (1996) Lignin and hydroxycinnamic acids in walls of brown midrib mutants of sorghum, pearl millet and maize stems. Journal of the Science of Food and Agriculture 71, 174–178.

630. Lambein, F., Haque, R., Khan, J.K., Kebede, N., and Kuo, Y.H. (1994) From soil to brain: zinc deficiency increases the neurotoxicity of *Lathyrus sativus* and may affect the susceptibility for the motorneurone disease neurolathyrism. Toxicon 32, 461–466.

631. Lancaster, M.C., Jenkins, F.P., and Philip, J.M. (1961) Toxicity associated with certain samples of groundnuts. Nature 192, 1095–1096.

632. Lange, T. (1998) Molecular biology of gibberellin synthesis. Planta 204, 409–419.

633. Langevin, S.A., Clay, K., and Grace, J.B. (1990) The incidence and effects of hybridization between cultivated rice and its related weed red rice (*Oryza sativa* L.). Evolution 44, 1000–1008.

634. Larkin, V.D. (1953) Favism-report of a case and brief review of the literature. Journal of Pediatrics 42, 453–456.

635. Latham, J.R., Wilson, A.K., and Steinbrecher, R.A. (2006) The mutational consequences of plant transformation. Journal of Biomedicine and Biotechnology 2, Article ID 25376.

636. Leah, J.M., Caseley, J.C., Riches, C.R., and Valverde, B. (1994) Association between elevated activity of aryl-acylamidase and propanil resistance in jungle rice *Echinochloa colona*. Pesticide Science 42, 281–289.

637. Lee, D.K., and Boe, A. (2005) Biomass production of switchgrass in central South Dakota. Crop Science 45, 2583–2590.

638. Lefol, E., Danielou, V., and Darmency, H. (1996) Predicting hybridization between transgenic oilseed rape and wild mustard. Field Crops Research 45, 153–161.

639. Lefol, E., Fleury, A., and Darmency, H. (1996) Gene dispersal from transgenic crops.2. Hybridization between oilseed rape and the wild heavy mustard. Sexual Plant Reproduction 9, 189–196.

640. Lentini, Z., and Mercedes Espinoza, A. (2005) Coexistence of weedy rice and rice in tropical America: gene flow and genetic diversity. In Crop Ferality and Volunteerism (ed. Gressel, J.), 305–322. CRC Press, Boca Raton, FL.

641. Levings, C.S.I., and Siedow, J.N. (1992) Molecular basis of disease susceptibility in the Texas cytoplasm of maize. Plant Molecular Biology 19, 135–147.

642. Levy, I., Shani, Z., and O., S. (2002) Modification of polysaccharides and plant cell wall by endo-1,4-α-glucanase and cellulose-binding domains. Biomolecular Engineering 19, 17–30.

643. Lewis, M.W. (1992) Green delusions: An environmentalist critique of radical environmentalism. Duke Univ. Press, Durham, NC.

644. Li, C., Zhou, A., and Sang, T. (2006) Rice domestication by reduced shattering. Science 311, 1936–1939.

645. Li, C., Zhou, A. and Sang, T. (2006) Genetic analysis of rice domestication syndrome with the wild annual species, Oryza nivara. New Phytologist 170, 185–194.

646. Li, H.Q., and Li, M.R. (2004) RecQ helicase enhances homologous recombination in plants. FEBS Letters 574, 151–155.

647. Li, L., Zhou, Y., Cheng, X.F., Sun, J., Marita, J.M., Ralph, J., and Chiang, V.L. (2003) Combinatorial modification of multiple lignin traits in trees through multigene cotransformation. Proceedings of the National Academy of Sciences USA 100, 4939–4944.

648. Li, Q., Robson, P.R.H., Bettany, A.J.E., Donnison, I.S., Thomas, H., and Scott, I.M. (2004) Modification of senescence in ryegrass transformed with IPT under the control of a monocot senescence-enhanced promoter. Plant Cell Reports 22, 816–821.

649. Li, S.Q., and Zhang, Q.H. (2001) Advances in the development of functional foods from buckwheat. Critical Reviews in Food Science and Nutrition 41, 451–464.

650. Li, W., and Gill, B.S. (2006) Multiple genetic pathways for seed shattering in the grasses. Functional and integrative Genomics 6, 300–309.

651. Li, Z., Zhao, L., Cui, C., Kai, G., Zhang, L., Sun, X. and Tang, K. (2005) Molecular cloning and characterization of an anti-bolting related gene (BrpFLC) from Brassica rapa ssp. Pekinensis. Plant Science 168, 407–413.

652. Liljegren, S.J., Ditta, G.S., Eshed, Y., Savidge, B., Bowman, J.L., and Yanofsky, M.F. (2000) SHATTERPROOF MADS-box genes control seed dispersal in Arabidopsis. Nature 404, 766–770.

653. Liljegren, S.J., Roeder, A.H.K., Kempin, S.A., Gremski, K., Ostergaard, L., Guimil, S., Reyes, D.K., and Yanofsky, M.F. (2004) Control of fruit patterning in Arabidopsis by INDEHISCENT. Cell 116, 843–853.

654. Linder, C.R., Taha, I., Seiler, G.J., Snow, A.A., and Rieseberg, L.H. (1998) Long-term introgression of crop genes into wild sunflower populations. Theoretical and Applied Genetics 96, 339–347.

655. Lines, R.E., Persley, D., Dale, J.L., Drew, R., and Bateson, M.F. (2002) Genetically engineered immunity to papaya ringspot virus in Australian papaya cultivars. Molecular Breeding 10, 119–129.

656. Ling-Hwa, T., and Morishima, H. (1997) Genetic characterization of weedy rices and the inference on their origins. Breeding Science 47, 153–160.

657. Linke, K.H., Abd El-Monein, A.M., and Saxena, M.C. (1993) Variation in resistance of some forage legumes species to Orobanche crenata Forsk. Field Crops Research 32, 277–285.

658. Liu, Q., Singh, S.P., and Green, A.G. (2002) High-stearic and high-oleic cottonseed oils produced by hairpin RNA-mediated post-transcriptional gene silencing. Plant Physiology 129, 1732–1743.

659. Llewellyn, D.J. (2000) Herbicide tolerant forest trees. in Molecular biology of woody plants (eds. Jain, S.M., and Minocha, S.C.), 439–466. Kluwer, Dordrecht, The Netherlands.

660. Loc, N.T., Tinjuangjun, P., Gatehouse, A.M.R., Christou, P., and Gatehouse, J.A. (2002) Linear transgene constructs lacking vector backbone sequences generate transgenic rice plants which accumulate higher levels of proteins conferring insect resistance. Molecular Breeding 9, 231–244.

661. Loret, E., and Hammock, B.D. (2001) Structure and neurotoxicity of venoms. In Scorpion Biology and Research (eds. Brownell, P., and Polis, G.), 204–233. Oxford University Press, New York.

662. Lubulwa, A.S.G., and Davis, J.S. (1995) Estimating the social costs of the impacts of fungi and aflatoxins in maize and peanuts. In Stored-Product Protection. (eds. Highley, E., Wright, E.J., Banks, H.J., and Champ, B.R.),1017–1042. CAB International, Wallingford, United Kingdom.

663. Lucas, J.S., Lewis, S.A., and Hourihane, J.O. (2003) Kiwi fruit allergy: A review. Pediatric Allergy and Immunology 14, 420–428.

664. Lunn, J.A., and Hughes, D.T.D. (1967) Pulmonary hypersensitivity to the grain weevil. British Journal of Industrial Medicine 24, 158–161.

665. Lutz, W., Sanderson, W., and Scherbov, S. (2001) The end of world population growth. Nature 412, 543–545.

666. Lynch, J.M., Benedetti, A., Insam, H., Nuti, M.P., Smalla, K., Torsvik, V., and Nannipieri, P. (2004) Microbial diversity in soil: ecological theories, the contribution of molecular techniques and the impact of transgenic plants and transgenic microorganisms. Biology and Fertility of Soils 40, 363–385.

667. Ma, J.F. (2004) Role of silicon in enhancing the resistance of plants to biotic and abiotic stresses. Soil Science and Plant Nutrition 50, 11–18.

668. Ma, J.F., Hiradate, S., and Matsumoto, H. (1998) High aluminum resistance in buckwheat. II. Oxalic acid detoxifies aluminum internally. Plant Physiology 117, 753–759.

669. Ma, J.F., Tamai, K., Ichii, M., and Wu, G.F. (2002) A rice mutant defective in Si uptake. Plant Physiology 130, 2111–2117.

670. Ma, J.F., Mitani, N., Nagao, S., Konishi, S., Tamai, K., Iwashita, T., and Yano, M. (2004) Characterization of the silicon uptake system and molecular mapping of the silicon transporter gene in rice. Plant Physiology 136, 3284–3289.

671. Ma, J.F., Tamai, K., Yamaji, N., Mitani, N., Konishi, S., Katsuhara, M., Ishiguro, M., Murata, Y., and Yano, M. (2006) A silicon transporter in rice. Nature 440, 688–691.

672. Machmuller, A. (2006) Medium-chain fatty acids and their potential to reduce methanogenesis in domestic ruminants. Agriculture, Ecosystems and the Environment 112, 107–114.

673. Machuka, J.S., Okeola, O.G., Chrispeels, M.J., and Jackai, L.E.N. (2000) The African yambean seed lectin affects the development of the cowpea weevil but does not affect the development of larvae of the legume pod borer. Phytochemistry 53, 667–674.

674. MacKay, J.J., O'Malley, D.M., Presnell, T., Booker, F.L., Campbell, M.M., Whetten, R.W.. and Sederoff, R.R. (1997) Inheritance, gene expression, and lignin characterization in a mutant pine deficient in cinnamyl alcohol dehydrogenase. Proceedings of the National Academy of Science USA 94, 8255–8260.

675. Mader, P., Fliessbach, A., Dubois, D., Gunst, L., Fried, P., and Niggli, U. (2002) The ins and outs of organic farming (authors' response). Science 298, 1889–1890.

676. Madsen, K.H., Valverde, B.E., and Jensen, J.E. (2002) Risk assessment of herbicide-resistant crops: A Latin American perspective using rice (*Oryza sativa*) as a model. Weed Technology 16, 215–223.

677. Madsen, K.H., Holm, P.B., Lassen, J., and Sandoe, P. (2002) Ranking genetically modified plants according to familiarity. Journal of Agricultural and Environmental Ethics 15, 267–278.

678. Mahon, R.E., Bateson, M.F., Chamberlain, D.A., Higgins, C.M., Drew, R.A., and Dale,. J.L. (1996) Transformation of an Australian variety of *Carica papaya* using microprojectile bombardment. Australian Journal of Plant Physiology 23, 679–685.

679. Mairesse, M., and Ledent, C. (2003) Nocturnal asthma due to buckwheat. Revue Francaise D Allergologie et d Immunologie Clinique 43, 527–529.

680. Maisadour. (2007) www.maisadour-semences.com/sunflower-hybrids-catalogue.htm.

681. Majumder, N.D., Ram, T., and Sharma, A.C. (1997) Cytological and morphological variation in hybrid swarms and introgressed population of interspecific hybrids (*Oryza rufipogon* Griff. x *Oryza sativa* L.) and its impact on evolution of intermediate types. Euphytica 94, 295–302.

682. Maliga, P. (2004) Plastid transformation in higher plants. Annual Review of Plant Biology 55, 289–313.

683. Malik, R.K., and Singh, S. (1995) Littleseed canarygrass (*Phalaris minor*) resistance to isoproturon in India. Weed Technology 9, 419–425.

684. Mallikarjuna, N., Kranthi, K.R., Jadhav, D.R., Kranthi, S., and Chandra, S. (2004) Influence of foliar chemical compounds on the development of *Spodoptera litura* (Fab.) in interspecific derivatives of groundnut. Journal of Applied Entomology 128, 321–328.

685. Mamo, T., Richter, C., and Hoppenstedt, A. (1996) Phosphorus response studies on some varieties of durum wheat (*Triticum durum* Desf) and tef (*Eragrostis tef* (Zucc) Trotter) grown in sand culture. Journal of Agronomy and Crop Science 176, 189–197.

686. Maoka, T., and Hataya, T. (2005) The complete nucleotide sequence and biotype variability of Papaya Leaf Distortion Mosaic virus. Phytopathology 95, 128–135.

687. Marasas, W.F.O. (2001) Discovery and occurrence of the fumonisins: A historical perspective. Environmental Health Perspectives 109, 239–243.

688. Marasas, W.F.O. (2001) *Fusarium*. In Foodborne Disease Handbook (ed. Hui, Y.H.), 535–580. Marcel Dekker, New York.

689. Marchais, L. (1994) Wild pearl millet population (*Pennisetum glaucum*, Poaceae) integrity in agricultural Sahelian areas. An example from Keita (Niger). Plant Systematics and Evolution 189, 233–245.

690. Mariam, A.L., Zakri, A.H., Mahani, M.C., and Normah, M.N. (1996) Interspecific hybridization of cultivated rice, *Oryza sativa* L. with the wild rice, *O. minuta* Presl. Theoretical and Applied Genetics 93, 664–671.

691. Mariani, C., Debeuckeleer, M., Truettner, J., Leemans, J., and Goldberg, R.B. (1990) Induction of male sterility in plants by a chimeric ribonuclease gene. Nature 347, 737–741.

692. Marillonet, S., Thoeringer, C., Kandzia, R., Klimyuk, V., and Gleba, Y. (2005) Systemic *Agrobacterium tumefaciens*-mediated transfection of viral replicons for efficient transient expression in plants. Nature Biotechnology 25, 718–723.

693. Marsaro, A.L., Lazzari, S.M.N., Figueira, E.L.Z., and Hirooka, E.Y. (2005) Amylase inhibitors in corn hybrids as a resistance factor to *Sitophilus zeamais* (Coleoptera: Curculionidae). Neotropical Entomology 34, 443–450.

694. Martens, S., and Mithofer, A. (2005) Flavones and flavone synthases. Phytochemistry 66, 2399–2407.

695. Martens, S., Forkmann, G., Britsch, L., Wellmann, F., Matern, U., and Lukacin, R. (2003) Divergent evolution of flavonoid 2-oxoglutarate-dependent dioxygenases in parsley. FEBS Letters 544, 93–98.

696. Martin, A.C., Zim, H.S., and Nelson, A.L. (1961) American Wildlife Plants. Dover Publications, New York.

697. Martineau, B. (2001) First Fruit: The Creation of the Flavr Savr™ Tomato, McGraw-Hill, New York.

698. Matsui, K., Tetsuka, T., and Hara, T. (2003) Two independent gene loci controlling non-brittle pedicels in buckwheat. Euphytica 134, 203–208.

699. Matsui, K., Nishio, T., and Tetsuka, T. (2004) Genes outside the S supergene suppress S Functions in buckwheat (*Fagopyrum esculentum*). Annals of Botany 94, 805–809.

700. Matsui, K., Kiryu, Y., Komatsuda, T., Kurauchi, N., Ohtani, T., and Tetsuka, T. (2004) Identification of AFLP makers linked to non-seed shattering locus (*sht*1) in buckwheat and conversion to STS markers for marker-assisted selection. Genome 47, 469–474.

701. Mattana, M., Biazzi, E., Consonni, R,. Locatelli, F., Vannini, C., Provera, S., and Coraggio, I. (2005) Overexpression of *Osmyb*4 enhances compatible solute accumulation and increases stress tolerance of *Arabidopsis thaliana*. Physiologia Plantarum 125, 212–223.

702. Matusova, R., Rani, K., Verstappen, F.W.A., Franssen, M.C.R., Beale, M.H., and Bouwmeester, H.J. (2005) The strigolactone germination stimulants of the plant-parasitic *Striga* and *Orobanche* spp. are derived from the carotenoid pathway. Plant Physiology 139, 920–934.

703. Mayer, A.M. (2006) Pathogenesis by fungi and parasitic plants: Similarities and differences. Phytoparasitica 34, 3–16.

704. McCloskey, M.C., Firbank, L.G., Watkinson, A.R., and Webb, D.J. (1998) Interactions between weeds of winter wheat under different fertilizer, cultivation and weed management treatments. Weed Research 38, 11–24.

705. McFarlane, J.A. (1988) Pest-management strategies for Prostephanus truncatus (Horn) (Coleoptera, Bostrichidae) as a pest of stored maize grain—Present status and prospects. Tropical Pest Management 34, 121–132.

706. McGlasson, W.B., Last, J.H., Shaw, K.J., and Meldrum, S.K. (1987) Influence of the non-ripening mutants rin and nor on the aroma of tomato fruit. Hortscience 22, 632–634.

707. McGlynn, K.A., Rosvold, E.A., Lustbader, E.D., Hu, Y., Clapper, M.L., Zhou, T., Wild, C.P., Xia, X.L., Baffoe-Bonnie, A., Ofori-Adjei, D., Chen, G.C., London, W.T., Shen, F.M., and Buetow, K.H. (1995) Susceptibility to hepatocellular carcinoma is associated with genetic variation in the enzymatic detoxification of aflatoxin B_1. Proceedings of the National Academy of Sciences USA 92, 2384–2387.

708. McKenzie, M.J., Jameson, P.E., and Poulter, R.T.M. (1994) Cloning an ipt gene from Agrobacterium tumefaciens: characterization of cytokinins in derivative transgenic plant tissue. Plant Growth Regulation 14, 217–228.

709. McKevith, B. (2005) Nutritional aspects of oilseeds. Nutrition Bulletin 30, 13–26.

710. McLaughlin, S.B., and Kszos, L.A. (2005) Development of switchgrass (Panicum virgatum) as a bioenergy feedstock in the United States. Biomass and Bioenergy 28, 515–535.

711. McPartlan, H.C., and Dale, P.J. (1994) An assessment of gene transfer by pollen from field-grown transgenic potatoes to non-transgenic potatoes and related species. Transgenic Research 3, 216–225.

712. McPherson, M.A., Good, A., and Hall, L.M. (2005) Safflower-ferality in a plant-made pharmaceutical platform. in Crop ferality and volunteerism (ed. Gressel, J.), 242–247. CRC Press, Boca Raton, FL.

713. McPherson, M.A., Good, A.G., Topinka, A.K.C., and Hall, L.M. (2004) Theoretical hybridization potential of transgenic safflower (Carthamus tinctorius L.) with weedy relatives in the New World. Canadian Journal of Plant Science 84, 923–934.

714. Meakin, P.J., and Roberts, J.A. (1990) Dehiscence of fruit in oilseed rape (Brassica napus L.).1. anatomy of pod dehiscence. Journal of Experimental Botany 41, 995–1002.

715. Mechin, V., Argillier, O., Rocher, F., Hebert, Y., Mila, I., Pollet, B., Barriere, Y., and Lapierre, C. (2005) In search of a maize ideotype for cell wall enzymatic degradability using histological and biochemical lignin characterization. Journal of Agricultural and Food Chemistry 53, 5872–5881.

716. Meier, C. (2005) Reporter system for plants. New Zealand Patent NZ536874.

717. Meier, C., Bouquin, T., Nielsen, M.E., Raventos, D., Mattsson, O., Rocher, A., Schomburg, F., Amasino, R.M., and Mundy, J. (2001) Gibberellin response mutants identified by luciferase imaging. Plant Journal 25, 509–519.

718. Mekbib, F., Mantell, S.H., and Buchanan Wollaston, V. (1997) Callus induction and in vitro regeneration of tef [Eragrostis tef (Zucc.) Trotter] from leaf. Journal of Plant Physiology 151, 368–372.

719. Melo, F.R., Sales, M.P., Pereira, L.S., Bloch, C. Jr, Franco, O.L., amd Ary,. M.B.. (1999) alpha-Amylase inhibitors from cowpea seeds. Protein and Peptide Letters 6, 385–390.

720. Mencuccini, M., Micheli, M., Angiolillo, A., and Baldoni, L. (1999) Genetic transformation of olive (Olea europaea L.) using Agrobacterium tumefaciens. Acta Horticulturae 474, 515–520.

721. Mendez, J., Comtois, P., and Iglesias, I. (2005) Betula pollen: One of the most important aeroallergens in Ourense, Spain. Aerobiological studies from 1993 to 2000. Aerobiologia 21, 115–123.

722. Mensink, R.P., Zock, P.L., Kester, A.D.M., and Katan, M.M. (2003) Effects of dietary fatty acids and carbohydrates on the ratio of serum total to HDL cholesterol and on serum lipids and

apolipoproteins: a meta-analysis of 60 controlled trials. American Journal of Clinical Nutrition 77, 1146–1155.

723. Menzel, G., Apel, K., and Melzer, S. (1996) Identification of two MADS box genes that are expressed in the apical meristem of the long-day plant *Sinapis alba* in transition to flowering. Plant Journal 9, 399–408.

724. Mercuri, A., Sacchetti, A., De Benedetti, L., Schiva, T., and Alberti, S. (2001) Green fluorescent flowers. Plant Science 161, 961–968.

725. Metcalf, R.L. (1986) Coevolutionary adaptations of rootworm beetles to cucurbitacins. Journal of Chemical Ecology 12, 1109–1124

726. Michiyama, H., Arikuni, M., Hirano, T., and Hayashi, H. (2003) Influence of day length before and after the start of anthesis on the growth, flowering and seed-setting in common buckwheat (*Fagopyrum esculentum* Moench). Plant Production Science 6, 235–242.

727. Mignouna, H.D., Njukeng, P., Abang, M.M., and Asiedu, R. (2001) Inheritance of resistance to Yam mosaic virus, genus Potyvirus, in white yam (*Dioscorea rotundata*). Theoretical and Applied Genetics 103, 1196–1200.

728. Millward, D.J. (1999) The nutritional value of plant-based diets in relation to human amino acid and protein requirements. Proceedings of the Nutrition Society 58, 249–260.

729. Milstein, O., Bechar, A., Shragina, L,. and Gressel, J. (1987) Solar pasteurization of straw for nutritional upgrading and as a substrate for ligninolytic organisms. Biotechnology Letters 9, 269–274.

730. Milstein, O., Vered, Y., Sharma, A., Gressel, J., and Flowers, H.M. (1986) Heat and microbial treatments for nutritional upgrading of wheat straw. Biotechnology and Bioengineering 28, 381–386.

731. Minami, M., Ujihara, A., and Campbell, C.G. (1999) Morphology and inheritance of dwarfism in common buckwheat line, G410, and its stability under different growth conditions. Breeding Science 49, 27–32.

732. Miquel, M.F., and Browse, J. (1994) High-oleate oilseeds fail to develop at low temperature. Plant Physiology 106, 421–427.

733. Miron, J., Zuckerman, E., Sadeh, D., Adin, G., Nikbachat, M., Yosef, E., Ben-Ghedalia, D., Carmi, A., Kipnis, T. and Solomon, R. (2005) Yield, composition and in vitro digestibility of new forage sorghum varieties and their ensilage characteristics. Animal Feed Science and Technology 120, 17–32.

734. Mishnah. (post-Biblical) Trumot (donations). 7, 11.

735. Missaoui, A.M., Paterson, A.H., and Bouton, J.H. (2005) Investigation of genomic organization in switchgrass (*Panicum virgatum* L.) using DNA markers. Theoretical and Applied Genetics 110, 1372–1383.

736. Miura, R., and Terauchi, R. (2005) Genetic control of weediness traits and the maintenance of sympatric crop-weed polymorphism in pearl millet (*Pennisetum glaucum*). Molecular Ecology 14, 1251–1261.

737. Moffett, L. (1991) Pignut tubers from a bronze-age cremation at Barrow Hills, Oxfordshire, and the importance of vegetable tubers in the prehistoric period. Journal of Archaeological Science 18, 187–191.

738. Mol, J., Cornish, E., Mason, J., and Koes, R. (1999) Novel coloured flowers. Current Opinion in Biotechnology 10, 198–201.

739. Moloney, M.M., Walker, J.M., and Sharma, K.K. (1989) High efficiency transformation of *Brassica napus* using *Agrobacterium* vectors. Plant Cell Reports 8, 238–242.

740. Mondal, K.A.M.S.H., and Parween, S. (2000) Insect growth regulators and their potential in the management of storage-product insect pests. Integrated Pest Management Reviews 5, 255–295.

741. Mongkolporn, O., Kadkol, G.P., Pang, E.C.K., and Taylor, P.W.J. (2003) Identification of RAPD markers linked to recessive genes conferring siliqua shatter resis-tance in *Brassica rapa*. Plant Breeding 122, 479–484.

742. Moody, M.E., and Mack, R.N. (1988) Controlling the spread of plant invasions: the importance of nascent foci. Journal of Applied Ecology 25, 1009–1021.

743. Mooney, C. (2005) The Republican War on Science. Basic Books, New York.

744. Moore, S., Vrebalov, J., Payton, P., and Giovannoni, J. (2002) Use of genomics tools to isolate key ripening genes and analyse fruit maturation in tomato. Journal of Experimental Botany 53, 2023–2030.

745. Morales-Payan, P.J., Ortiz, J.R., Cicero, J., and Taveras, F. (2002) *Digitaria exilis* as a crop in the Dominican Republic. In Trends in New Crops and New Uses (eds. Janick, J. and Whipkey, A.), S1S3. ASHS Press, Alexandria, VA.

746. Moreno, A.B., Penas, G., Rufat, M., Bravo, J.M., Estopa, M., Messeguer, J., and San Segundo, B. (2005) Pathogen-induced production of the antifungal AFP protein from *Aspergillus giganteus* confers resistance to the blast fungus *Magnaporthe grisea* in transgenic rice. Molecular Plant-Microbe Interactions 18, 960–972.

747. Morgan, C.L., Bruce, D.M., Child, R., Ladbrooke, Z.L., and Arthur, A.E. (1998) Genetic variation for pod shatter resistance among lines of oilseed rape developed from synthetic *B. napus.* Field Crops Research 58, 153–165.

748. Morgan, C.L., Ladbrooke, Z.L., Bruce, D.M., Child, R., and Arthur, A.E. (2000) Breeding oilseed rape for pod shattering resistance. Journal of Agricultural Science 135, 347–359.

749. Morohoshi, N., and Kajita, S. (2001) Formation of a tree having a low lignin content. Journal of Plant Research 114, 517–523.

750. Morrell, P.L., Williams-Coplin, T.D., Lattu, A.L., Bowers, J.E., Chandler, J.M., and Paterson, A.H. (2005) Crop-to-weed introgression has impacted allelic composition of johnsongrass populations with and without recent exposure to cultivated sorghum. Molecular Ecology 14, 2143–2154.

751. Morris, C.E., and Sands, D.C. (2006) The breeder's dilemma-yield or nutrition. Nature Biotechnology 24, 1078–1080.

752. Morris, R., and Sears, E.R. (1976) The cytogenetics of wheat and its relatives. In Wheat and Wheat Improvement (ed. Quisenberry, K.S.), 19–88. American Society of Agronomy, Madison, WI.

753. Morris, W.F., Kareiva, P.M., and Raymer, P.L. (1994) Do barren zones and pollen traps reduce gene escape from transgenic crops. Ecological Applications 4, 157–165.

754. Morton, R.L., Schroeder, H.E., Bateman, K.S., Chrispeels, M.J., Armstrong, E., and Higgins, T.J.V. (2000) Bean alpha-amylase inhibitor 1 in transgenic peas (*Pisum sativum*) provides complete protection from pea weevil (*Bruchus pisorum*) under field conditions. Proceedings of the National Academy of Sciences USA 97, 3820–3825.

755. Mothes, N., Horak, F., and Valenta, R. (2004) Transition from a botanical to a molecular classification in tree pollen allergy: Implications for diagnosis and therapy. International Archives of Allergy and Immunology 135, 357–373.

756. Moyes, C.L., Lilley, J.M., Casais, C.A., Cole, S.G., Haeger, P.D., and Dale, P.J. (2002) Barriers to gene flow from oilseed rape (*Brassica napus*) into populations of *Sinapis arvensis*. Molecular Ecology 11, 103–112.

757. Muangprom, A., Mauriera, I., and Osborn, T.C. (2006) Transfer of a dwarf gene from *Brassica rapa* to oilseed *B. napus*, effects on agronomic traits, and development of a 'perfect' marker for selection. Molecular Breeding 17, 101–110.

758. Muangprom, A., Thomas, S.G., Sun, T.-P., and Osborn, T.C. (2005) A novel dwarfing mutation in a green revolution gene from *Brassica rapa*. Plant Physiology 137, 931–938.

759. Mudalige, R.G., and Kuehnle, A.R. (2004) Orchid biotechnology in production and improvement. HortScience 39, 11–17.

760. Mugo, S., De Groote, H., Bergvinson, D., Mulaa, M., Songa, J., and Gichuki, S. (2005) Developing Bt maize for resource-poor farmers-Recent advances in the IRMA project. African Journal of Biotechnology 4, 1490–1504.

761. Muir, J.P., Sanderson, M.A., Ocumpaugh, W.R., Jones, R.M., and Reed, R.L. (2001) Biomass production of 'Alamo' switchgrass in response to nitrogen, phosphorus, and row spacing. Agronomy Journal 93, 896–901.
762. Mulcahy, D.L., and Mulcahy, G.B. (1987) The effects of pollen competition. American Scientist 75, 44–50.
763. Muller, H., Jordal, O., Kierulf, P., Kirkhus, B., and Pedersen, J.I. (1998) Replacement of partially hydrogenated soybean oil by palm oil in margarine without unfavorable effects on serum lipoproteins. Lipids 33, 879–887.
764. Mundy, P.J. (2000) Red-billed queleas in Zimbabwe. in Workshop on research priorities for migrant pests of agriculture in southern Africa (eds. Cheke, R.A., Rosenberg, L.J., and Kieser, M.E.), 97–102. Natural Resources Institute, Chatham, United Kingdom.
765. Munimbazi, C., and Bullerman, L.B. (1996) Molds and mycotoxins in foods from Burundi. Journal of Food Protection 59, 869–875.
766. Munkvold, G.P. (2003) Changing genetic approaches to managing mycotoxins in maize. Annual Review of Phytopathology 41, 99–116.
767. Munkvold, G.P., Hellmich, R.L., and Rice, L.G. (1999) Comparison of fumonisin concentrations in kernels of transgenic Bt maize hybrids and nontransgenic hybrids. Plant Disease 83, 130–138.
768. Murray, J.D. (1987) Modeling the spread of rabies. American Scientist 75, 280–284.
769. Murthy, H.N., Jeong, J.H., Choi, Y.E., and Paek, K.Y. (2003) Agrobacterium-mediated transformation of niger [Guizotia abyssinica (L. f.) Cass.] using seedling explants. Plant Cell Reports 21, 1183–1187.
770. Mutengwa, C.S., Tongoona, P.B., and Sithole-Niang, I. (2005) Genetic studies and a search for molecular markers that are linked to Striga asiatica resistance in sorghum. African Journal of Biotechnology 4, 1355–1361.
771. Mäder, P., Fliessbach, A., Dubois, D., Gunst, L., Fried, P., and Niggli, U. (2002) Soil fertility and biodiversity in organic farming. Science 296, 1694–1696.
772. Nakagahra, M., Okuno, K., and Vaughan, D. (1997) Rice genetic resources: history, conservation, investigative characterization and use in Japan. Plant Molecular Biology 35, 69–77.
773. Nakagawa, Y., Lee, Y.M., Lehmberg, E., Herrmann, R., Herrmann, R., Moskowitz, H., Jones, A.D., and Hammock, B.D. (1997) New anti-insect toxin (AaIT5) from Androctonus australis. European Journal of Biochemistry 246, 496–501.
774. Nakagawa, Y., Sadilek, M., Lehmberg, E., Herrmann, R., Moskowitz, H., Lee, Y.M., Thomas, B.A., Shimizu, R., Kuroda, M., Jones, A.D., and Hammock, B.D. (1998) Rapid purification and molecular modeling of AaIT peptides from venom of Androctonus australi. Archives of Insect Biochemistry and Physiology 38, 53–65.
775. Nan, G.L., Tang, C.S., Kuehnle, A.R., and Kado, C.I. (1997) Dendrobium orchids contain an inducer of Agrobacterium virulence genes. Physiological and Molecular Plant Pathology 51, 391–399.
776. Narayanan, N.N., Baisakh, N., Oliva, N.P., Vera Cruz, C.M., Gnanamanickam, S., Datta, K., and. Datta, S.K. (2004) Molecular breeding: marker-assisted selection combined with biolistic transformation for blast and bacterial blight resistance in Indica rice (cv. CO39). Molecular Breeding 14, 61–71.
777. Narula, A., Kumar, S., and Srivastava, P.S. (2005) Abiotic metal stress enhances diosgenin yield in Dioscorea bulbifera L. cultures. Plant Cell Reports 24, 250–254.
778. Nassar, N.M.A. (1980) Attempts to hybridize wild Manihot species with cassava. Economic Botany 34, 13–15.
779. Nassar, N.M.A. (1984) Natural hybridization between Manihot reptans Pax and M. alutacea Rogers & Appan. Canadian Journal of Plant Science 64, 423–425.
780. Natilla, A., Piazzolla, G., Nuzzaci, M., Saldarelli, P., Tortorella, C., Antonaci, S., and Piazzolla, P. (2004) Cucumber mosaic virus as carrier of a hepatitis C virus-derived epitope. Archives of Virology 149, 137–154.

781. Naylor, R.L., Falcon, W.P., Goodman, R.M., Jahn, M.M., Sengooba, T., Tefera, H., and Nelson, R.J. (2004) Biotechnology in the developing world: a case for increased investments in orphan crops. Food Policy 29, 15–44.

782. Nelson, O.E., Mertz, E.T., and Bates, L.S. (1965) Second mutant gene affecting the amino acid pattern of maize endosperm proteins. Science 150, 1469.

783. Nentwig, W. (1999) Weedy plant species and their beneficial arthropods: potential for manipulation in field crops. in Enhancing Biological Control (eds. Pickett, C.H., and Bugg, R.L.), 49–72. University of California Press, Berkeley.

784. Ness, J.E., Del Cardayre, S.B., Minshull, J., and Stemmer, W.P.C. (2000) Mo-lecular breeding: the natural approach to protein design. Advances in Protein Chemistry 55, 261–292.

785. Newcomb, R.D., Campbell, P.M., Ollis, D.L., Cheah, E., Russell, R.J., and Oakeshott, J.G. (1997) A single amino acid substitution converts a carboxylesterase to an organophosphorus hydrolase and confers insecticide resistance to blowfly. Proceedings of the National Academy of Science USA 94, 7464–7468.

786. Newhouse, K., Smith, W., Starrett, M., Schaefer, T., and Singh, B. (1989) Tolerance to imidazolinone herbicides in wheat. Plant Physiology 100, 882–886.

787. Ni, W., Paiva, N.L., and Dixon, R.A. (1994) Reduced lignin in transgenic plants containing a caffeic acid O-methyltransferase antisense gene. Trangenic Research 3, 120–126.

788. Nielsen, P.K., Bonsager, B.C., Fukuda, K., and Svensson, B. (2004) Barley alpha-amylase/subtilisin inhibitor: structure, biophysics and protein engineering. Biochimica et Biophysica Acta-Proteins and Proteomics 1696, 157–164.

789. Nigatu, A., and Gashe, B.A. (1994) Inhibition of spoilage and food-borne pathogens by lactic-acid bacteria isolated from fermenting tef (*Eragrostis tef*) Dough. Ethiopian Medical Journal 32, 223–229.

790. Nigatu, A., and Gashe, B.A. (1998) Effect of heat treatment on the antimicrobial properties of tef dough, injera, kocho and aradisame and the fate of selected pathogens. World Journal of Microbiology and Biotechnology 14, 63–69.

791. Nir, N., Canoles, M., Beaudry, R., Baldwin, E., and Mehla, C.P. (2004) Inhibiting tomato ripening with 1-methylcyclopropene. Journal of the American Society for Horticultural Science 129, 112–120.

792. Nishijima, T., Katsura, N., Koshioka, M., Yamazaki, H., Nakayama, M., Yamane, H. Yamaguchi, I., Yokota, T., Murofushi, N., Takahashi, N., and Nonaka, M. (1998) Effects of gibberellins and gibberellin-biosynthesis inhibitors on stem elongation and flowering of *Raphanus sativus* L. Journal of the Japanese Society for Horticultural Science 67, 325–330.

793. Nishimura, A., Ashikari, M., Lin, S., Takashi, T., Angeles, E.R., Yamamoto, T., and Matsuoka, M. (2005) Isolation of a rice regeneration quantitative trait loci gene and its application to transformation systems. Proceedings of the National Academy of Science USA 102, 11940–11944.

794. Nishizawa, Y., Nishio, Z., Nakazono, K.. Soma, M., Nakajima, E., Ugaki, M., and Hibi, T. (1999) Enhanced resistance to blast (*Magnaporthe grisea*) in transgenic Japonica rice by constitutive expression of rice chitinase. Theoretical and Applied Genetics 99, 383–390.

795. Njapau, H., Muzungaile, E.M., and Changa, R.C. (1998) The effect of village processing techniques on the contents of aflatoxins in corn and peanuts. Journal of Science of Food in Agriculture 76, 450–456.

796. Noguchi, T., Fujioka, S., Choe, S., Takatsuto, S., Yoshida, S., Yuan, H., Feldmann, K.A., and Tax, F.E. (1999) Brassinosteroid-insensitive dwarf mutants of *Arabidopsis* accumulate brassinosteroids. Plant Physiology 121, 743–752.

797. Noldin, J.A., Chandler, J.M., and McCauley, G.N. (1999) Red rice (*Oryza sativa*) biology. I. Characterization of red rice ecotypes. Weed Technology 13, 12–18.

798. Noldin, J.A., Chandler, J.M., Ketchersid, M.L., and McCauley, G.N. (1999) Red rice (*Oryza sativa*) biology. II. Ecotype sensitivity to herbicides. Weed Technology 13, 19–24.

799. O'Donovan, J.T., and McClay, A.S. (2002) Relationship between relative time of emergence of Tartary buckwheat (*Fagopyrum tataricum*) and yield loss of barley. Canadian Journal of Plant Science 82, 861–863.

800. O'Keefe, D.P., Tepperman, J.M., Dean, C., Leto, K.J., Erbes, D.L., and Odell, J.T. (1994) Plant expression of a bacterial cytochrome-P450 that catalyzes activation of a sulfonylurea proherbicide. Plant Physiology 105, 473–482.

801. O'Rourke, M.K., and Buchmann, S. (1986) Pollen yield from olive tree cvs. Manzanillo and Swan Hill in closed urban environments. Journal of the American Society for Horticultural Science 111, 980–984.

802. Oard, J., Cohn, M.A., Linscombe, S., Gealy, D., and Gravois, K. (2000) Field evaluation of seed production, shattering, and dormancy in hybrid populations of transgenic rice (*Oryza sativa*) and the weed, red rice (*Oryza sativa*). Plant Science 157, 13–22.

803. Oard, J.H., Linscombe, S.D., Braverman, M.P., Jodari, F., Blouin, D.C., Leech, M., Kholi, A., Vain, P., Cooley, J.C., and Christou, P. (1996) Development, field evaluation and agronomic performance of transgenic herbicide resistant rice. Molecular Breeding 2, 359–368.

804. Obilana, A.B. (2003) Overview: importance of millets in Africa. In AFRIPRO Workshop on the Proteins of Sorghums and Millets: Enhancing the Nutritional and Funtional Properties for Africa (eds. Belton, P.S., and Taylor, J.R.N.). Online www.afripro.org.uk/papers/Paper02Obilana.pdf, Pretoria.

805. Ochatt, S.J., Durieu, P., Jacas, L., and Pontécaille, C. (2001) Protoplast, cell and tissue cultures for the biotechnological breeding of grasspea (*Lathyrus sativus* L.). Lathyrus Lathyrism Newsletter 2, 35–38.

806. Ogunwolu, E.O., and Odunlami, A.T. (1996) Suppression of seed bruchid (*Callosobruchus maculatus* F.) development and damage on cowpea (*Vigna unguiculata* (L.) Walp.) with *Zanthoxylum zanthoxyloides* (Lam.) Waterm. (Rutaceae) root bark powder when compared to neem seed powder and pirimiphos-methyl. Crop Protection 15, 603–607.

807. Ohlrogge, J.B., Mhaske, V.B., Beisson, F., and Ruuska, S. (2004) Genomics approaches to lipid biosynthesis. In New Directions for a Diverse Planet. www.cropscience.org.au, on CD-ROM, Brisbane.

808. Ohmiya, Y., Nakai, T., Park, Y.W., Aoyama, T., Oka, A., Sakai, F., and Hayashi, T. (2003) The role of PopCel1 and PopCel2 in poplar leaf growth and cellulose biosynthesis. Plant Journal 33, 1087–1097.

809. Ohnishi, O. (1990) Analyses of genetic variants in common buckwheat *Fagopyrum esculentum* Moench: A review. Fagopyrum 10, 12–22.

810. Ohnishi, O. (1994) Buckwheat in Karakora and the Hindukush. Fagopyrum 14, 17–25.

811. Oikawa, T., Koshioka, M., Kojima, K., Yoshida, H., and Kawata, M. (2004) A role of *OsGA20ox1*, encoding an isoform of gibberellin 20-oxidase, for regulation of plant stature in rice. Plant Molecular Biology 55, 687–700.

812. Okamoto, M. (1957) Asynaptic effect of chromosome V. Wheat Information Service 5, 6–8.

813. Okeola, O.G., and Machuka, J. (2001) Biological effects of African yam bean lectins on *Clavigralla tomentosicollis* (Hemiptera: Coreidae). Journal of Economic Entomology 94, 724–729.

814. Okuley, J., Lightner, J., Feldmann, K., Yadav, N., Lark, E., and Browse, J. (1994) *Arabidopsis FAD2* gene encodes the enzyme that is essential for polyunsaturated lipid synthesis. Plant Cell 6, 147–158.

815. Okuzaki, A., and Toriyama, K. (2004) Chimeric RNA/DNA oligonucleotide-directed gene targeting in rice. Plant Cell Reports 22, 509–512.

816. Oleson, J.D., Park, Y.L., Nowatzki, T.M., and Tollefson, J.J. (2005) Node-injury scale to evaluate root injury by corn rootworms (Coleoptera: Chrysomelidae). Journal of Economic Entomology 98, 1–8.

817. Oliver, M.J., Quisenberry, J.E., Trolinder, N.L.G., and Keim, D.L. (1998) Control of plant gene expression. U.S. Patent 5,723,765.

818. Oliver, M.J., Luo, H., Kausch, A., and Collins, H. (2004) Seed-based strategies for transgene containment. In 8th International Symposium on the Biosafety of Genetically Modified Organisms, 154–161. INRA, Montpellier.

819. Olmo, H.P. (1995) Grapes. In Evolution of Crop Plants (eds. Smartt, J., and Simmonds, N.W.), 485–490. Longman, Harlow, United Kingdom.

820. Olsson, G. (1960) Species cross within the genus *Brassica*. II. Artificial *Brassica napus* L. Hereditas 46, 351–386.

821. Omitogun, O.G., Jackai, L.E.N., and Thottappilly, G. (1999) Isolation of insecticidal lectin-enriched extracts from African yam bean (*Sphenostylis stenocarpa*) and other legume species. Entomologia Experimentalis et Applicata 90, 301–311.

822. Ono, E.Y.S., Sasaki, E.Y., Hashimoto, E.H., Hara, L.N., Corrêa, B., Itano, E.N., Sugiura, T., Ueno, Y., and Hirooka, E.Y. (2002) Post-harvest storage of corn: effect of beginning moisture content on mycoflora and fumonisin contamination. Food Additives and Contaminants 19, 1081–1090.

823. Onyeike, E.N., Abbey, B.W., and Anosike, E.O. (1991) Kinetics of heat-inactivation of trypsin-inhibitors from the African yam bean (*Sphenostylis stenocarpa*). Food Chemistry 40, 9–23.

824. Oraby, H.F., Ransom, C.B., Kravchenko, A.N., and Sticklen, M.B. (2005) Barley *HVA1* gene confers salt tolerance in R3 transgenic oat. Crop Science 45, 2218–2227.

825. Orlikowska, T.K., Cranston, H.J., and Dyer, W.E. (1995) Factors influencing *Agrobacterium tumefaciens*-mediated transformation and regeneration of the safflower cultivar 'Centennial'. Plant Cell, Tissue and Organ Culture 40, 85–91.

826. Ortiz-Garcia, S., Ezcurra, E., Schoel, B., Acevedo, F., Soberon, J., and Snow, AA. (2005) Absence of detectable transgenes in local landraces of maize in Oaxaca, Mexico (2003–2004). Proceedings of the National Academy of Sciences USA 102, 12338–12343.

827. Osborne, T.B., and Mendel, L.B. (1914) Amino acids in nutrition and growth. Journal of Biological Chemistry 17, 325–349.

828. Osman, S.F., Herb, S.F., Fitzpatrick, T.J., and Schmiediche, P. (1978) Glycoalkaloid composition of wild and cultivated tuber-bearing *Solanum* species of potential value in potato breeding programs. Journal of Agricultural and Food Chemistry 26, 1246–1248.

829. Ostergaard, L., Kempin, S.A., Bies, D., Klee, H.J., and Yanofsky, M.F. (2006) Pod shatter resistant *Brassica* fruit produced by ectopic expression of the *FRUITFULL* gene. Plant Biotechnology Journal 4, 45–51.

830. Otsuki, T., Wilson, J.S., and Sewadeh, M. (2001) Saving two in a billion: quantifying the trade effect of European food safety standards on African exports. Food Policy 26, 495–514.

831. Outchkourov, N.S., de Kogel, W.J., Wiegers, G.L., Abrahamson, M., and Jongsma, M.A. (2004) Engineered multidomain cysteine protease inhibitors yield resistance against western flower thrips (*Franklinielia occidentalis*) in greenhouse trials. Plant Biotechnology Journal 2, 449–458.

832. Owen, M.D.K., and Gressel, J. (2001) Non-traditional concepts of $ynergy for evaluating integrated weed management. In Pesticide Biotransformations in Plants and Microorganisms: Similarities and Divergences (eds. Hall, J.C., Hoagland, R., and Zablotowicz, R.), 376–396. American Chemical Society, Washington, DC.

833. Ozkan, H., and Feldman, M. (2001) Genotypic variation in tetraploid wheat affecting homoeologous pairing in hybrids with *Aegilops peregrina*. Genome 44, 1000–1006.

834. Pahkala, K., and Sankari, H. (2001) Seed loss as a result of pod shatter in spring rape and spring turnip rape in Finland. Agricultural and Food Science in Finland 10, 209–216.

835. Papanikolaou, I., Barderas, R., Thibaudon, M., and Pauli, G. (2005) Ash tree pollinosis: botanical aspects, description of allergens and cross-reactivities. Revue Francaise d Allergologie et d Immunologie Clinique 45, 395–405.

836. Parveez, G.K., Masri, M.M., Zainal, A., Majid, N.A., Yunus, A.M., Fadilah, H.H., Rasid, O., and Cheah, S.C. (2000) Transgenic oil palm: production and projection. Biochemical Society Transactions 28, 969–972.

837. Patkar, R.N., and Chattoo, B.B. (2006) Transgenic indica rice expressing ns-LTP-Like protein shows enhanced resistance to both fungal and bacterial pathogens. Molecular Breeding 17, 159–171.

838. Payne, T., Johnson, S.D., and Koltunow, A.M. (2004) *KNUCKLES* (*KNU*) encodes a C2H2 zinc-finger protein that regulates development of basal pattern elements of the *Arabidopsis* gynoecium. Development 131, 3737–3749.

839. Pedersen, J.F., Marx, D.B., and Funnell, D.L. (2003) Use of A(3) cytoplasm to reduce risk of gene flow through sorghum pollen. Crop Science 43, 1506–1509.

840. Pedersen, J.F., Vogel, K.P., and Funnell, D.L. (2005) Impact of reduced lignin on plant fitness. Crop Science 45, 812–819.

841. Pedersen, M.B., Kjaer, C., and Elmegaard, N. (2000) Toxicity and bioaccumulation of copper to black bindweed (*Fallopia convolvulus*) in relation to bioavailability and the age of soil contamination. Archives of Environmental Contamination and Toxicology 39, 431–439.

842. Pedersen, M.B., Axelsen, J.A., Strandberg, B., Jensen, J., and Attrill, M.J. (1999) The impact of a copper gradient on a microarthropod field community. Ecotoxicology 8, 467–483.

843. Peng, J., Richards, D.E., Hartley, N.M., Murphy, G.P., Devos, K.M., Flintham, J.E., Beales, J., Fish, L.J., Worland, A.J., Pelica, F., Sudhakar, D., Christou, P., Snape, J.W., Gale, M.D., and Harberd, N.P. (1999) 'Green revolution' genes encode mutant gibberellin response modulators. Nature 400, 256–261.

844. Pennycooke, J.C., Jones, M.L., and Stushnoff, C. (2003) Down-regulating alpha-galactosidase enhances freezing tolerance in transgenic petunia. Plant Physiology 133, 901–909.

845. Peter, L.J. (1969) The Peter Principle. William Morrow, New York.

846. Petersen, M., Sander, L., Child, R., van Onckelen, H., Ulvskov, P., and Borkhardt, B. (1996) Isolation and characterisation of a pod dehiscence zone-specific polygalacturonase from *Brassica napus*. Plant Molecular Biology 31, 517–527.

847. Pfiffner, L., and Mader, P. (1997) Effects of biodynamic, organic and conventional production systems on earthworm populations. Biological Agriculture & Horticulture 15, 3–10.

848. Piazzolla, G., Nuzzaci, M., Tortorella, C., Panella, E., Natilla, A., Boscia, D., De Stradis, A., Piazzolla, .P, and Antonaci, S. (2005) Immunogenic properties of a chimeric plant virus expressing a hepatitis C virus (HCV)-derived epitope: New prospects for an HCV vaccine. Journal of Clinical Immunology 25, 142–152.

849. Pichon, F.J., Uquillas, J.E., and Frechione, J. (1999) Traditional and Modern Natural Resource Management in Latin America. University of Pittsburgh Press, Pittsburgh, PA.

850. Pilon-Smits, E., and Pilon, M. (2002) Phytoremediation of metals using transgenic plants. Critical Reviews in Plant Sciences 21, 439–456.

851. Pipithsangchan, S., and Morallo-Rejesus, B. (2005) Insecticidal activity of diosgenin isolated from three species of grape ginger (*Costus* spp.) on the diamondback moth, *Plutella xylostella* (L.). Philippine Agricultural Scientist 88, 317–327.

852. Pirttila, A.M., McIntyre, L.M., Payne, G.A., and Woloshuk, C.P. (2004) Expression profile analysis of wild-type and fec1 mutant strains of *Fusarium verticillioides* during fumonisin biosynthesis. Fungal Genetics and Biology 41, 647–656.

853. Pistrick, K. (2002) Notes on neglected and underutilized crops—Current taxonomical overview of cultivated plants in the families Umbelliferae and Labiatae. Genetic Resources and Crop Evolution 49, 211–225.

854. Pitt, J.I. (2000) Toxigenic fungi and mycotoxins. British Medical Bulletin 56, 184–192.

855. Placinta, C.M., D'Mello, J.P.F., and Macdonald, A.M.C. (1999) A review of worldwide contamination of cereal grains and animal feed with *Fusarium* mycotoxins. Animal Feed Science and Technology 78, 21–37.

856. Plattner, R.D., Norred, W.P., Bacon, C.W., Voss, K.A., Peterson, R., Shackelford, D.D., and Weisleder, D. (1990) A method of detection of fumonisins in corn samples associated with field cases of equine leukoencephalomalacia. Mycologia 82, 698–702.

857. Ploetz, R.C. (2001) Black Sigatoka of Banana. The Plant Health Instructor www.apsnet.org/education/feature/banana/.

858. Pogue, G.P., Lindbo, J.A., Garger, S.J., and Fitzmaurice, W.P. (2002) Making an ally from an enemy: Plant virology and the new agriculture. Annual Review of Phytopathology 40, 45–74.

859. Popelka, J.C., Terryn, N., and Higgins, T.J.V. (2004) Gene technology for grain legumes: can it contribute to the food challenge in developing countries? Plant Science 167, 195–206.

860. Popov, V.N., Orlova, I.V., Kipaikina, N.V., Serebriiskaya, T.S., Merkulova, N.V., Nosov, A.M., Trunova, T.I., Tsydendambaev, V.D., and Los, D.A. (2005) The effect of tobacco plant transformation with a gene for acyl-lipid delta 9-desaturase from *Synechococcus vulcanus* on plant chilling tolerance. Russian Journal of Plant Physiology 52, 664–667.

861. Porat, R., Reiss, N., Atzorn, R., Halevy, A.H., and Borochov, A. (1995) Examination of the possible involvement of lipoxygenase and jasmonates in pollination-induced senescence of *Phalaenopsis* and *Dendrobium* orchid flowers. Physiologia Plantarum 94, 205–210.

862. Porat, R., Nadeau, J.A., Kirby, J.A., Sutter, E.G., and O'Neill, S.D. (1998) Characterization of the primary pollen signal in the postpollination syndrome of *Phalaenopsis* flowers. Plant Growth Regulation 24, 109–117.

863. Potter, D., and Doyle, J.J. (1992) Origins of the African yam bean (*Spheno-stylis stenocarpa*, Leguminosae)-evidence from morphology, isozymes, chloroplast DNA, and linguistics. Economic Botany 46, 276–292.

864. Powell, M. (1997) Science in sanitary and phytosanitary dispute resolution. Discussion Paper, Resources for the Future, Washington, DC.

865. Prakash, S., and Chopra, V.L. (1990) Reconstruction of allopolyploid Brassicas through nonhomologous recombination-introgression of resistance to pod shatter in *Brassica napus*. Genetical Research 56, 1–2.

866. Prakesh, S., Misra, B.K., Adsule, R.N., and Barat, G.K. (1977) Distribution of β-N-oxalyl-L. α,β diaminopropionic acid in different tissues of aging *Lathyrus sativus* plant. Biochemie Physiologie Pflanzen 171, 369–374.

867. Prescott, V.E., Campbell, P.M., Moore, A., Mattes, J., Rothenberg, M.E., Foster, P.S., Higgins, T.J., and Hogan, S.P. (2005) Transgenic expression of bean alpha-amylase inhibitor in peas results in altered structure and immunogenicity. Journal of Agricultural and Food Chemistry 53, 9023–9030.

868. Preston, C., and Powles, S.B. (1998) Amitrole inhibits diclofop metabolism and synergises diclofop-methyl in a diclofop-methyl-resistant biotype of *Lolium rigidum*. Pesticide Biochemistry and Physiology 62, 179–189.

869. Price, J.S., Hobson, R.N., Nealle, M.A., and Bruce, D.M. (1996) Seed losses in commercial harvesting of oilseed rape. Journal of Agricultural Engineering Research 65, 183–191.

870. Protalix. (2006) www.protalix.com.

871. Puchta, H. (2002) Gene replacement by homologous recombination in plants. Plant Molecular Biology 48, 173–182.

872. Puhakainen, T., Hess, M.W., Makela, P., Svensson, J., Heino, P., and Palva, E.T.. (2004) Overexpression of multiple dehydrin genes enhances tolerance to freezing stress in *Arabidopsis*. Plant Molecular Biology 54, 743–753.

873. Punnonen, J., Whalen, R., Patten, P.A., and Stemmer, W. (2000) Molecular breeding by DNA shuffling. Science and Medicine 7, 38–47.

874. Purvis, A., and Hector, A. (2000) Getting the measure of biodiversity. Nature 405, 212–219.

875. Pusztai, A., Bardocz, G.G., Alonso, R., Chrispeels, M.J., Schroeder, H.E., Tabe, L.M., and Higgins, T.J. (1999) Expression of the insecticidal bean alpha-amylase inhibitor transgene has minimal detrimental effect on the nutritional value of peas fed to rats at 30% of the diet. Journal of Nutrition 129, 1597–1603.

876. Quan, R.D., Shang, M., Zhang, H., Zhao, Y.X., and Zhang, J.R. (2004) Improved chilling tolerance by transformation with *bet*A gene for the enhancement of glycinebetaine synthesis in maize. Plant Science 166, 141–149.

877. Quist, D., and Chapela, I.H. (2001) Transgenic DNA introgressed into traditional maize landraces in Oaxaca, Mexico. Nature 414, 541–543.

878. Rachmiel, M., Verleger, H., Waisel, Y., Keynan, N., Kivity, S., and Katz, Y. (1996) The importance of the pecan tree pollen in allergic manifestations. Clinical and Experimental Allergy 26, 323–329.

879. Radhakrishnan, P., and Srivastava, V. (2005) Utility of the FLP-FRT recombination system for genetic manipulation of rice. Plant Cell Reports 23, 721–726.

880. Raikhel, N., and Minorsky, P. (2001) Celebrating plant diversity. Plant Physiology 127, 1325–1327.
881. Rains, D.W., Epstein, E., Zasoski, R.J., and Aslam, M. (2006) Active silicon uptake by wheat. Plant and Soil 280, 223–228.
882. Raja, N., Jeyasankar, A., Venkatesan, S.J., and Ignacimuthu, S. (2005) Efficacy of *Hyptis suaveolens* against Lepidopteran pests. Current Science 88, 220–222.
883. Rajani, S., and Sundaresan, V. (2001) The *Arabidopsis myc/bHLH* gene *ALCATRAZ* enables cell separation in fruit dehiscence. Current Biology 11, 1914–1922.
884. Rajguru, S.N., Burgos, N.R., Shivrain, V.K., and Stewart, J.M. (2005) Mutations in the red rice *ALS* gene associated with resistance to imazethapyr. Weed Science 53, 567–577.
885. Ralph, J., Guillaumie, S., Grabber, J.H., Lapierre, C., and Barriere, Y. (2004) Genetic and molecular basis of grass cell-wall biosynthesis and degradability. III. Towards a forage grass ideotype. Comptes Rendus Biologies 327, 467–479.
886. Ramadan, M.F., and Morsel, J.T. (2004) Oxidative stability of black cumin (*Nigella sativa* L.), coriander (*Coriandrum sativum* L.) and niger (*Guizotia abyssinica* Cass.) crude seed oils upon stripping. European Journal of Lipid Science and Technology 106, 35–43.
887. Rank, C., Rasmussen, L.S., Jensen, S.R., Pierce, S., Press, M.C., and Scholes, J.D. (2004) Cytotoxic constituents of *Alectra* and *Striga* species. Weed Research 44, 265–270.
888. Ravishankar, G.A., and Grewal, S. (1991) Development of media for growth of *Dioscorea deltoidea* cells and in vitro diosgenin production-influence of media constituents and nutrient stress. Biotechnology Letters 13, 125–130.
889. Rawal, K.M. (1975) Natural hybridization among wild, weedy and cultivated *Vigna unguiculata* (L.) Walp. Euphytica 24, 699–707.
890. Reiss, B. (2003) Homologous recombination and gene targeting in plant cells. International Review of Cytology 228, 85–139.
891. Rekha, M.R., Sasikiran, K., and Padmaja, G. (2004) Inhibitor potential of protease and alpha-amylase inhibitors of sweet potato and taro on the digestive enzymes of root crop storage pests. Journal of Stored Products Research 40, 461–470.
892. Rhew, R.C., Ostergaard, L., Saltzman, E.S., and Yanofsky, M.F. (2003) Genetic control of methyl halide production in *Arabidopsis*. Current Biology 13, 1809–1813.
893. Ribeiro, A.P.O., Pereira, E.J.G., Galvan, T.L., Picanco, M.C., Picoli, E.A.T., da Silva, D.J.H., Fari, M.G., and Otoni, W.C. (2006) Effect of eggplant transformed with oryzacystatin gene on *Myzus persicae* and *Macrosiphum euphorbiae*. Journal of Applied Entomology 130, 84–90.
894. Rich, P.J., Grenier, C., and Ejeta, G. (2004) *Striga* resistance in the wild relatives of sorghum. Crop Science 44, 2221–2229.
895. Richards, H.A., Rudas, V.A., Sun, H., McDaniel, J.K., Tomaszewski, Z., and Conger, B.V. (2001) Construction of GFP-BAR plasmid and its use for switchgrass transformation. Plant Cell Reports 20, 48–54.
896. Richardson, C.P., and Bonmati, E. (2005) 2,4,6-trinitrotoluene transformation using *Spinacia oleracea*: Saturation kinetics of the nitrate reductase enzyme. Journal of Environmental Engineering 131, 800–809.
897. Rick, C.M., and Yoder, J.I. (1988) Classical and molecular genetics of tomato: highlights and perspectives. Annual Review of Genetics 22, 281–300.
898. Rieger, M.A., Lamond, M., Preston, C., Powles, S.B., and Roush, R.T. (2002) Pollen-mediated movement of herbicide resistance between commercial canola fields. Science 296, 2386–2388.
899. Riepe, M., Spencer, P.S., Lambein, F., Ludolph, A.C., and Allen, C.N. (1995) In vitro toxicological investigations of isoxazolinone amino acids of *Lathyrus sativus*. Natural Toxins 3, 58–64.
900. Rieseberg, L.H., Whitton, J., and Gardner, K. (1999) Hybrid zones and the genetic architecture of a barrier to gene flow between two wild sunflower species. Genetics 152, 713–727.
901. Rieseberg, L.H., Kim, M.J., and Seiler, G.J. (1999) Introgression between the cultivated sunflower and a sympatric wild relative, *Helianthus petiolaris* (Asteraceae). Intlernational Journal of Plant Science 160, 102–108.

902. Riley, R., and Chapman, V. (1958) Genetic control of the cytologically diploid behavior of hexaploid wheat. Nature 182, 713–715.

903. Riley, R., Campbell, C.S., Young, R.M., and Belfield, A.M. (1966) Control of meiotic chromosome pairing by the chromosomes of homoeologous group 5 of *Triticum aestivum*. Nature 212, 1475–1477.

904. Riopel, J.L., and Timko, M.P. (1995) Haustorial initiation and differentiation. in Parasitic plants (eds. Press, M.C., and Graves, J.D.), 39–79. Chapman and Hall, London.

905. Rissler, J., and Mellon, M. (1993) Perils amidst the promise-ecological risks of BD-HRCs in a global market. Union of Concerned Scientists, Cambridge MA.

906. Roach, B.T. (1995) Sugar canes. In Evolution of Crop Plants (eds. Smartt, J., and Simmonds, N.W.), 160–166. Longman, Harlow, United Kingdom.

907. Roberts, J.A., Elliott, K.A., and Gonzalez-Carranza, Z.H. (2002) Abscission, dehiscence, and other cell separation processes. Annual Review of Plant Biology 53, 131–158.

908. Robertson, G.H., Doyle, L.R., Sheng, P., Pavlath, A.E., and Goodman, N. (1989) Diosgenin formation by freely suspended and entrapped plant-cell cultures of *Dioscorea deltoidea*. Biotechnology and Bioengineering 34, 1114–1125.

909. Robson, P.R.H., McCormac, A.C., Irvine, A.S., and Smith, H. (1996) Genetic engineering of harvest index in tobacco through overexpression of a phytochrome gene. Nature Biotechnology 14, 995–998.

910. Robson, P.R., Donnison, I.S., Wang, K., Frame, B., Pegg, S.E., Thomas, A., and Thomas, H. (2004) Leaf senescence is delayed in maize expressing the *Agrobacterium IPT* gene under the control of a novel maize senescence-enhanced promoter. Plant Biotechnology Journal 2, 101–112.

911. Rodenburg, J., Bastiaans, L., Weltzien, E., and Hess, D.E. (2005) How can field selection for *Striga* resistance and tolerance in sorghum be improved. Field Crops Research 93, 34–50.

912. Rodrigues, F.A., Jurick, W.M., Datnoff, L.E., Jones, J.B., and Rollins, J.A. (2005) Silicon influences cytological and molecular events in compatible and incompatible rice—*Magnaporthe grisea* interactions. Physiological and Molecular Plant Pathology 66, 144–159.

913. Rodriguez, R., Villalba, M., Batanero, E., Gonzalez, E.M., Monsalve, R.I., Huecas, S., Tejera, M.L., and Ledesma, A. (2002) Allergenic diversity of the olive pollen. Allergy 57, 6–16.

914. Roeder, A.H., Ferrandiz, C., and Yanofsky, M.F. (2003) The role of the REPLUMLESS homeodomain protein in patterning the *Arabidopsis* fruit. Current Biology 13, 1630–1635.

915. Rohini, V.K., and Rao, K.S. (2000) Embryo transformation, a practical approach for realizing transgenic plants of safflower (*Carthamus tinctorius* L.). Annals of Botany 86, 1043–1049.

916. Rojas, R., Alba, J., Magana-Plaza, I., Cruz, F., and Ramos-Valdivia, A.C. (1999) Stimulated production of diosgenin in *Dioscorea galeottiana* cell suspension cultures by abiotic and biotic factors. Biotechnology Letters 21, 907–911.

917. Rokem, J.S., Schwarzberg, J., and Goldberg, I. (1984) Autoclaved fungal mycelia increase diosgenin production in cell-suspension cultures of *Dioscorea deltoidea*. Plant Cell Reports 3, 159–160.

918. Romeis, J., Meissle, M., and Bigler, F. (2006) Transgenic crops expressing *Bacillus thuringiensis* toxins and biological control. Nature Biotechnology 24, 63–71.

919. Rong, J., Lu, B.R., Song, Z., Su, J., Snow, A.A., Zhang, X., Sun, S., Chen, R., and Wang, F. (2007) Dramatic reduction of crop-to-crop gene flow within a short distance from transgenic rice fields. New Phytologist 173, 346–353.

920. Rotter, R.G., Marquardt, R.R., and Campbell, C.G. (1991) The nutritional-value of low lathyrogenic lathyrus (*Lathyrus sativus*) for growing chicks. British Poultry Science 32, 1055–1067.

921. Roy, D.N. (1980) Trypsin inhibitor from *Lathyrus sativus* seeds: final purification, separation of protein components, properties, and characterization. Journal of Agricultural and Food Chemistry 28, 48–54.

922. Roy, D.N., and Rao, S.P. (1971) Evidence, isolation, purification and some properties of a trypsin inhibitor in *Lathyrus sativus*. Journal of Agricultural and Food Chemistry 19, 257–259.

923. Roy, P.K., Ali, K., Gupta, A., Barat, G.K., and Mehta, S.L. (1993) Beta-N-oxalyl-L-alpha, beta-diaminopropionic acid in somaclones derived from internode explants of *Lathyrus sativus*. Journal of Plant Biochemistry and Biotechnology 2, 9–13.

924. Rudolph, R., Stresemann, E., Stresemann, B., and Haupthof, M. (1987) Sensitations against *Tribolium confusum* Du Val in patients with occupational and non-occupational exposure. Experientia Supplement 51, 177–182.

925. Rudra, M.R., Junaid, M.A., Jyothi, P., and Rao, S.L. (2004) Metabolism of dietary ODAP in humans may be responsible for the low incidence of neurolathyrism. Clinical Biochemistry 37, 318–322.

926. Ruel, K., Montiel, M.D., Goujon, T., Jouanin, L., Burlat, V., and Joseleau, J.P. (2002) Interrelation between lignin deposition and polysaccharide matrices during the assembly of plant cell walls. Plant Biology 4, 2–8.

927. Ruestow, E.G. (1984) Leeuwenhoek and the campaign against spontaneous generation. Journal of the History of Biology 17, 225–248.

928. Ruiter, R., van den Brande, I., Stals, E., Delaure, S., Cornelissen, M., and D'Halluin, K. (2003) Spontaneous mutation frequency in plants obscures the effect of chimeraplasty. Plant Molecular Biology 53, 715–729.

929. Ruiz, O.N., and Daniell, H. (2005) Engineering cytoplasmic male sterility via the chloroplast genome by expression of beta-ketothiolase. Plant Physiology 138, 1232–1246.

930. Ryals, J. (1996) Agricultural Biotechnology '96. Molecular Breeding 2, 91–93.

931. Saha, B.C., Iten, L.B., Cotta, M.A., and Wu, Y.V. (2005) Dilute acid pretreatment, enzymatic saccharification and fermentation of wheat straw to ethanol. Process Biochemistry 40, 3693–3700.

932. Saini, H.S., Attieh, J.M., and Hanson, A.D. (1995) Biosynthesis of halomethanes and methanethiol by higher-plants via a novel methyltransferase reaction. Plant Cell and Environment 18, 1027–1033.

933. Sakakibara, H. (2005) Cytokinin biosynthesis and regulation. in Plant hormones: Biosynthesis, signal transduction, action (ed. Davies, P.J.), 95–114. Springer, Dordrecht, The Netherlands.

934. Sakin, M., and Yildirim, A. (2004) Induced mutations for yield and its components in durum wheat (*Triticum durum* Desf.). Food Agriculture and Environment 2, 285–290.

935. Sala, F., Rigano, M.M., Barbante, A., Basso, B., Walmsley, A.M., and Castiglione, S. (2003) Vaccine antigen production in transgenic plants: strategies, gene constructs and perspectives. Vaccine 21, 803–808.

936. Sambrook, J., Fritsch, E.F., and Maniatis, T. (1989) Molecular Cloning: A Laboratory Manual. Cold Spring Harbor, New York.

937. Samimy, C., Bjorkman, T., Siritunga, D., and Blanchard, L. (1996) Overcoming the barrier to interspecific hybridization of *Fagopyrum esculentum* with *Fagopyrum tataricum*. Euphytica 91, 323–330.

938. Sankula, S., Braverman, M.P., and Oard, J.H. (1998) Genetic analysis of glufosinate resistance in crosses between transformed rice (*Oryza sativa*), and red rice (*Oryza sativa*). Weed Technology 12, 209–214.

939. Sankula, S., Braverman, M.P., Jodari, F., Linscombe, S.D., and Oard, J.H. (1997) Evaluation of glufosinate on rice (*Oryza sativa*) transformed with the *bar*-gene and red rice (*Oryza sativa*). Weed Technology 11, 70–75.

940. Sarmah, B.K., Moore, A., Tate, W., Molvig, L., Morton, R.L., Rees, D.P., Chiaese, P., Chrispeels, M.J., Tabe, L.M., and Higgins, T.J.V. (2004) Transgenic chickpea seeds expressing high levels of a bean alpha-amylase inhibitor. Molecular Breeding 14, 73–82.

941. Sartelet, H., Serghat, S., Lobstein, A., Ingenbleek, Y., Anton, R., Petitfrere, E., Aguie-Aguie, G., Martiny, L., and Haye, B. (1996) Flavonoids extracted from fonio millet (*Digitaria exilis*) reveal potent antithyroid properties. Nutrition 12, 100–106.

942. Sato, S., Kominato, M., and Sayama, H. (1992) Inoculation of plant virus. Japanese Patent 4,330,005.

943. Sato, S., Kominato, M., Sayama, H., and Ishimura, E. (2002) Inoculation of plant virus. Japanese Patent 2000,2201,25 35.

944. Savant, N.K., Snyder, G.H., and Datnoff, L.E. (1997) Silicon management and sustainable rice production. Advances in Agronomy 58, 151–199.

945. Sawada, K., Hasegawa, M., Tokuda, L., Kameyama, J., Kodama, O., Kohchi, T., Yoshida, K., and Shinmyo, A. (2004) Enhanced resistance to blast fungus and bacterial blight in transgenic rice constitutively expressing *OsSBP*, a rice homologue of mammalian selenium-binding proteins. Bioscience Biotechnology and Biochemistry 68, 873–880.

946. Schaefer, D.G. (2002) A new moss genetics: targeted mutagenesis in *Physcomitrella patens*. Annual Review of Plant Biology 53, 477–501.

947. Schaffrath, U., Mauch, F., Freydl, E., Schweizer, P., and Dudler, R. (2000) Constitutive expression of the defense-related *Rir1b* gene in transgenic rice plants confers enhanced resistance to the rice blast fungus *Magnaporthe grisea*. Plant Molecular Biology 43, 59–66.

948. Schaller, H., Bouvier-Navé, P., and Benveniste, P. (1998) Overexpression of an Arabidopsis cDNA encoding a sterol-C24-methyltransferase in tobacco modifies the ratio of 24-methyl cholesterol to sitosterol and is associated with growth reduction. Plant Physiology 18, 461–469.

949. Schernthaner, J.P., Fabijanski, S.F., Arnison, P.G., Racicot, M., and Robert, L.S. (2003) Control of seed germination in transgenic plants based on the segregation of a two-component genetic system. Proceedings of the National Academy of Sciences USA 100, 6855–6859.

950. Schlapfer, F., Pfisterer, A.B., and Schmid, B. (2005) Non-random species extinction and plant production: implications for ecosystem functioning. Journal of Applied Ecology 42, 13–24.

951. Schlechter, M., Marasas, W.F.O., Sydenham, E.W., Stockenstrom, S., Vismer, H.F., and Rheeder, J.P. (1998) Incidence of *Fusarium moniliforme* and fumonisins in commercial maize products, intended for human consumption, obtained from retail outlets in the United States and South Africa. South African Journal of Science 94, 185–187.

952. Schmer, M.R., Vogel, K.P., Mitchell, R., Moser, L.E., Eskridge, K.M., and Perrin, P.K. (2006) Establishment stand thresholds for switchgrass grown as a bioenergy crop. Crop Science 46, 157–161.

953. Schmidt-Dannert, C., Umeno, D., and Arnold, F.H. (2000) Molecular breeding of carotenoid biosynthetic pathways. Nature Biotechnology 18, 750–753.

954. Schnepf, E., Crickmore, N., Van Rie, J., Lereclus, D., Baum, J., Feitelson, J., Zeigler, D.R., and Dean, D.H. (1998) *Bacillus thuringiensis* and its pesticidal crystal proteins. Microbiology and Molecular Biology Reviews 62, 775–806.

955. Schouten, H.J., van Tongeren, C.A.M., and van den Bulk, R.W. (2002) Fitness effects of *Alternaria dauci* on wild carrot in The Netherlands. Environmental Biosafety Research 1, 39–47.

956. Schouten, H.J., Krens, F.A., Jacobsen, K., and Jacobsen, E. (2006) Cisgenic plants are similar to traditionally bred plants-International regulations for genetically modified organisms should be altered to exempt cisgenesis. EMBO Reports 7, 750–753.

957. Schroeder, H.E., Gollasch, S., Moore, A., Tabe, L.M., Craig, S., Hardie, D.C., Chrispeels, M.J., Spencer, D., and Higgins, T. (1995) Bean alpha-amylase inhibitor confers resistance to the pea weevil (*Bruchus pisorum*) in transgenic peas (*Pisum sativum* L). Plant Physiology 107, 1233–1239.

958. Schuch, W., Kanczler, J., Robertson, D., Hobson, G., Tucker, G., Grierson, D., Bright, S.W.J., and Bird, C. (1991) Fruit quality characteristics of transgenic tomato fruit with altered polygalacturonidase activity. HortScience 26, 1517–1520.

959. Schulman, R.N., Salt, D.E., and Raskin, I. (1999) Isolation and partial characterization of a lead-accumulating *Brassica juncea* mutant. Theoretical and Applied Genetics 99, 398–404.

960. Schultz, J.C., Schonrogge, K., and Lichtenstein, C.P. (2001) Plant response to bruchins. Trends in Plant Science 6, 406.

961. Scott, P.M. (1996) Mycotoxins transmitted into beer from contaminated grain during brewing. Journal of the AOAC International 79, 875–882.

962. Sears, E.R. (1976) Genetic control of chromosome pairing in wheat. Annual Review of Genetics 10, 31–51.

963. Sears, E.R. (1984) Mutations in wheat that raise the level of meiotic chromosome pairing.In Gene Manipulation in Plant Improvement (ed. Gustafson, J.P.), 295–300. Plenum Press, New York.

964. Sedlacek, J.D., Komaravalli, S.R., Hanley, A.M., Price, B.D., and Davis, P.M. (2001) Life history attributes of Indian meal moth (Lepidoptera: Pyralidae) and Angoumois grain moth (Lepidoptera: Gelechiidae) reared on transgenic corn kernels. Journal of Economic Entomology 94, 586–592.

965. Seefeldt, S.S., Zemetra, R., Young, F.L., and Jones, S.S. (1998) Production of herbicide-resistant jointed goatgrass (*Aegilops cylindrica*) x wheat (*Triticum aestivum*) hybrids in the field by natural hybridization. Weed Science 46, 632–634.

966. Setamou, M., Cardwell, K.F., Schulthess, F., and Hell, K. (1998) Effect of insect damage to maize ears, with special reference to *Mussidia nigrivenella* (Lepidoptera: Pyralidae), on *Aspergillus flavus* (Deuteromycetes: Monoliales) infection and aflatoxin production in maize before harvest in the Republic of Benin. Journal of Economic Entomology 91, 433–438.

967. Shaked, H., Melamed-Bessudo, C., and Levy, A.A. (2005) High-frequency gene targeting in *Arabidopsis* plants expressing the yeast *RAD54* gene. Proceedings of the National Academy of Sciences USA 102, 12265–12269.

968. Shani, Z., Dekel, M., Tzbary, T., Goren, R., and Shoseyov, O. (2004) Growth enhancement of transgenic poplar plants by overexpression of *Arabidopsis thaliana* endo-1,4-beta-glucanase (*cel1*). Molecular Breeding 14, 321–330.

969. Shani, Z., Dekel, M., Tsabary, G., Jensen, C.S., Tzfira, T., Goren, R., Altman, A., and Shoseyov, O. (1999) Expression of *Arabidopsis thaliana* endo-1,4-beta-glucanase (*cel1*) in transgenic plants. in Plant biotechnology and in vitro biology in the 21st century (eds. Altman, A., Ziv, M., and Izhar, S.), 209–212. Kluwer Academic Publishers, Dordrecht, The Netherlands.

970. Shani, Z., Shpigel, E., Roiz, L., Goren, R., Vinocur, B., Tzfira, T., Altman, A., and Shoseyov, O. (1999) Cellulose binding domain, increases cellulose synthase activity in *Acetobacter xylinum*, and biomass of transgenic plants. in Plant biotechnology and in vitro biology in the 21st century (eds. Altman, A., Ziv, M., and Izhar, S.), 213–218. Kluwer Academic Publishers, Dordrecht, The Netherlands.

971. Sheahan, J.J. (1996) Sinapate esters provide greater UV-B attenuation than flavonoids in *Arabidopsis thaliana* (Brassicaceae). American Journal of Botany 83, 679–686.

972. Shen, R.F., Iwashita, T., and Ma, J.F. (2004) Form of Al changes with Al concentration in leaves of buckwheat. Journal of Experimental Botany 55, 131–136.

973. Shephard, G.S., van der Westhuizen, L., Thiel, P.G., Gelderblom, W.C.A., Marasas, W.F.O., and van Schalkwyk, D.J. (1996) Disruption of sphingolipid metabolism in non-human primates consuming diets of fumonisin-containing *Fusarium moniliforme* culture material. Toxicon 34, 527–534.

974. Shi, X.H., Qin, C.D., Song, J.L., Xia, X.P., and Wang, H. (2005) Immunoblot detection of silica-binding protein in rice and other graminaceous plants. Progress in Biochemistry and Biophysics 32, 371–376.

975. Shiboleth, Y.M., Arazi, T., Wang, Y., and Gal-On, A. (2001) A new approach for weed control in a cucurbit field employing an attenuated potyvirus-vector for herbicide resistance. Journal of Biotechnology 92, 37–46.

976. Shimamura, C., Ohno, R., Nakamura, C., and Takumi, S. (2006) Improvement of freezing tolerance in tobacco plants expressing a cold-responsive and chloroplast-targeting protein WCOR15 of wheat. Journal of Plant Physiology 163, 213–219.

977. Shorter, R., Gibson, P., and Frey, K.J. (1978) Outcrossing rates in oat species crosses (*Avena sativa* L X *Avena sterilis* L). Crop Science 18, 877–878.

978. Shou, H.X., Bordallo, P., Fan, J.-B., Yeakley, J.M., Bibikova, M., Sheen, J.. and Wang, K. (2004) Expression of an active tobacco mitogen-activated protein kinase kinase kinase enhances freezing tolerance in transgenic maize. Proceedings of the National Academy of Sciences USA 101, 3298–3303.

979. Siame, B.A., Weersuriya, Y., Wood, K., Ejeta, G., and Butler, L.G. (1993) Isolation of strigol, a germination stimulant for *Striga asiatica,* from host plants. Journal of Agricultural and Food Chemistry 41, 1486–1491.

980. Siame, B.A., Mpuchane, S.F., Gashe, B.A., Allotey, J., and Teffera, G. (1998) Occurrence of aflatoxins, fumonisin B_1, and zearalenone in foods and feeds in Botswana. Journal of Food Protection 61, 1670–1673.

981. Sill, W.H.J., Lowe, A.E., Bellingham, R.C., and Fellows, H. (1954) Transmission of wheat streak-mosaic virus by abrasive leaf contacts during strong winds. Plant Disease Reporter 38, 445–447.

982. Sillero, J.C., Moreno, M.T., and Rubiales, D. (1996) in Advances in Parasitic Weed Research (eds. Moreno, M.T. et al.), 659–664. Junta de Andalucia, Cordoba, Spain.

983. Rubiales, D., Perez-de-Luque, A., Fernandez-Aparico, M., Sillero, J.C., Roman, B., Kharrat, M., Khalil, S., Joel, D.M., and Riches, C. (2006) Screening techniques and sources of resistance against parasitic weeds in grain legumes Euphytica 147, 187–199.

984. Sillero, J.C., Cubero, J.I., Fernández-Aparicio, M., and Rubiales, D. (2005) Search for resistance to crenate broomrape (*Orobanche crenata*) in *Lathyrus*. Lathyrus Lathyrism Newsletter 4, 7–9.

985. Sindel, B.M. (1997) Outcrossing of transgenes to weedy relatives. in Risk, benefit and trade considerations commercialisation of BD-HRCs (eds. McLean, G.D., Waterhouse, P.M., Evans, G., and Gibbs, M.J.), 43–81. Research Center Plant Science and Bureau of Resource Science, Canberra.

986. Singh, R.J., and Hymowitz, T. (1989) The genomic relationships among *Glycine soja* Sieb. and Zucc., *G. max* (L.) Merr. and '*G. gracilis*' Skvortz. Plant Breeding 103, 171–173.

987. Singh, S., Kirkwood, R., and Marshall, G. (1999) Biology and control of *Phalaris minor* Retz. (littleseed canarygrass) in wheat. Crop Protection 18, 1–16.

988. Singh, S., Thomaeus, S., Lee, M., Stymne, S., and Green, A. (2001) Transgenic expression of a Delta 12-epoxygenase gene in *Arabidopsis* seeds inhibits accumulation of linoleic acid. Planta 212, 872–879.

989. Singh, S.P., Zhou, X.R., Liu, Q., Stymne, S., and Green, A.G. (2005) Metabolic engineering of new fatty acids in plants. Current Opinion in Plant Biology 8, 197–203.

990. Sisler, E.C., and Serek, M. (1997) Inhibitors of ethylene responses in plants at the receptor level: recent developments. Physiologia Plantarum 100, 577–582.

991. Sivakumar, S., Franco, O.L., Tagliari, P.D., Bloch, C. Jr, Mohan, M., and Thayumanavan, B. (2005) Screening and purification of a novel trypsin inhibitor from *Prosopis juliflora* seeds with activity toward pest digestive enzymes. Protein and Peptide Letters 12, 561–565.

992. Skinner, J.S., Meilan, R., Ma, C.P., and Strauss, S.H. (2003) The *Populus PTD* promoter imparts floral-predominant expression and enables high levels of floral-organ ablation in *Populus, Nicotiana* and *Arabidopsis*. Molecular Breeding 12, 119–132.

993. Skogsmyr, I. (1994) Gene dispersal from tansgenic potatoes to con-specifics: A field trial. Theoretical and Applied Genetics 88, 770–774.

994. Smalla, K., and Fry, J. (2002) The contribution of mobile genetic elements in bacterial adaptability and diversity. Editorial. FEMS Microbiology Ecology 42, 163.

995. Smalla, K., and Sobecky, P.A. (2002) The prevalence and diversity of mobile genetic elements in bacterial communities of different environmental habitats: insights gained from different methodological approaches. FEMS Microbiology Ecology 42, 165–175.

996. Smeda, R.J., Currie, R.S., and Rippee, J.H. (2000) Fluazifop-P resistance expressed as a dominant trait in sorghum (*Sorghum bicolor*). Weed Technology 14, 397–401.

997. Smith, H., and Whitelam, G.C. (1997) The shade avoidance syndrome: Multiple responses mediated by multiple phytochromes. Plant Cell and Environment 20, 840–844.

998. Smith, M.C., Holt, J.S., and Webb, M. (1993) Population model of the parasitic weed *Striga hermonthica* (Scrophulariaceae) to investigate the potential of *Smicronyx umbrinus* (Coleoptera: Curculionidae) for biological control in Mali. Crop Protection 12, 473–476.

999. Smith, N.J.H., Williams, J.T., Plucknett, D.L., and Talbot, J.P. (1992) Oil palm. In Tropical Forests and Their Crops, 231–251. Cornell Univ. Press, Ithaca, NY.

1000. Smith, R.A.H., and Bradshaw, A.D. (1979) Use of metal tolerant plant populations for the reclamation of metalliferous wastes. Journal of Applied Ecology 16, 595–603.

1001. Sneller, C.H. (2003) Impact of transgenic genotypes and subdivision on diversity within the North American soybean germplasm. Crop Science 43, 409–414.

1002. Sohn, M.H., Lee, S.Y., and Kim, K.E. (2003) Prediction of buckwheat allergy using specific IgE concentrations in children. Allergy 58, 1308–1310.

1003. Somers, D.A., Rines, H.W., Gu, W., Kaeppler, H.F., and Bushnell, W.R. (1992) Fertile transgenic oat plants. Bio/Technology 10, 1589–1594.

1004. Somleva, M.N., Tomaszewski, Z., and Conger, B.V. (2002) *Agrobacterium*-mediated genetic transformation of switchgrass. Crop Science 42, 2080–2087.

1005. Spence, J., Vercher, Y., Gates, P. and Harris, N. (1996) 'Pod shatter' in *Arabidopsis thaliana*, *Brassica napus* and *B. juncea*. Journal of Microscopy-Oxford 181, 195–203.

1006. Spennemann, D.H.R., and Allen, L.R. (2000) Feral olives (*Olea europaea*) as future woody weeds in Australia: a review. Australian Journal of Experimental Agriculture 40, 889–901.

1007. Srnic, W.P. (2005) Technical performance of some commercial glyphosate-resistant crops. Pest Management Science 61, 225–234.

1008. Staniforth, A.R. (1982) Straw for Fuel, Feed and Fertilizer. Farming Press, Ipswich, UK.

1009. Steber, C.M., Cooney, S.E., and McCourt, P. (1998) Isolation of the GA-response mutant sly1 as a suppressor of ABI1–1 in *Arabidopsis thaliana*. Genetics 149, 509–521.

1010. Steinfeld, H., Gerber, P., Wassenaar, T., Castel, V., Rosales, M., and de Haan, C. (2006) Livestock's Long Shadow—Environmental Issues and Options. FAO, Rome.

1011. Stemmer, W.P. (1994) DNA shuffling by random fragmentation and reassembly: in vitro recombination for molecular evolution. Proceedings of the National Academy of Science USA 91, 10747–10751.

1012. Stender, S., and Dyerberg, J. (2004) Influence of trans fatty acids on health. Annals of Nutrition and Metabolism 48, 61–66.

1013. Stephenson, A.G., Winsor, J.A., Schlichting, C.D., and Davis, L.E. (1988) Pollen competition: nonrandom fertilization and progeny fitness: a reply to Charlseworth. American Naturalist 132, 303–308.

1014. Sterling, A., Baker, C.J., Berry, P.M., and Wade, A. (2003) An experimental investigation of the lodging of wheat. Agricultural and Forest Meteorology 119, 149–165.

1015. Stevens, V.L., and Tang, J. (1997) Fumonisin B1-induced sphingolipid depletion inhibits vitamin uptake via the GPI-anchored folate receptor. Journal of Biological Chemistry 272, 18020–18025.

1016. Stewart, C.N., Richards, H.A., and Halfhill, M.D. (2000) Transgenic plants and biosafety: Science, misconceptions and public perceptions. Biotechniques 29, 832.

1017. Stewart, C.N., Halfhill, M.D., and Warwick, S.I. (2003) Transgene introgression from genetically modified crops to their wild relatives. Nature Reviews Genetics 4, 806–817.

1018. Stewart, C.N., Jr., Millwood, R.J., Halfhill, M.D., Ayalew, M., Cardoza, V., Kooshki, M., Capelle, G.A., Kyle, K.R., Piaseki, D., McCrum, G., and Di Benedetto, J. (2005) Laser-induced fluorescence imaging and spectroscopy of GFP transgenic plants. Journal of Fluorescence 15, 699–705.

1019. Stewart, C.N., Jr. (2004) Genetically Modified Planet. Oxford University Press, New York.

1020. Stokstad, E. (2006) Agriculture-A kinder, gentler Jeremy Rifkin endorses biotech, or does he? Science 312, 1586–1587.

1021. Stoughton, R.B. (2005) Applications of DNA microarrays in biology. Annual Review of Biochemistry 74, 53–82.

1022. Suarez, L., Hendricks, K.A., Cooper, S.P., Sweeney, A.M., Hardy, R.J., and Larsen, R.D. (2000) Neural tube defects among Mexican Americans living on the US-Mexico border: Effects of folic acid and dietary folate. American Journal of Epidemiology 152, 1017–1023.

1023. Subramanian, V. (2002) DNA shuffling to produce nucleic acids for mycotoxin detoxification. U.S. Patent 6,500,639.

1024. Suh, H.S., Sato, Y.I., and Morishima, H. (1997) Genetic characterization of weedy rice (*Oryza sativa*) based on morpho-physiology, isozymes, and RAPD markers. Theoretical and Applied Genetics 94, 316–321.

1025. Sukopp, H., and Sukopp, U. (1993) Ecological long-term effects of cultigens becoming feral and of naturalization of nonnative species. Experientia 49, 210–218.

1026. Sukopp, U., Pohl, M., Driessen, S., and Bartsch, D. (2005) Feral beets-with help from the maritime wild? in Crop Ferality and Volunteerism (ed. Gressel, J.), 45–57. CRC Press, Boca Raton, FL.

1027. Sun, L., Ghosh, I., Paulus, H., and Xu, M.Q. (2001) Protein trans-splicing to produce herbicide-resistant acetolactate synthase. Applied and Environmental Microbiology 67, 1025–1029.

1028. Sun, M., and Corke, H. (1992) Population genetics of colonizing success of weedy rye in northern California. Theoretical and Applied Genetics 83, 321–329.

1029. Sun, Q.X., Ni, Z.F., Liu, Z.Y., Gao, J.W., and Huang, T.C. (1998) Genetic relationships and diversity among Tibetan wheat, common wheat, and European spelt wheat revealed by rapid markers. Euphytica 99, 205–211.

1030. Sundram, K., French, M.A., and Clandinin, M.T. (2003) Exchanging partially hydrogenated fat for palmitic acid in the diet increases LDL-cholesterol and endogenous cholesterol synthesis in normocholesterolemic women. European Journal of Nutrition 42, 188–181.

1031. Surov, T., Aviv, D., Aly, R., Joel, D.M., Goldman-Guez, T., and. Gressel, J. (1998) Generation of transgenic asulam-resistant potatoes to facilitate eradication of parasitic broomrapes (*Orobanche spp.*) with the *sul* gene as the selectable marker. Theoretical and Applied Genetics 96, 132–137.

1032. Suzuki, H., Achnine, L., Xu, R., Matsuda, S.P.T., and Dixon, R.A. (2002) A genomics approach to the early stages of triterpene saponin biosynthesis in *Medicago truncatula*. Plant Journal 32, 1033–1048.

1033. Suzuki, T., Honda, Y., Funatsuki, W., and Nakatsuka, K. (2004) In-gel detection and study of the role of flavon 3-glucosidase in the bitter taste generation in tartary buckwheat. Journal of the Science of Food and Agriculture 84, 1691–1694.

1034. Swaminathan, M.S. (1993) Perspectives for crop protection in sustainable agriculture. Ciba Foundation Symposia 177, 257–272.

1035. Sydenham, E.W., Shephard, G.S., Thiel, P.G., Marasas, W.F.O., and Stockenstrom, S. (1991) Fumonisin contamination of commercial corn-based human foodstuffs. Journal of Agricultural and Food Chemistry 39, 2014–2018.

1036. Sydenham, E.W., Thiel, P.G., Marasas, W.F.O., Shephard, G.S., van Schalkwyk, D.J., and Koch, K.R. (1990) Natural occurrence of some *Fusarium* mycotoxins in corn from low and high esophageal cancer prevalence areas of the Transkei, Southern Africa. Journal of Agricultural and Food Chemistry 38, 1900–1903.

1037. Szeinberg, A., Sheba, C., Hirshorn, N., and Bodonyi, E. (1957) Studies on erythrocytes in cases with past history of favism and drug-induced acute hemolytic anemia. Blood 12, 603–613.

1038. S rensen, M. (1996) Yam bean (*Pachyrhizus* DC.). Promoting the Conservation and Use of Underutilized and Neglected Crops. 2, 143 pp., International Plant Genetic Resources Institute, Rome.

1039. Tabashnik, B.E., Fabrick, J.A., Henderson, S., Biggs, R.W., Yafuso, C.M., Nyboer, M.E., Manhardt, N.M., Coughlin, L.A., Sollome, J., Carriere, Y., Dennehy, T.J., and Morin, S. (2006) DNA screening reveals pink bollworm resistance to Bt cotton remains rare after a decade of exposure. Journal of Economic Entomology 99, 1525–1530.

1040. Tadesse, A., and Basedow, T. (2005) Laboratory and field studies on the effect of natural control measures against insect pests in stored maize in Ethiopia. Journal of Plant Diseases and Protection 112, 156–172.

1041. Tadesse, Y., Sagi, L., Swennen, R., and Jacobs, M. (2003) Optimisation of transformation conditions and production of transgenic sorghum (*Sorghum bicolor*) via microparticle bombardment. Plant Cell Tissue and Organ Culture 75, 1–18.

1042. Takahashi, E., Ma, J.F., and Miyake, Y. (1990) The possibility of silicon as an essential element for higher plants. Agricultural and Food Chemistry 2, 99–122.

1043. Tal, B., Gressel, J., and Goldberg, I. (1982) The effect of medium constituents on growth by *Dioscorea deltoidea* cells grown in batch cultures. Planta Medica 44, 111–115.

1044. Tal, B., Rokem, J.S., Gressel, J., and Goldberg, I. (1984) The effect of chlorophyll-bleaching herbicides on growth, carotenoid and diosgenin levels in cell suspension cultures of *Dioscorea deltoidea*. Phytochemistry 23, 1333–1335.

1045. Tamai, K. and Ma, J.F. (2003) Characterization of silicon uptake by rice roots. New Phytologist 158, 431–436.

1046. Tanaka, Y., Katsumoto, Y., Brugliera, F., and Mason, J. (2005) Genetic engineering in floriculture. Plant Cell Tissue and Organ Culture 80, 1–24.

1047. Tarimo, T.M.C. (2000) Cyanogenic glycoside dhurrin as a possible cause of bird-resistance in Ark-3048 sorghum. In Workshop on Research Priorities for Migrant Pests of Agriculture in Southern Africa (eds. Cheke, R.A., Rosenberg, L.J., and Kieser, M.E.), 103–111. Natural Resources Institute, Chatham, United Kingdom.

1048. Tavassoli, A. (1986) The cytology of *Eragrostis tef* with special reference to *E. tef* and its relatives. PhD Thesis, University of London, London.

1049. Taylor, D.P., and Obendorf, R.L. (2001) Quantitative assessment of some factors limiting seed set in buckwheat. Crop Science 41, 1792–1799.

1050. Teklu, Y., and Tefera, H. (2005) Genetic improvement in grain yield potential and associated agronomic traits of tef (*Eragrostis tef*). Euphytica 141, 247–254.

1051. Tennant, P., Fermin, G., Fitch, M.M., Manshardt, R.M., Slightom, J.L., and Gonsalves, D. (2001) Papaya ringspot virus resistance of transgenic Rainbow and SunUp is affected by gene dosage, plant development, and coat protein homology. European Journal of Plant Pathology 107, 645–653.

1052. Tennant, P., Souza, M.T., Jr., Gonsalves, D., Fitch, M.M., Manshardt, R.M., and Slightom, J.L. (2005) Line 63-1: A new virus-resistant transgenic papaya. Hortscience 40, 1196–1199.

1053. Thebault, E., and Loreau, M. (2005) Trophic interactions and the relationship between species diversity and ecosystem stability. American Naturalist 166, E95E114.

1054. Thelen, J.J., and Ohlrogge, J.B. (2002) Metabolic engineering of fatty acid biosynthesis in plants. Metabolic Engineering 4, 12–21.

1055. Theurer, C., Gruetzner, K., Freeman, S., and Koetter, U. (1997) In vitro phototoxicity of hypericin, fagopyrin rich, and fagopyrin free buckwheat herb extracts. Pharmaceutical and Pharmacological Letters 7, 113–115.

1056. Thill, D.C., and Mallory-Smith, C.A. (1997) The nature and consequence of weed spread in cropping systems. Weed Science 45, 337–342.

1057. Thomas, H. (1995) Oats. in Evolution of Crop Plants (eds. Smartt, J., and Simmonds, N.W.), 132–137. Longman, Harlow, United Kingdom.

1058. Thomson, J.A. (2004) The status of plant biotechnology in Africa. AgBioForum 7, 9–12. Online www.agbioforum.org.

1059. Thorsoe, K.S., Bak, S., Olsen, C.E., Imberty, A., Breton, C., Moller ,BL. (2005) Determination of catalytic key amino acids and UDP sugar donor specificity of the cyanohydrin glycosyltransferase UGT85B1 from *Sorghum bicolor*. Molecular modeling substantiated by site-specific mutagenesis and biochemical analyses. Plant Physiology 139, 664–673.

1060. Tillman, D.A. (2000) Biomass cofiring: the technology, the experience, the combustion consequences. Biomass and Bioenergy 19, 365–384.

1061. Tilman, D. (2000) Causes, consequences and ethics of biodiversity. Nature 405, 208–211.

1062. Tilman, D., Polasky, S., and Lehman, C. (2005) Diversity, productivity and temporal stability in the economies of humans and nature. Journal of Environmental Economics and Management, 49, 405–4126.

1063. Timmons, A.M., Obrien, E.T., Charters, Y.M., Dubbels, S.J,. and Wilkinson, M.J. (1995) Assessing the risks of wind pollination from fields of genetically-modified *Brassica napus* ssp *oleifera*. Euphytica 85, 417–423.

1064. Timmons, A.M., Charters, Y.M., Crawford, J.W., Burn, D., Scott, S.E., Dubbels, S.J., Wilson, N.J., Robertson, A., O'Brien, E.T., Squire, G.R., and Wilkinson, M.J. (1996) Risks from transgenic crops. Nature 380, 487–487.

1065. Tobias, C.M., and Chow, E.K. (2005) Structure of the cinnamyl-alcohol dehydrogenase gene family in rice and promoter activity of a member associated with lignification. Planta 220, 678–688.

1066. Tomasino, S.F., Leister, R.T., Dimock, M.B., Beach, R.M., and Kelly, J.L. (1995) Field performance of *Clavibacter xyli* subsp. *cynodontis* expressing the insecticidal protein gene *cryIA(c)* of *Bacillus thuringiensis* against European corn borer in field corn. Biological Control 5, 442–448.

1067. Tor, M., Mantell, S.H., and Ainsworth, C. (1992) Endophytic bacteria expressing beta-glucuronidase cause false positives in transformation of *Dioscorea* species. Plant Cell Reports 11, 452–456.

1068. Tor, M., Ainsworth, C., and Mantell, S.H. (1993) Stable transformation of the food yam *Dioscorea alata* L by particle bombardment. Plant Cell Reports 12, 468–473.

1069. Tor, M., Twyford, C.T., Funes, I., Boccon-Gibod, J., Ainsworth, C.C., and Mantell, S.H. (1998) Isolation and culture of protoplasts from immature leaves and embryogenic cell suspensions of *Dioscorea* yams: tools for transient gene expression studies. Plant Cell Tissue and Organ Culture 53, 113–125.

1070. Torgersen, H. (1996) Risk assessment in transgenic plants: what can we learn from the ecological impacts of traditional crops? BINAS News 2: (3&4). www.binas. unido.org/binas/News/96issue34/risk.html.

1071. Torii, K.U., McNellis, T.W., and Deng. X.-W. (1998) Functional dissection of Arabidopsis COP1 reveals specific roles of its three structural modules in light control of seedling development. EMBO Journal 17, 5577–5587.

1072. Tourte, Y., and Tourte, C. (2005) Genetic Engineering and Biotechnology: Concepts, Methods and Agronomic Applications. Science Publishers, Enfield, NH.

1073. Travis, A.J., Murison, S.D., Hirst, D.J., Walker, K.C., and Chesson, A.C. (1996) Comparison of the anatomy and degradability of straw from varieties of wheat and barley that differ in susceptibility to lodging. Journal of Agricultural Science 127, 1–10.

1074. Tripathi, S., Bau, H.J., Chen, L.F., and Yeh, S.D. (2004) The ability of papaya ringspot virus strains overcoming the transgenic resistance of papaya conferred by the coat protein gene is not correlated with higher degrees of sequence divergence from the transgene. European Journal of Plant Pathology 110, 871–882.

1075. Tripathi, S.C., Sayre, K.D., and Kaul, J.N. (2003) Fibre analysis of wheat genotypes and its association with lodging: Effects of nitrogen levels and ethephon. Cereal Research Communications 31, 429–436.

1076. Tsai, C.-J., Popko, J.L., Mielke, M.R., Hu, W.J., Podila, G.K., and Chiang, V.L. (1998) Suppression of O-methyltransferase gene by homologous sense transgene in quaking aspen causes red-brown wood phenotypes. Plant Physiology 117, 101–112.

1077. Tsanuo, M.K., Hassanali, A., Hooper, A.M., Khan, Z., Kaberia, F., Pickett, J.A., and Wadhams, L.J. (2003) Isoflavanones from the allelopathic aqueous root exudate of *Desmodium uncinatum*. Phytochemistry 64, 265–273.

1078. Tsybina, T., Dunaevsky, Y., Musolyamov, A., Egorov, T., Larionova, N., Popykina, N., and Belozersky, M. (2004) New protease inhibitors from buckwheat seeds: properties, partial amino acid sequences and possible biological role. Biological Chemistry 385, 429–434.

1079. Turner, P.C., Moore, S.E., Hall, A.J., Prentice, A.M., and Wild, C.P. (2003) Modification of immune function through exposure to dietary aflatoxin in Gambian children. Environmental Health Perspectives 111, 217–220.

1080. Turner, P.C., Sylla, A., Diallo, M.S., Castegnaro, J.J., Hall, A.J., and Wild, C.P. (2002) The role of aflatoxins and hepatitis viruses in the etiopathogenesis of hepatocellular carcinoma: A basis for primary prevention in Guinea-Conakry, West Africa. Journal of Gastroenterology and Hepatology 17, S441S448.

1081. U, N. (1935) Genome-analysis in *Brassica* with special reference to the experimental formation of *B. napus* and peculiar mode of fertilization. Japanese Journal of Botany 7, 390–452.

1082. Ubbink, J.B., Christianson, A., Bester, M.J., Van Allen, M.I., Venter, P.A., Delport, R., Blom, H.J., van der Merwe, A., Potgieter, H., and Vermaak, W.J. (1999) Folate status, homocysteine metabolism, and methylene tetrahydrofolate reductase genotype in rural South African blacks with a history of pregnancy complicated by neural tube defects. Metabolism-Clinical and Experimental 48, 269–274.

1083. Ueguchi-Tanaka, M., Ashikari, M., Nakajima, M., Itoh, H., Katoh, E., Kobayashi, M., Chow, T.Y., Hsing, Y.I.C., Kitano, H., Yamaguchi, I., and Matsuoka, M. (2005) *GIBBERELLIN INSENSITIVE DWARF*1 encodes a soluble receptor for gibberellin. Nature 437, 693–698.

1084. Ullstrup, A.J. (1972) The impacts of the southern leaf blight epidemics of 1970–1971. Annual Review of Phytopathology 10, 37–50.

1085. UNIDO. (2007) Decision support system for safety assessment of genetically modified crop plants. http://binas.unido.org/xdtree.

1086. Urga, K., and Narashima, H.V. (1997) Effect of natural fermentation on the HCl-extractability of minerals from tef (*Eragrostis tef*). Bulletin of the Chemical Society of Ethiopia 11, 3–10.

1087. Ursin, V.M. (2003) Modification of plant lipids for human health: development of functional land-based omega-3 fatty acids. Journal of Nutrition 133, 4271–4274.

1088. Uvere, P.O., Adenuga, O.D., and Mordi, C. (2000) The effect of germination and kilning on the cyanogenic potential, amylase and alcohol levels of sorghum malts used for burukutu production. Journal of the Science of Food and Agriculture 80, 352–358.

1089. Valverde, B., and Itoh, K. (2001) World rice and herbicide resistance. in Herbicide resistance in world grains (eds. Powles, S.B., and Shaner, D.), 195–250. CRC Press, Boca-Raton, FL.

1090. Valverde, B.E. (2005) The damage by weedy rice-can feral rice remain undetected? In Crop Ferality and Volunteerism (ed. Gressel, J.), 279–294. CRC Press, Boca Raton, FL.

1091. Valverde, B.E., Riches, C.R., and Caseley, J.C. (2000) Prevention and management of herbicide resistant weeds in rice: Experiences from Central America with *Echinochloa colona*. Cámara de Insumos Agropecuarios, Costa Rica.

1092. Van Bueren, E.T.L., Struik, P.C., and Jacobsen, E. (2002) Ecological concepts in organic farming and their consequences for an organic crop ideotype. Netherlands Journal of Agricultural Science 50, 1–26.

1093. van Bueren, E.T.L., Struik, P.C., and Jacobsen, E. (2003) Organic propagation of seed and planting material: an overview of problems and challenges for research. Njas-Wageningen Journal of Life Sciences 51, 263–277.

1094. Van der Schaar, W., Alonso-Blanco, C., Leon-Kloosterziel, K.M., Jansen, R.C., van Ooijen, J.W., and Koornneef, M. (1997) QTL analysis of seed dormancy in *Arabidopsis* using recombinant inbred lines and MQM mapping. Heredity 79, 190–200.

1095. van der Walt, E. (2000) Research at PPRI on environmental effects of quelea control operations. In Workshop on Research Priorities for Migrant Pests of Agriculture in Southern Africa (eds. Cheke, R.A., Rosenberg, L.J., and Kieser, M.E.), 91–95. Natural Resources Institute, Chatham, United Kingdom.

1096. Van Slageren, M.W. (1994) Wild wheats: A monograph of *Aegilops* L. and *Amblyopyrum* (Jaub. & Spach) Eig (Poaceae). Wageningen Agricultural University, Wageningen, The Netherlands.

1097. Van Staden, J., and Davey, J.E. (1980) Effect of silver thiosulfate on the senescence of emasculated orchid (*Cymbidium*) flowers. South African Journal of Science 76, 314–315.

1098. van Valen, L. (1973) A new evolutionary law. Evolutionary Theory 1, 1–30.

1099. Vanjildorj, E., Bae, T.W., Riu, K.Z., Kim, S.Y., and Lee, H.Y. (2005) Overexpression of *Arabidopsis ABF3* gene enhances tolerance to drought and cold in transgenic lettuce (*Lactuca sativa*). Plant Cell Tissue and Organ Culture 83, 41–50.

1100. Vareschi, V. (1980) Vegetations ökologie der Tropen. Ulmer, Stuttgart.

1101. Varga, E., Schmidt, A.S., Reczey, K., and Thomsen, A.B. (2003) Pretreatment of corn stover using wet oxidation to enhance enzymatic digestibility. Applied Biochemistry and Biotechnology 104, 37–50.

1102. Vaughan, D.A. (1994) The wild relatives of rice. International Rice Research Institute, Manila.

1103. Vaughan, D.A., Sanchez, P.L., Ushiki, J., Kaga, A., and Tomooka, N. (2005) Asian rice and weedy rice-evolutionary perspectives. In Crop Ferality and Volunteerism (ed. Gressel, J.), 257–277. CRC Press, Boca Raton, FL.

1104. Vaughan, L.K., Ottis, V., Prazak-Havey, A.M., Conaway-Bormans, C.A., Sneller, C.H., Chandler, J.M., and Park, W.D. (2001) Is all red rice found in commercial rice really *Oryza sativa?* Weed Science 49, 468–476.

1105. Vaughn, D. (2005) *Vigna* in Asia. In Crop Ferality and Volunteerism (ed. Gressel, J.), 235–236. CRC Press, Boca Raton, FL.

1106. Vavilov, N.I. (1935) Origin and Geography of Cultivated Plants. In The Phytogeographical Basis for Plant Breeding, 316–366. Cambridge Univ. Press, Cambridge, United Kingdom.

1107. Vaz-Patto, M.C., Skiba, B., Pang, E.C.K., Ochatt, S.J., Lambein, F., and Rubiales, D. (2006) *Lathyrus* improvement for resistance against biotic and abiotic stresses: From classical breeding to marker assisted selection. Euphytica 147, 133–147.

1108. Verhoog, H., Matze, M., Van Bueren, E.L., and Baars, T. (2003) The role of the concept of the natural (naturalness) in organic farming. Journal of Agricultural & Environmental Ethics 16, 29–49.

1109. Vietmeyer, N.D. (ed.) (1996) Fonio. In Lost Crops of Africa, Vol. 1 Grains, National Academy Press, Washington DC.

1110. Vodouhe, S.R., and Achigan Dako, E. (eds.) (2001) Renforcement de la contribution du Fonio a la securite alimentaire et aux revenus des paysans en Afrique de l'Oueste. IPGRI, Rome.

1111. Voelker, T.A., and Kinney, A.J. (2001) Variations in the biosynthesis of seed-storage lipids. Annual Review of Plant Physiology and Plant Molecular Biology 52, 335–361.

1112. Vogel, K.P., and Jung, H.J.G. (2001) Genetic modification of herbaceous plants for feed and fuel. Critical Reviews in Plant Sciences 20, 15–49.

1113. Vogler, R.K., Ejeta, G., and Butler, L.G. (1996) Inheritance of low production of *Striga* germination stimulant in sorghum. Crop Science 36, 1185–1191.

1114. Vollbrecht, E., Springer, P.S., Goh, U., Buckler, E.S., and Martienssen, R. (2005) Architecture of floral branch systems in maize and related grasses. Nature 436, 1119–1126.

1115. Vos, P., Hogers, R., Bleeker, M., Reijans, M., van de Lee, T., Hornes, M., Frijters, A., Pot, J., Peleman, J., Kuiper, M., and Zabeau, M. (1995) AFLP-A new technique for DNA-fingerprinting. Nucleic Acids Research 23, 4407–4414.

1116. Vrebalov, J., Ruezinsky, D., Padmanabhan, V., White, R., Medrano, D., Drake, R., Schuch, W., and Giovannoni, J.. (2002) A MADS-box gene necessary for fruit ripening at the tomato ripening-inhibitor (*Rin*) locus. Science 343, 343–346.

1117. Vurro, M., Gressel, J., Butt, T., Harman, G.E., Pilgeram, A., St. Leger, R.J., and Nuss, D.L. (eds.) (2001) Enhancing biocontrol agents and handling risks, IOS Press, Amsterdam.

1118. Wakabayashi, K. (2000) Changes in cell wall polysaccharides during fruit ripening. Journal of Plant Research 113, 231–237.

1119. Walker, B.H. (1992) Biodiversity and ecological redundancy. Conservation Biology 6, 18–23.

1120. Walker, B.H., and Langridge, J.L. (2002) Measuring functional diversity in plant communities with plant life forms: A problem of hard and soft attributes. Ecosystems 5, 529–538.

1121. Wall, D.A., and Smith, M.A.H. (1999) Weed management in common buckwheat (*Fagopyrum esculentum*). Canadian Journal of Plant Science 79, 455–461.

1122. Wall, E.M., Lawrence, T.S., Green, M.J., and Rott, M.E. (2004) Detection and identification of transgenic virus resistant papaya and squash by multiplex PCR. European Food Research and Technology 219, 90–96.

1123. Wang, J.-S., Shen, X., He, X., Zhu, Y.R., Zhang, B.C., Wang, J.B., Qian, G.S., Kuang, S.Y., Zarba, A., Egner, P.A., Jacobson, L.P., Munoz, A., Helzlsouer, K.J., Groopman, J.D., and Kensler, T.W. (1999) Protective alterations in phase I and phase 3 metabolism of aflatoxin B1 by oltipraz in residents of Qidong, Peoples Republic of China. Journal of the National Cancer Institute 91, 347–353.

1124. Wang, N.N., Yang, S.F., and Charng, Y.Y. (2001) Differential expression of 1-aminocyclo-propane-1-carboxylate synthase genes during orchid flower senescence induced by the protein phosphatase inhibitor okadaic acid. Plant Physiology 126, 253–260.

1125. Wang, Q., Zhao, X.P., Wu, C.X., Dai, F., Wu, L.Q., Xu, H., Li, M.S., and He, J.H.. (2001) Occurrence of weeds in wet-seeded rice fields and application of Nominee (bispyribac sodium) in Zhejianng, China. In Proceedings 18th Asian-Pacific Weed Science Society Conference-Beijing, 251–257.

1126. Wang, T., Li, Y., Shi, Y., Reboud, X., Darmency, H., and Gressel, J. (2004) Low frequency transmission of a plastid encoded trait in *Setaria italica*. Theoretical and Applied Genetics 108, 315–320.

1127. Wang, Y., Scarth, R., and Campbell, C. (2005) Interspecific hybridization between diploid *Fagopyrum esculentum* and tetraploid *F. homotropicum*. Canadian Journal of Plant Science 85, 41–48.

1128. Wang, Y.J., Scarth, R., and Campbell, G.C. (2005) Inheritance of seed shattering in interspecific hybrids between *Fagopyrum esculentum* and *F. homotropicum*. Crop Science 45, 693–697.

1129. Wang, Y.J., Scarth, R., and Campbell, C. (2005) S^h and S^c-two complementary dominant genes that control self-compatibility in buckwheat. Crop Science 45, 1229–1234.

1130. Wang, Z.N., Zemetra, R.S., Hansen, J., and Mallory-Smith, C.A. (2001) The fertility of wheat x jointed goatgrass hybrid and its backcross progenies. Weed Science 49, 340–345.

1131. Wang, Z.Y., Zhang, K.W., Sun, X.F., Tang, K.X., and Zhang, J.R. (2005) Enhancement of resistance to aphids by introducing the snowdrop lectin gene *gna* into maize plants. Journal of Biosciences 30, 627–638.

1132. Warchalewski, J.R., Gralik, J., Winiecki, Z., Nawrot, J., and Piasecka-Kwiatkowska, D. (2002) The effect of wheat alpha-amylase inhibitors incorporated into wheat-based artificial diets on development of *Sitophilus granarius* L., *Tribolium confusum* Duv, and *Ephestia kuehniella* Zell. Journal of Applied Entomology 126, 161–168.

1133. Warwick, S.I., and Stewart C.N. Jr. (2005) Crops come from wild plants: How domestication, transgenes, and linkage together shape ferality. In Crop Ferality and Volunteerism (ed. Gressel, J.), 9–30. CRC Press, Boca Raton, FL.

1134. Warwick, S.I., Simard, M.J., Legere, A., Beckie, H.J., Braun, L., Zhu, B., Mason, P., Seguin-Swartz, G., and Stewart, C.N. Jr. (2003) Hybridization between transgenic *Brassica napus* L. and its wild relatives: *B. rapa* L., *Raphanus raphanistrum* L., *Sinapis arvensis* L., and *Erucastrum gallicum* (Willd.) O.E. Schulz. Theoretical and Applied Genetics. 107, 528–539.

1135. Waterhouse, P.M., and Helliwell, C.A. (2003) Exploiting plant genomes by RNA-induced gene silencing. Nature Review Genetics 4, 29–38.

1136. Webb, S.E., Appleford, N.E.J., Gaskin, P., and Lenton, J.R. (1998) Gibberellins in internodes and ears of wheat containing different dwarfing alleles. Phytochemistry 47, 671–677.

1137. Webster, R., and Hulse, D. (2005) Controlling avian flu at the source. Nature 435, 415–416.

1138. Wei, W., Schuler, T.H., Clark, S.J., Stewart, C.N., and Poppy, G.M. (2005) Age-related increase in levels of insecticidal protein in the progenies of transgenic oilseed rape and its efficacy against a susceptible strain of diamondback moth. Annals of Applied Biology 147, 227–234.

1139. Weissmann, S., Feldman, M., and Gressel, J. (2005) Sporadic inter-generic DNA introgression from wheat into wild *Aegilops* spp. Molecular Biology and Evolution 22, 2055–2062.

1140. Weissmann, S., Feldman, M., and Gressel, J. (2007) Can real chaperons prevent promiscuous transgene flow from polyploid crops to their wild/weedy relatives? Manuscript in preparation.

1141. Weller, G.L., Cassells, J.A., and Banks, H.J. (1998) The Grainscanner —Could it play a role in the control of insects in bulk grain? Australian Postharvest Technical Conference, 328–333.

1142. Wen, Y., Hatabayashi, H., Arai, H., Kitamoto, H.K., and Yabe, K. (2005) Function of the *cypX* and *moxY* genes in aflatoxin biosynthesis in *Aspergillus parasiticus*. Applied and Environmental Microbiology 71, 3192–3198.

1143. Wendorf, F., Close, A.E., Schild, R., Wasylikowa, K., Housley, R.A., Harlan, J.R., and Króiik, H. (1992) Saharan exploitation of plants 8,000 years BP. Nature 359, 721–724.

1144. Werner, T., Motyka, V., Laucou, V., Smets, R., Van Onckelen, H., and Schmülling, T. (2003) Cytokinin-deficient transgenic *Arabidopsis* plants show multiple developmental alterations indicating opposite functions of cytokinins in the regulation of shoot and root meristem activity. Plant Cell 15, 2532–2550.

1145. Westwood, J. (2000) Characterization of the *Orobanche-Arabidopsis* system for studying parasite-host interactions. Weed Science 48, 742–748.

1146. Wheeler, J.L., Mulcahy, C., Walcott, J.J., and Rapp, G.G. (1990) Factors affecting the hydrogen-cyanide potential of forage sorghum. Australian Journal of Agricultural Research 41, 1093–1100.

1147. Whitham, S.A., Yamamoto, M.L., and Carrington, J.C. (1999) Selectable viruses and altered susceptibility mutants in *Arabidopsis thaliana*. Proceedings of the National Academy of Science USA 96, 772–777.

1148. Whitham, T.G. et al. (1999) Plant hybrid zones affect biodiversity: Tools for a genetic-based understanding of community structure. Ecology 80, 416–428.

1149. WHO. (2002) Evaluation of certain mycotoxins in food. Fifty-sixth report of the joint FAO/WHO expert committee on food additives (JECFA). Technical Report 906, World Health Organization, Geneva.

1150. Widstrom, N.W., Hanson, W.D., and Redlinger, L.M. (1975) Inheritance of maize weevil resistance in maize. Crop Science 15, 467–470.

1151. Wiechert, R. (1970) Modern steroid problems. Angewandte Chemie International Edition-English 9, 321–332.

1152. Wielopolska, A., Townley, H., Moore, I., Waterhouse, P., and Helliwell, P. (2005) A high throughput inducible RNAi vector for plants. Plant Biotechnology Journal 3, 583–590.

1153. Williams, M.E. (1995) Genetic engineering for pollination control. Trends in Biotechnology 13, 344–349.

1154. Williams, R.J., Johnson, A.C., Smith, J.J.L., and Kanda, R. (2003) Steroid estrogens profiles along river stretches arising from sewage treatment works discharges. Environmental Science and Technology 37, 1744–1750.

1155. Williamson, M. (1993) Invaders, weeds, and risks from genetically manipulated organisms. Experientia 49, 219–224.

1156. Wilson, J.D., Morris, A.J., Arroyo, B.E., Clark, S.C., and Bradbury, R.B. (1999) A review of the abundance and diversity of invertebrate and plant foods of granivorous birds in northern Europe in relation to agricultural change. Agriculture Ecosystems and Environment 75, 13–30.

1157. Wilson, J.S., and Otsuki, T. (2001) Global Trade and Food Safety, Winners and Losers in a Fragmented System. World Bank, Washington DC.

1158. Wilson, S., Blaschek, K., and de Mejia, E.G. (2005) Allergenic proteins in soybean: Processing and reduction of P34 allergenicity. Nutrition Reviews 63, 47–58.

1159. Winston, J. (2004) Some history of the treatment of epidemics with homeopathy by homeopathy. www.homeopathic.org/crhistJW2.htm.

1160. Wisniewski, J.P., Frangne, N., Massonneau, A., and Dumas, C. (2002) Between myth and reality: genetically modified maize, an example of a sizeable scientific controversy. Biochimie 84, 1095–1103.

1161. Wolfe, M.S., Brandle, U., Koller, B., Limpert, E., McDermott, J.M., Mfiller, K., and Schaffner, D. (1992) Barley mildew in Europe-population biology and host-resistance. Euphytica 63, 125–139.

1162. Wolski, W.E., Lalowski, M., Jungblut, P., and Reinert, K. (2005) Calibration of mass spectrometric peptide mass fingerprint data without specific external or internal calibrants. BMC Bioinformatics 6, 203. www.biomedcentral.com/content/pdf/1471–21[OSX]05–6–203.pdf.

1163. Woo, S.H., Adachi, T., Jong, S.K., and Campbell, C.G. (1999) Inheritance of self-compatibility and flower morphology in an inter-specific buckwheat hybrid. Canadian Journal of Plant Science 79, 483–490.

1164. Wyman, C.E., Spindler, D.D., and Grohmann, K. (1992) Simultaneous saccharification and fermentation of several lignocellulosic feedstocks to fuel ethanol. Biomass and Bioenergy 3, 301–307.

1165. Xuan, T.D., and Tsuzuki, E. (2004) Allelopathic plants: Buckwheat (*Fagopyrum* spp.). Allelopathy Journal 13, 137–148.

1166. Yamada, T., Moriyama, R., Hattori, K., and Ishimoto, M. (2005) Isolation of two alpha-amylase inhibitor genes of tepary bean (*Phaseolus acutifolius* A. Gray) and their functional characterization in genetically engineered azuki bean. Plant Science 169, 502–511.

1167. Yamaguchi, S., Sun, T.P., Kawaide, H., and Kamiya, Y. (1998) The GA2 locus of *Arabidopsis thaliana* encodes ent-kaurene synthase of gibberellin. Plant Physiology 116, 1271–1278.

1168. Yamane, K., Yasui, Y., and Ohnishi, O. (2003) Intraspecific cpDNA variations of diploid and tetraploid perennial buckwheat, *Fagopyrum cymosum* (Polygonaceae). American Journal of Botany 90, 339–346.

1169. Yamane, K., Yamaki, Y., and Fujishige, N. (2004) Effects of exogenous ethylene and 1-MCP on ACC oxidase activity, ethylene production and vase life in *Cattleya* alliances. Journal of the Japanese Society for Horticultural Science 73, 128–133.

1170. Yanai, O., Shani, E., Dolezal, K., Tarkowski, P., Sablowski, R., Sandberg, G., Samach, A., and Ori, N. (2005) *Arabidopsis* KNOXI proteins activate cytokinin biosynthesis. Current Biology 15, 1566–1571.

1171. Yang, J.S., and Ye, C.A. (1992) Plant-regeneration from petioles of in vitro regenerated papaya (*Carica papaya* L) shoots. Botanical Bulletin of Academia Sinica 33, 375–381.

1172. Yang, Z.Y., Shim, W.B., Kim, J.H., Park, S.J., Kang, S.J., Nam, B.S., and Chung, D.H. (2004) Detection of aflatoxin-producing molds in Korean fermented foods and grains by multiplex PCR. Journal of Food Protection 67, 2622–2626.

1173. Yanofsky, M.F., and Liljegren, S.J. (2001) Seed plants characterized by delayed seed dispersal. U.S. Patent 6,288,305.

1174. Yanofsky, M.F., and Liljegren, S.J. (2005) Control of fruit dehiscence in plants by indehiscent 1 genes. U.S. Patent Application 20050,120,41 7.

1175. Yao, J.H., Zhao, X.Y., Qi, H.X., Wan, B.L., Chen, F., Sun, X.F., Yu, S.Q., and Tang, K.X. (2004) Transgenic tobacco expressing an *Arisaema heterophyllum* agglutinin gene displays enhanced resistance to aphids. Canadian Journal of Plant Science 84, 785–790.

1176. Yasui, Y., Wang, Y.J., Ohnishi, O., and Campbell, C.G. (2004) Amplified fragment length polymorphism linkage analysis of common buckwheat (*Fagopyrum esculentum*) and its wild self-pollinated relative *Fagopyrum homotropicum*. Genome 47, 345–351.

1177. Ye, B., and Gressel, J. (2000) Transient, oxidant-induced antioxidant transcript and enzyme levels correlate with greater oxidant-resistance in paraquat-resistant *Conyza bonariensis*. Planta 211, 50–61.

1178. Yen, M.L., Su, J.L., Chien, C.L., Tseng, K.W., Yang, C.Y., Chen, W.F., Chang, C.C., and Kuo, M.L. (2005) Diosgenin induces hypoxia-inducible factor-1 activation and angiogenesis through estrogen receptor-related phosphatidylinositol 3-kinase/Akt and p38 mitogen-activated protein kinase pathways in osteoblasts. Molecular Pharmacology 68, 1061–1073.

1179. Yi, S.Y., Kim, J.H., Joung, Y.H., Lee, S., Kim, W.T., Yu, S.H., and Choi, D. (2004) The pepper transcription factor *CaPF*1 confers pathogen and freezing tolerance in *Arabidopsis*. Plant Physiology 136, 2862–2874.

1180. Yin, F., Giuliano, A.E., Law, R.E., and Van Herle, A.J. (2001) Apigenin inhibits growth and induces G2/M arrest by modulating cyclin-CDK regulators and ERK MAP kinase activation in breast carcinoma cells. Anticancer Research 21, 413–420.

1181. Ying, M.C., Dyer, W.E., and Bergman, J.W. (1992) *Agrobacterium tumefaciens*-mediated transformation of safflower (*Carthamus tinctorius* L.) cv. 'Centennial'. Plant Cell Reports 11, 581–585.

1182. Yizengaw, T., and Verheye, W. (1994) Modeling production potentials of tef (*Eragrostis tef*) in the Central Highlands of Ethiopia. Soil Technology 7, 269–277.

1183. York, J., and Daniel, L. (1991) Release statement of Ark-3048 and six other inbred grain sorghum lines. University of Arkansas Agricultural Experiment Station, Fayetteville, AR.

1184. Yoshioka, H., Ohmoto, T., Urisu, A., Mine, Y., and Adachi, T. (2004) Expression and epitope analysis of the major allergenic protein Fag e 1 from buckwheat. Journal of Plant Physiology 161, 761–767.

1185. Yu, J.J., Chang, P-K., Ehrlich, K.C., Cary, J.W., Bhatnagar, D., Cleveland, T.E., Payne, G.A., Linz, J.E., Woloshuk, C.P., and Bennett, J.W. (2004) Clustered pathway genes in aflatoxin biosynthesis. Applied and Environmental Microbiology 70, 1253–1262.

1186. Yu, J.N., Zhang, J.S., Shan, L., and Chen, S.Y. (2005) Two new group 3 *LEA* genes of wheat and their functional analysis in yeast. Journal of Integrative Plant Biology 47, 1372–1381.

1187. Yusibov, V., Hooper, D.C., Spitsin, S.V., Fleysh, N., Kean, R.B., Mikheeva, T., Deka, D., Karasev, A., Cox, S., Randall, J., and Koprowski, H. (2002) Expression in plants and immunogenicity of plant virus-based experimental rabies vaccine. Vaccine 20, 3155–3164.

1188. Zacchini, M., and De Agazio, M. (2004) Micropropagation of a local olive cultivar for germplasm preservation. Biologia Plantarum 48, 589–592.

1189. Zarba, A. et al. (1992) Aflatoxin-M1 in human breast-milk from the Gambia, west Africa, quantified by combined monoclonal-antibody immunoaffinity chromatography and HPLC. Carcinogenesis 13, 891–894.

1190. Zarkadas, M., and Case, G. (2005) Celiac disease and the gluten free diet. Topics in Clinical Nutrition 20, 127–138.

1191. Zegeye, A. (1997) Acceptability of injera with stewed chicken. Food Quality and Preference 8, 293–295.

1192. Zemetra, R.S., Hansen, J., and Mallory-Smith, C.A. (1998) Potential for gene transfer between wheat (*Triticum aestivum*) and jointed goatgrass (*Aegilops cylindrica*). Weed Science 46, 313–317.

1193. Zhang, W., Linscombe, S.D., Webster, E., Tan, S., and Oard, J. (2006) Risk assessment of the transfer of imazethapyr herbicide tolerance from Clearfield rice to red rice (*Oryza sativa*). Euphytica 152, 75–86.

1194. Zhao, L., Weiner, D.P., and Hickle, L. (2004) Transaminases, deaminases and aminomutases and compositions and methods for enzymatic detoxification. World Patent WO2004085624 .

1195. Zheng, S.J., Ma, J.F., and Matsumoto, H. (1998) High aluminum resistance in buckwheat. I. Al-induced specific secretion of oxalic acid from root tips. Plant Physiology 117, 745–751.

1196. Zhong, R.W., Morrison H. III, Negrel, J, and Ye, Z.H. (1998) Dual methylation pathways in lignin biosynthesis. Plant Cell 10, 2033–2046.

1197. Zhou, H., Arrowsmith, J.W., Fromm, M.E., Hironaka, C.M., Taylor, M.L., Rodriguez, D., Pajeau, M.E., Brown, S.M., Santino, C.G., and Fry, J.E. (1995) Glyphosate-tolerant CP4 and GOX genes as a selectable marker in wheat transformation. Plant Cell Reports 15, 159–163.

1198. Zhu, T., Mettenburg, K., Peterson, D.J., Tagliani, L., and Baszczynski, C.L. (2000) Engineering herbicide-resistant maize using chimeric RNA/DNA oligonucleotides. Nature Biotechnology 18, 555–558.

1199. Zhu, Y., Wang, Y., Chen, H., and Lu, B.R. (2003) Conserving traditional rice varieties through management for crop diversity. BioScience 53, 158–162.

1200. Zhu, Y.J., Agbayani, R., Jackson, M.C., Tang, C.S., and Moore, P.H. (2004) Expression of the grapevine stilbene synthase gene VST1 in papaya provides increased resistance against diseases caused by *Phytophthora palmivora*. Planta 220, 241–250.

1201. Zhu, Y.L., Pilon-Smits, E.A.H., Tarun, A., Weber, S.U., Jouanin, L., and Terry, N. (1999) Cadmium tolerance and accumulation in Indian mustard is enhanced by overexpressing gamma-glutamylcysteine synthetase. Plant Physiology 121, 1169–1177.

1202. Zhu-Salzman, K., Koiwa, H., Salzman, R.A., Shade, R.E., and Ahn, J.E. (2003) Cowpea bruchid *Callosobruchus maculatus* uses a three-component strategy to overcome a plant defensive cysteine protease inhibitor. Insect Molecular Biology 12, 135–145.

1203. Zohary, D. (2000) Domestication of Plants in the Old World: The Origin and Spread of Cultivated Plants in West Asia, Europe, and the Nile Valley. Oxford University Press, Oxford.

1204. Zuo, J.R., and Chua, N.H. (2000) Chemical-inducible systems for regulated expression of plant genes. Current Opinion in Biotechnology 11, 146–151.

Index